AF361657

Performance-Based Design in Structural Fire Engineering

Performance-Based Design in Structural Fire Engineering

Editor

Maged A. Youssef

MDPI • Basel • Beijing • Wuhan • Barcelona • Belgrade • Manchester • Tokyo • Cluj • Tianjin

Editor
Maged A. Youssef
Civil and Environmental
Engineering
Western University
London
Canada

Editorial Office
MDPI
St. Alban-Anlage 66
4052 Basel, Switzerland

This is a reprint of articles from the Special Issue published online in the open access journal *Fire* (ISSN 2571-6255) (available at: www.mdpi.com/journal/fire/special_issues/Structural_Fire_Engineering).

For citation purposes, cite each article independently as indicated on the article page online and as indicated below:

LastName, A.A.; LastName, B.B.; LastName, C.C. Article Title. *Journal Name* **Year**, *Volume Number*, Page Range.

ISBN 978-3-0365-4340-6 (Hbk)
ISBN 978-3-0365-4339-0 (PDF)

Contents

About the Editor

Maged A. Youssef

Dr. Maged A. Youssef is a Professor of Civil and Environmental Engineering at Western University, Canada. He received his PhD from McMaster University, Canada, in 2000 and then joined Murray Engineering P.C., New York City, as a senior structural engineer. Dr. Youssef joined Western University in August of 2002. He was promoted to the rank of Professor in 2015. He also acted as the Associate Chair for Undergraduate Affairs from 2012 to 2015. Dr. Youssef has also received the 2007 R. Mohan Mathur Award for Excellence in Teaching. In addition to university teaching, he is actively involved in the professional development of practicing engineers. Dr. Youssef's research focuses on improving safety of our civil structures to natural and human-made disasters. His research covers areas in earthquake engineering, structural fire engineering, and smart materials and systems. He was a co-publisher of three standards, and he has also published 2 book chapters, 70 journal articles, 65 conference papers, and several technical reports. He has presented his research as a keynote/expert speaker in national and international conferences and was also invited to present his research at Tohoku University (Japan), Kyoto University (Japan), and South China University of Technology (China). Dr. Youssef supervised 12 PhD students and 10 Master's students to graduation. He is currently supervising 10 PhD students and 2 Master's students. He is an active member of the professional committees of the Canadian Society of Civil Engineers and the American Concrete Institute. He is a licensed professional engineer and is actively involved in consulting activities for structural engineering.

Preface to "Performance-Based Design in Structural Fire Engineering"

The performance-based design of structures in fire is gaining growing interest as a rational alternative to the traditionally adopted prescriptive code approach. This interest has led to its introduction in different codes and standards around the world. Although engineers widely use performance-based methods to design structural components in earthquake engineering, the adoption of such methods in fire engineering is still very limited. This Special Issue addresses this shortcoming by providing engineers with the needed knowledge and recent research activities addressing performance-based design in structural fire engineering, including the use of hotspot analysis to estimate the magnitude of risk to people and property in urban areas; simulations of the evacuation of large crowds; and the identification of fire effects on concrete, steel, and special structures.

Maged A. Youssef
Editor

Article

Hotspot Analysis of Structure Fires in Urban Agglomeration: A Case of Nagpur City, India

Priya P. Singh *[id], Chandra S. Sabnani and Vijay S. Kapse

Department of Architecture and Planning, Visvesvaraya National Institute of Technology, Nagpur 440010, India; csabnani@arc.vnit.ac.in (C.S.S.); vskapse@arc.vnit.ac.in (V.S.K.)
* Correspondence: priya.work.singh@gmail.com or priyasingh@students.vnit.ac.in

Abstract: Fire Service is the fundamental civic service to protect citizens from irrecoverable, heavy losses of lives and property. Hotspot analysis of structure fires is essential to estimate people and property at risk. Hotspot analysis for the peak period of last decade, using a GIS-based spatial analyst and statistical techniques through the Kernel Density Estimation (KDE) and Getis-Ord Gi* with Inverse Distance Weighted (IDW) interpolation is performed, revealing fire risk zones at the city ward micro level. Using remote sensing, outputs of hotspot analysis are integrated with the built environment of Land Use Land Cover (LULC) to quantify the accurate built-up areas and population density of identified fire risk zones. KDE delineates 34 wards as hotspots, while Getis-Ord Gi* delineates 17 wards within the KDE hotspot, the central core areas having the highest built-up and population density. A temporal analysis reveals the maximum fires on Thursday during the hot afternoon hours from 12 noon to 5 p.m. The study outputs help decision makers for effective fire prevention and protection by deploying immediate resource allocations and proactive planning reassuring sustainable urban development. Furthermore, updating the requirement of the National Disaster Management Authority (NDMA) to build urban resilient infrastructure in accord with the Smart City Mission.

Keywords: fire incidence; hotspot analysis; KDE; Getis-Ord Gi*; IDW interpolation; fire risk zones; built-up areas; temporal analysis; sustainable development

Citation: Singh, P.P.; Sabnani, C.S.; Kapse, V.S. Hotspot Analysis of Structure Fires in Urban Agglomeration: A Case of Nagpur City, India. *Fire* **2021**, *4*, 38. https://doi.org/10.3390/fire4030038

Academic Editor: Maged A. Youssef

Received: 23 June 2021
Accepted: 19 July 2021
Published: 21 July 2021

1. Introduction

Structure fires are the fires involving the structural components of various types of residential, commercial, educational, or industrial buildings. Structure fires have substantial consequences adversely affecting urban sustainable development threatening life safety, property protection, continuity of operations, environmental protection, and heritage conservation. As per the International Association of Fire and Rescue Services, India has accounted the average fire rate as 1.18 per 1000 inhabitants per year with an average fire death of 1.04 per 100 fires for a period of five years from 2014 to 2018 [1]. The Indian Risk Survey Report 2018 has listed fire as the third of the top five identified risks, with the increased vulnerability causing tremendous losses to physical assets over the last three years, and in 2019, fire risk was on the tenth rank. Hence, fire is of major concern [2]. The National Crime Records Bureau 2019 data of India has accounted for a total of 11,037 accidental fires, with 69% of these fires being in the structures of schools or commercial, residential, and governmental buildings, and a total of 10,915 deaths, with 62% in structure fires, and a total of 441 persons injured, with 78% in structure fires [3]. Hence, it is evident that structure fires have an adverse impact on the sustainability of an urban built environment, affecting and disrupting the urban functionality with heavy losses of property and lives [4]. Therefore, the fire incidence pattern of the urban agglomeration is to be assessed delineating the hotspot area along with statistically significant fire risk areas for effective and efficient mitigation [5].

Urban agglomeration is an inevitable phenomenon in the process of urbanization sheltering nearly 68% of the Earth's population by 2050 and with its center of gravity in the Asian cities of China and India [6,7]. Mckinsey and company have predicted the probability of Indian urban agglomeration with an intense rise in population density accounting for nearly thirteen cities with a population of more than four million and a million-plus population in sixty-eight cities by 2030 [8]. With the rapid urban agglomeration, cities are leading to have compact development and expansion by urban sprawl development with land-use transformations resulting in a multitude of challenges by increased and new fire risks [9–11]. In addition, the increasing demography increases the vulnerability of fire risk, demanding a significant availability of fire service provisions for efficient and reliable fire safety management [12]. As the urban growth develops, the provision of fire service facilities becomes a priority to cope up with the alarming demand for fire safety [13] and has to be strengthened by comprehensive and accurate information for balanced decision making with an emergent response [14]. Therefore, understanding the fire incidence pattern with its severity, particularly structure fires in the context of sustainable urban development, is of great significance for the implementation through systematic risk assessment by mitigating measures [15]. The planning of preparedness of fire service on the basis of risk assessment can improve the emergency response and thus enhance the efficiency of fire service. It is, therefore, recognized as an essential part of fire prevention and signs to assess the fire risk zones, delineating hotspots based on the historical fire incidences to understand the fire incidence pattern at the specific geographical location, as the geographical characteristics vary globally.

Geospatial tools comprising Geographic Information Systems (GIS) and Remote Sensing (RS) are powerful tools to evaluate the spatial fire distribution patterns integrating the temporal data [16] and are widely adopted as an analysis system for urban infrastructures [17]. The spatial and temporal patterns of structure fires are of interest, integrating the potential dimensions of space and time [15–21]. The fire distribution often has a wide variation over space and time, and it is critical to categorize fire distributing under uniform or random patterns with the changing challenges of the urban agglomerating space [22,23]. The fire distribution pattern has a close association with human activities and the surrounding built environment, as well as the demographic and socioeconomic factors [11,24]. Built environments with high population densities reflect high human activities with an increased risk of structure fires [15,20]. Fire risk has been researched in residential fires associated with varied socioeconomic aspects [25–28]. The impact of fire incidences was revealed with high risk to very young children and very old residents of Canada [29]. Structure fire studies have analyzed the various causes of fire incidents, integrating time in months and hours [30]. The temporal data analyses of fire incidences of previous studies in Australia revealed maximum fire incidence frequencies on weekdays and school holidays, establishing the close association between fire incidences and the socioeconomic conditions of the urban areas [20,28], with an increased rate over the pace of time [11]. RS integrated with GIS has many applications in the various fields of weather, forestry, agriculture, surface changes, biodiversity, and many more [30]. In the urban planning context for fire services, RS technologies can be used for detecting land use and land cover (LULC) with active fires (hotspots) determining the physical properties of land with accuracy and precision [11,31,32], quantifying the built-up land for allocation of fire service resources and enhancing the efficiency of emergency responses with sustainability.

Previous studies quantified the fire risk correlating the various aspects of the socioeconomic characteristics of neighborhoods in developed countries at macro-level spatial units such as countries, states, and census tracks [15–18,33]. In South Asia, Indian cities have undergone a rapid decadal transformation of the built environment, changing the urban landscape with social structures accommodating the increasing population and resulting in the urban agglomeration of a developing country [34,35], which are comparatively less researched. The fire incidence pattern in urban areas at the micro-level of urban agglomeration in developed countries is a research topic of great interest and an emergent need as well [24,36]. Research on fire severity quantifying the losses to assess the impacts of fire and identifying the fire risk areas for strategic interventions has become increasingly popular in recent years [37]. Population density as well can be the final output and target for resource allocation.

Hotspot analysis determines the dense concentration of events within a limited geographical area. Numerous statistical models such as descriptive statistics, Poisson regression, binomial regression, and Bayesian network models for hotspot analysis were adopted in varied disciplines, dealing with the randomness of events in space and time [38–40]. Being statistical, these methods do not consider the spatial characteristics of the events. GIS tools have advanced techniques to estimate and quantify hotspots identifying the high concentration of events to detect areas with active fires inferring as high fire risk zones, referred to as a hotspot, represented by cartographic maps for visualization [41]. In addition, Kernel Density Estimation (KDE) and Hotspot Analysis (Getis-Ord Gi*) HA(GOG*) with Inverse Distance Weighted (IDW) interpolation are widely applied in varied disciplines of geography, traffic safety management, and crime [42–44].

In India, a developing country, cities have undergone fundamental transformations in an urban landscape and social structures, and the process of urbanization has increased fire incidences and intensified consequences [2,35]. Fire service is a part of the responsibilities of Urban Local Bodies (ULB) to provide fire safety of urban areas [45] and plays a significant role in the success of all schemes by the Government of India carried out for the betterment of citizen's life and infrastructures integrating sustainable urbanization in the cities such Atal Mission for Rejuvenation and Urban Transformation (AMRUT) [46], Pradhan Mantri Awas Yojna (PMAY) [47], Urban Livelihood Mission, and Heritage City Development and Augmentation Yojna (HRIDAY) [48]. The National Disaster Management Authority (NDMA) has listed above 95% deficiencies in fire services throughout the country in 2012, with updating requirements of later date, which is still awaiting [49]. The spatial accessibility of fire vehicles for emergency response was a major consideration for assessing the deficiencies. Thus, the questions raised for considering the fire incidence pattern of urban areas for updating and strengthening the deficiencies are as follows:

- Are fire incidences evenly distributed?
- Are the fire risk areas identified and quantified?
- Are the fire occurrences analyzed on the temporal scale?
- Are the causes of fire incidences assessed for the identified fire risk areas?
- Are urban and human activities responsible for fire occurrences? How so?

Therefore, to address the research questions, the objective of the study is hotspot analysis delineating the fire risk zones to understand the fire pattern on a spatial and temporal scale with cause-wise analysis on historical fire incidences. The results of the study are to be integrated in the reassessment and restructuring of the fire service building community in order to achieve a resilient and sustainable built environment. The continual reassessment and restructuring of fire service provisions are essential in reducing fire severity in terms of fire deaths, injuries, and property damage [50]. Resource allocations for fire service involve a heavy budget investment and hence a long-term peak period has to be assessed when updating the requirements. The study aims to assess structure fire patterns in the urban agglomeration for a decadal period of historical fire incidences from 2011 to 2020, delineating the hotspot areas with the quantification of the built-up areas and population density under significant fire risk zones for effective and efficient mitigation with proactive planning during the peak period. The study has the potential to inform policy makers of other ULBs of similar cities to reassess and restructure fire services, integrating Smart City Mission, assuring sustainability [51], overlapping with the Sustainable Development Goals (SDG-11) [52] to develop a sustainable city, state, and nation.

2. Materials and Methods

2.1. Study Area

The urban agglomeration of Nagpur city of Maharashtra state is centrally located in India with a zero-mile location at 21°9′ N and 79°5′ E coordinates. The population of the city is 2.45 million as per the 2011 census and is ranked as the third most populated urban centre in the state and thirteenth in the country, with an average population density of 10,873 persons per km², covering an area of 225.08 km², merging two census towns of Narsala and Hudkeshwar [53], and subdivided into 138 wards as shown in Figure 1. Nagpur has a tropical savannah climate (Aw in Koppen climate classification) with dry conditions throughout the year, where summer temperatures intensify up to 47.8 °C, making it the hottest place in India [54] and suitable for the escalation of the fire frequency. The winter temperature declines to 10° to 12 °C and has the average annual rainfall of 1161.54 mm [55].

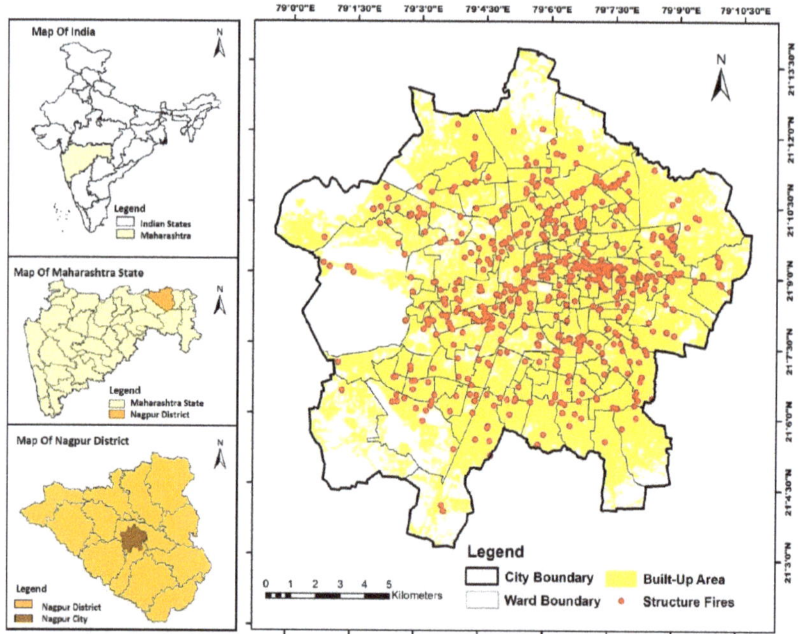

Figure 1. The geographical location of the study area with structure fires across the study period.

Structures refer to the urban functional confined spaces that have a significant impact on human life and daily activities [56]. According to the Census 2011, 594,272 buildings were housed for various purposes covering 52% of the built up land area of the city, expanding to 73% in 2020. Out of these buildings, 82.24% are used as a residence; 2.79% for residence-cum-other use; 8.60% for shops and offices; 0.31% for schools and colleges, 0.26% for hotels, lodges, guest houses, etc.; 0.48% as hospital, dispensary, etc.; 0.90% as factory, workshop, work shed, etc.; 0.58% as a place of worship; and 3.85% as other non-residential use.

Census 2011 has accounted for the structures of the residences and residence-cum-other by proportional building materials listed under roof, walls, and floors. The material of walls comprised a maximum of burnt bricks with 65.5%, followed with concrete—11.2%, Mud/unburnt—10.6, stoned packed—4.3%, stone not packed—3%, grass/thatch—2.2%, wood—1.3%, plastic/polythene—0.5%, and any other—0.2%. Material of roofs comprised a maximum of concrete with 62.6%, followed with galvanized iron/metals/asbestos—12.9%, machine-made 11.8%, hand-made—5.7%, stone/slate—2.5%, grass/thatch—2%, burnt brick—1.1%, plastic/polythene—1%, and any other—0.3%. Material of floors comprised a maximum of cement with 42.5%, followed by mosaic/floor tiles—40.3%, mud—8.6%, stone—6%, burnt brick—1.5%, any other material—0.9%, and wood/bamboo—0.2%.

2.1.1. Data and Sources

Nagpur Municipal Corporation (NMC) administers the urban centre of Nagpur, and the administrative data with regard to ward limits and the population was provided by NMC in kml format. The decadal population growth rate for 2011 was 19.3% and a growth of 20.9% is projected for 2021, 21.2% for 2031, and 20.7% for 2041 [55]. The yearly population size from 2011 to 2020 was procured from World Urbanization Prospects 2018 [57] to analyze on a yearly basis.

Fire incidence data for the period of a decade from 2011 to 2020 was procured from the Fire Service Department of NMC on a yearly, monthly, and daily basis, revealing the rise of fire incidents with the population growth. The maximum incidences were observed in the hottest month of May with a highest z-score value of 2.01 > 1.96 at a 95% confidence level as shown in Figure 2, indicating the impact of climatic conditions on fire incidence frequency. Therefore, the dataset for the hottest month of May with maximum fire occurrences is researched cumulatively from 2011 to 2020 for hotspot analysis of structure fires.

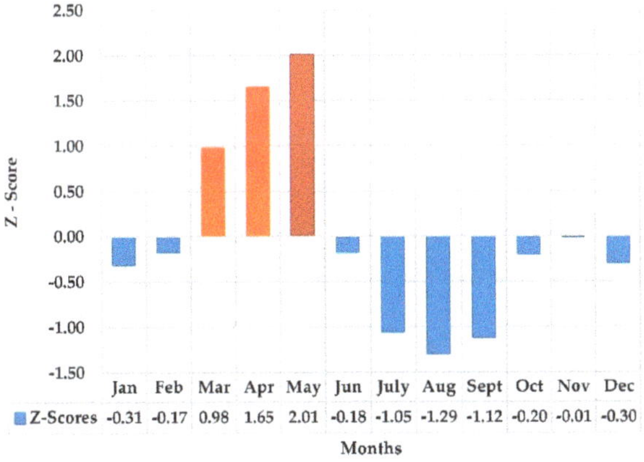

Figure 2. Monthly fire incidences.

Structure fire incident data of the city were considered for the study, with a cumulative count of 570 fire events for the hottest month of May with maximum fire frequency from 2011 to 2020. The daily fire incidence data was in regional language in the daily call register, including the addresses, date, call time, cause of the fire, and brief of occupancy type for each fire event (such as house fire, shop fire, hospital fire, etc., for structure fires). For the study, structure fires are grouped and classified by occupancy type based on the National Building Code of India (NBCI), 2016 [58] and the proportion of each structure fire reveals that the Residential fires have the highest proportion of 48% with next mercantile with 32%, while industrial of 7%, business of 4%, assembly of 3%, storage of 3%, educational of 2%, and institutional of 2%. The material of construction was not mentioned in the register and hence the Census 2011 building material data is considered with the maximum of framed structures with concrete and masonry.

2.1.2. Land Cover Data

Landsat-5 and Landsat-8 satellite images from the USGS Earth Explorer website were procured with a minimum cloud cover of less than 5% (details described in Table 1) for the years 2011 and 2020 to understand the urban expansion with built infrastructure accommodating the increased population. The images were classified using the false-color bands of 7, 6, and 4 to develop the LULC maps of the city, and the land cover was classified as built-up area, vegetation, fallow land, and water body. For analysis, only built-up areas were delineated comprising the structures and other land covers were excluded.

Table 1. Details of Landsat-8 satellite imaginary used in the study.

Landsat Satellite	Sensor	Scene ID	Path/Row	Acquisition Date
Landsat-5	TM	LT51440452011162KHC04	144/45	11 June 2011
Landsat-8	OLI_TIRS	LC81440452020139LGN00	144/45	18 May 2020

2.2. Methods

The hotspot analysis of structure fires is performed with two spatial analyst techniques of KDE and HA(GOG*) with IDW interpolation to arrive at fire risk zones. Administrative data with population density are joined with the ward areas, and furthermore, the results from these two methods are compared to assess the significant fire risk zones. The RS tool is used to acquire the built-up areas to quantify the fire risk zones. Furthermore, population density is estimated for each fire risk zones and ranked to reveal the fire risk due to structure fires in the city for implications. The temporal analysis on yearly, weekdays, and time, along with cause-wise analysis was also performed to reveal the association of urban and human activities with the fire occurrences. The study is performed as represented in Figure 3.

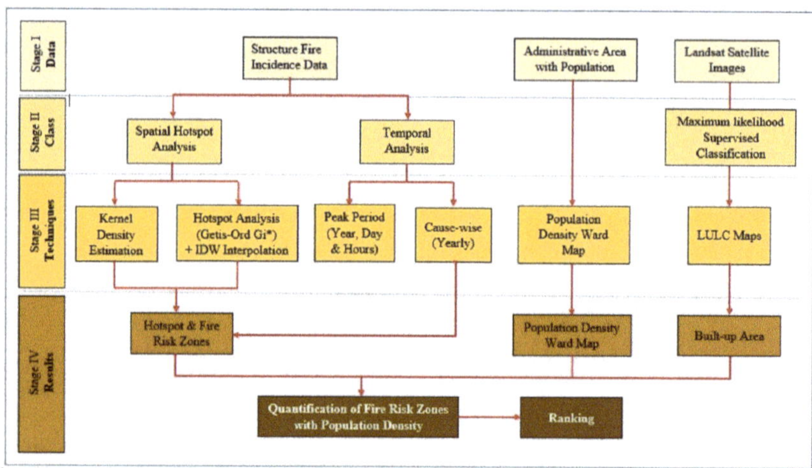

Figure 3. Methodological flow chart of the study.

2.2.1. Data Pre-Processing

The fire incidences were tagged in Google Earth Pro after acquiring the coordinates physically using Epicollect5 and imported in ArcGIS for analysis. Before performing the hotspot analysis, the data are combined using the 'collect event' tool to combine all the events in the same geographical location with the new point feature class of 'ICOUNT' and used as input for both methods.

2.2.2. Kernel Density Estimation (KDE)

The KDE identifies the dense areas based on the total count of the geographical events over time and is helpful to rectify the spatial pattern with classified intensities of density estimate values [56,59]. The KDE technique was adopted in the developed countries to reveal the occurrence of fire foci under different land uses in the State of Amazonas during 2005 [60] and to explore the spatial and temporal dynamics of fire incidents in South Wales, UK [18]. The KDE method statistically represents the spatial smooth continuous surface for intensities of the geographical event points over the space of the study area [61]. The kernel is the circular area of the defined bandwidth radius around each event, indicating the surrounding area under influence with a statistical value indicating the density per unit area, and adding of all the values at all places gives a surface of density estimates. ArcGIS 10.6 is used for evaluation considering the default search radius (bandwidth) based on the spatial configuration and number of input points [30]. The kernel density tool calculates a magnitude per unit area from a point using a kernel function to fit a smoothly tapered surface to each point. The surface value is highest at the location and diminishes at the edge of the surface radius considering the distance decay effect [62]. KDE is performed by the mathematical formula as in Equation (1) [61] as follows:

$$\hat{f}(s,b) = \frac{1}{nb^2} \sum_{i=1}^{n} K\left(\frac{s-s_j}{b}\right) \tag{1}$$

where n = total number of observations; b = smoothing parameter (bandwidth); s = coordinate vector that indicates where the function is being estimated; sj = coordinate vector representing each event point; and K = density function that satisfies the following condition given by Equation (2):

$$\int K(s)ds = 1 \tag{2}$$

2.2.3. Hotspot Analysis (Getis-Ord Gi*)–HA(GOG*)

A hotspot analysis by Gi* statistics was introduced by Getis and Ord for identifying statistically significant spatial clusters of each area at the local level with clusters of high values as Hotspots and low values as cold spots [41,63,64]. The HA(GOG*) technique is widely adopted in varied disciplines of geography, traffic safety management, and crime [56,64,65]. The vulnerable areas associated with high crime rate along with fire were revealed for targeted fire prevention in the city of Surrey, British Columbia, Canada [27]. The G_i^* statistic is a z-score at a local level, calculated by the mathematical formula as expressed in Equation (3) [41,66]. A high positive GiZ score indicates hotspots while negative low GiZ scores indicate cold spots and values near to zero indicates a random distribution of clusters with significance, as follows:

$$G_i^* = \frac{\sum_{j=1}^{n} w_{i,j}x_j - \overline{X}\sum_{j=1}^{n} w_{i,j}}{S\sqrt{\frac{\left[n\sum_{j=1}^{n} w_{i,j}^2 - \left(\sum_{j=1}^{n} w_{i,j}\right)^2\right]}{n-1}}} \tag{3}$$

where x_j is the attribute value for event j, $w_{i,j}$ is the spatial weight between event i and j, n is the total number of events, $\overline{X}$ = mean is calculated by Equation (4), and S = standard deviation is calculated by Equation (5).

$$\overline{X} = \frac{\sum_{j=1}^{n} x_j}{n} \tag{4}$$

$$S = \sqrt{\frac{\sum_{j=1}^{n} x_j^2}{n} - \left(\overline{X}\right)^2} \tag{5}$$

Hotspot analysis technique in ArcGIS is used to conduct a Gi* statistical significance test identifying the clusters with high concentration values surrounded by high concentration values indicating the clusters to be a hotspot and the low concentration values are surrounded by low concentration values indicating the clusters to be a cold spot [67,68].

Identification of hotspots is of particular operational interest, and with this goal in mind, the count itself was an appropriate measure [44,65,69,70].

2.2.4. Inverse Distance Weighted (IDW)

The IDW interpolation method estimates the measured values by hotspot analysis surrounding the prediction location [30]. The measured values closest to the prediction location have more influence on the predicted value than those farther away. IDW assumes that each measured point has a local influence that diminishes with distance. It gives greater weights to points closest to the prediction location, and the weights diminish as a function of distance, hence the name inverse distance weighted [30,42].

The IDW interpolation method is employed for demarcation of significant fire risk areas representing the spatial distribution in the study area. The IDW interpolation output was spatially joined with the census ward level of the city using a zonal spatial analyst tool to estimate the significant fire risk zonal areas and population under the risk areas. Furthermore, the IDW interpolation is used for predicting the high fire risk zonal areas with the population under risk integrating urbanization for implications.

2.2.5. Built-Up Area Estimation from LULC

Land cover has been conducted in ArcGIS 10.6 employing maximum likelihood with an image classification tool to measure the built-up area [71]. The land cover was classified into four major classes of built-up area (structures, roads, and small open spaces), vegetation (trees in forest areas, large open spaces, and wetland vegetations), fallow land (remaining open or unutilized land), and water bodies (lakes and ponds) [72]. An accuracy assessment was conducted for each classified image with ground truth data from Google Earth Pro using overall accuracy (OA) and kappa coefficient (K) as shown in Equations (6) and (7) [73]:

$$OA = \frac{CD}{TP} \times 100 \tag{6}$$

$$K = \frac{P_{(o)} - P_{(e)}}{1 - P_{(e)}} \tag{7}$$

where CD = a total number of reference samples chosen; TP = total number of correctly classified samples; $P_{(o)}$ is the observed proportion of agreement; and $P_{(e)}$ is the proportion expected by chance [74].

The LULC map was used to assess the actual built-up areas for quantification of population density under the risk of GiZ-score fire types. The built-up density is calculated by using Equation (8), and population density by Equation (9) as follows:

$$BUD = \frac{\sum_{i=1}^{n} TB_i}{\sum_{i=1}^{n} TA_i} \tag{8}$$

$$PD = \frac{\sum_{i=1}^{n} TP_i}{\sum_{i=1}^{n} TB_i} \tag{9}$$

where BUD = Built-up Density; TB = Sum of the total built-up area of all wards; TA = sum of the total area of all wards; PD = Population Density; and TP = sum of the total urban population of the wards.

2.2.6. Temporal and Cause-Wise Analysis

Temporal data statistical analysis significantly evaluates the association with the correlated variables responsible for fire incidences through the period of years, months, weekdays, and hourly events [36]. The statistical methodology helps in the interpretation of the parameters and, due to its simple application, is extensively applied at urban scale by the decision makers for econometrics, financial inferences, and planning disciplines [75,76]. Temporal analysis on a yearly basis for the selected month is performed statistically, fire index for the time series of the year is calculated using Equation (10), z-score for each year is calculated by using Equation (11), and the probability evaluation is performed using z-score value to reveal the highest fire incidences. The ratio between fire incidences and population size of the year is evaluated to understand the impact of urban activities with population growth on fire incidences. Furthermore, the fire incidences are segregated on basis of weekdays and hours to arrive at the maximum frequency by circular statistics represented in radar charts. Pearson's correlation is evaluated for the hourly fire frequencies using Equation (12) and the sensitivity analysis is performed by Durbin–Watson (DW) statistic using Equation (13) [77]:

$$FI = \frac{\sum_{i=1}^{n} X}{\overline{X}} \tag{10}$$

$$Z = \frac{x - \mu}{\sigma} \tag{11}$$

$$R = \frac{\sum(x - \overline{x})(y - \overline{y})}{\sqrt{\sum(x - \overline{x})^2 \sum(y - \overline{y})^2}} \tag{12}$$

$$d = \frac{\sum_{t=2}^{T}(e_t - e_{t-1})^2}{\sum_{t=1}^{T} e_t^2} \tag{13}$$

where FI is the fire index for the time series of a year, X is the number of fire incidents for the time series of a year, and $\overline{X}$ is the mean value of the time series of a year. Z = standard score, x = observed value, μ = mean of sample, and σ = standard deviation of the sample. R = Pearson's correlation coefficient and x and y are variables. T = Number of observations, e_t = residual given by et = $\rho e_{t-1} + v_t$, and ρ = null hypothesis.

Human and urban activities causing fire incidences on a yearly temporal scale were investigated to relate with the fire severity [36,78,79]. Cause-wise categorization of the fire incidences is performed, and percentage calculation helps to understand the fire cause pattern due to urban and human activities in the identified fire risk zones for implications.

3. Results

3.1. Hotspot Spatial Analysis

3.1.1. Kernel Density Estimation Result

The KDE analysis is performed by applying a spatial analyst tool. The maps generated are spatially joined with the ward map as represented in Figure 4. The output is classified into four groups at the geometric interval as listed in Table 2. The intense area in red color is the identified hotspot area categorized with a very high fire risk zone, with the highest value comprising 9% of the total area and with the highest population density located in the central core and extending towards the northern part of the city with high rise development. The ranking of population density reveals the very high fire risk zone with a maximum population under threat on the least percentage area, while the high fire risk zone has the lowest population density covering the maximum of 54% of the city area with a maximum of 33% of population size, spreading majorly towards the fringe area of the city with low rise development

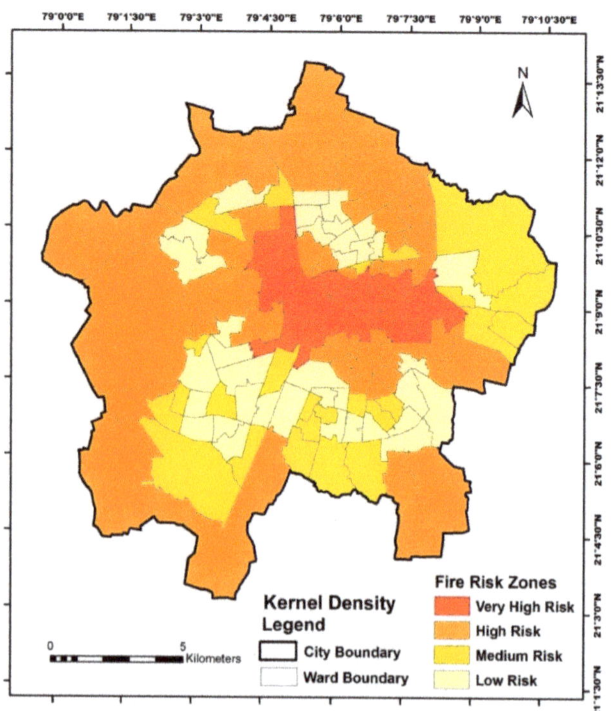

Figure 4. Hotspot from the Kernel Density Estimation.

Table 2. Fire risk zone classification for KDE outputs.

Geometric Interval Values	Fire Risk Zones	Number of Wards	Percentage Area	Percentage Population	Population Density per km^2	Ranking Based on Population Density
3.43–24.48	Very High	34	9	20	29,860	1
−0.50–3.42	High	43	54	33	8339	4
−1.23−−0.51	Medium	25	21	22	14,268	3
−5.17−−1.24	Low	36	17	26	21,078	2

3.1.2. Hotspot Analysis (Getis-Ord Gi*)–HA(GOG*) Result

The local level HA(GOG*) was performed to identify the statistical spatial distribution pattern after the process of collected events in ArcGIS to include all overlapping events, and the evaluated GiZ-score varies from 3.07216 to −1.07437 as represented in Figure 5. The results reveal that the hotspots with higher values of 3.07216 exceed the critical value of 1.96 at a 95% confidence level, indicating clusters with high fire event values are surrounded by high fire event values. The lower value of −1.07437 > −1.96 is within the 95% confidence level, indicating no cold spots in the spatial distribution pattern.

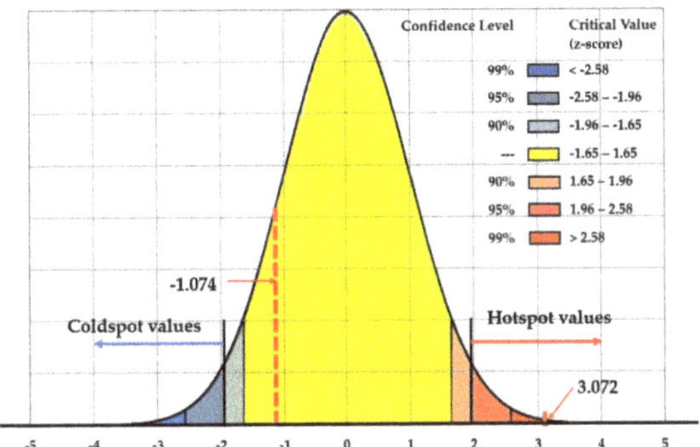

Figure 5. Hotspot (Getis-Ord Gi*) results.

3.1.3. Inverse Distance Weighted (IDW) Interpolation Result

The IDW Interpolation techniques are extensively performed for the generation of HA(GOG*) outputs on spatial dimensions [30,42]. The hotspot analysis results were spatially joined by a zonal statistics tool to the census ward map after IDW interpolation to identify administrative areas for estimation and implications, as represented in Figure 6. Fire disasters are not planned events and any fire incidence can be disastrous with heavy losses. Therefore, every fire event has to be considered with significant risk. Therefore, all GiZ scores at 95% confidence level are categorized under the four fire risk types as listed in Table 3 and reveal the highest population density in the hotspot zones.

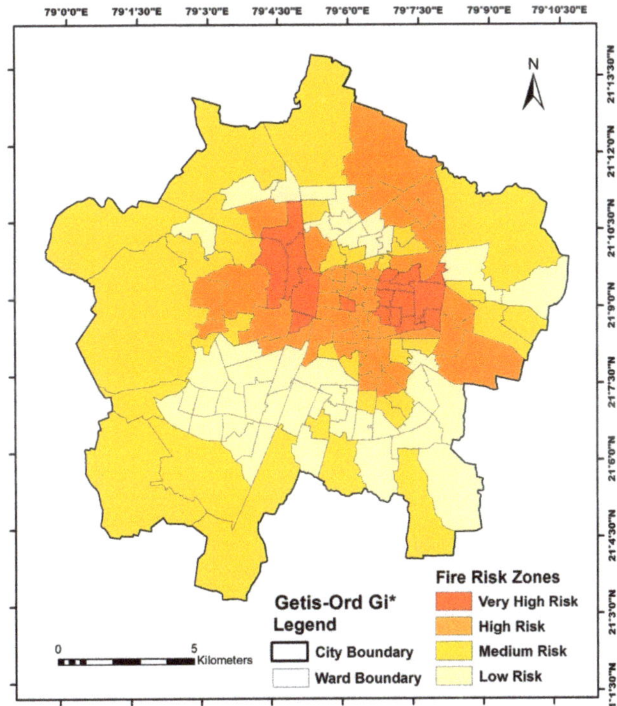

Figure 6. IDW Interpolation of hotspot analysis (Getis-Ord Gi*) joined to the census ward map.

Table 3. Fire risk zone classification for hotspot analysis (Getis-Ord Gi*) of census wards.

Z-Score	Type	Number of Wards	Percentage Area	Percentage Population	Population Density per km^2	Ranking Based on Population Density
>1.96	Very High	17	5	10	25,526	1
1.96 to 0.65	High	43	19	28	20,313	2
0.65 to −0.65	Medium	34	52	29	7655	4
−0.65 to −1.96	Low	44	24	33	18,857	3

3.1.4. Built-Up Area Estimation from LULC

The Landsat Satellite images from the USGS Earth Explorer website were classified through the maximum likelihood supervised classification tool in ArcGIS 10.3 to assess the urban built environment of 2011 and 2020 to then estimate the significant fire risk areas. Figure 7 represents the built-up areas of the city with structure fires from 2011 to 2020, comprising 52% in 2011 and increasing to 73% in 2020, which is an increase of 21%. The accuracy assessment of both the LULC maps has been performed using the ground truth data marked in Figures A2 and A1 and as per the findings listed in Tables A1 and A2. The overall accuracy for both years is 91%, and the Kappa coefficient is 85% in 2011 and 86% in 2020, satisfying the 85% limit for minimum accuracy [74].

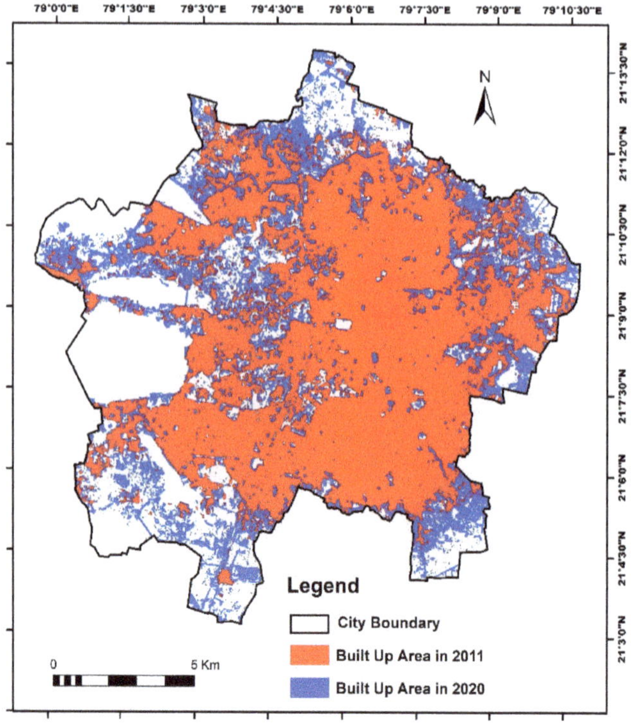

Figure 7. Built-up areas from 2011 to 2020.

The quantification of the built environment with classified fire types is performed, and the results are listed in Table 4. The spatial distribution for visualization is represented in Figure 8, which reveals that the highest percentage of actual built-up areas is in the Medium category with 46%, the highest built-up density percentage is in high fire risk category with 91%, revealing the highest development in the built environment and the highest population density in very high fire risk category, while the lowest are in the low category, indicating the impact of urbanization.

Table 4. Estimation of fire risk zones with significant z-scores.

Z-Score	Type	Actual Built-Up Area Percentage	Built-Up Density Percentage	Population Density per km^2 of Built-Up Area	Ranking Based on Population Density
>1.96	Very High	7	91	27,994	1
1.96 to 0.65	High	22	86	23,659	2
0.65 to −0.65	Medium	42	60	12,750	4
−0.65 to −1.96	Low	29	90	20,904	3
	Total		73	18,508	

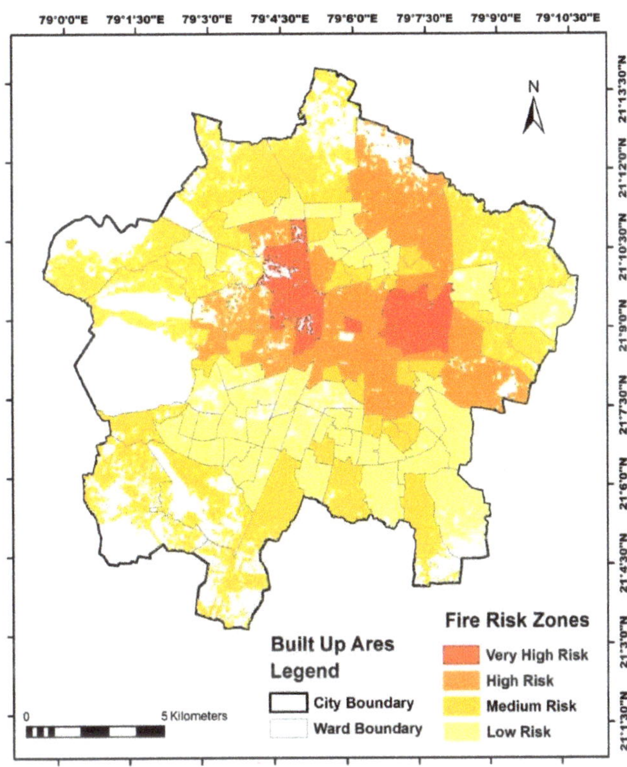

Figure 8. Built-up areas with structure fire risk zones.

Table 5 has the classified structures as per the NBCI, representing the percentage of fires to understand the human activity from occupancy fires. The hotspot area with a very high-risk zone in the central part of the city has a maximum of 39% fires in mercantile occupancy, 35% in residential, 12% in industrial, and 8% in the business occupancy. While other risk zones have maximum fires in residential class following with mercantile class of structures.

Table 5. Percentage of fires based on occupancy types in classified fire risk zones.

Occupancy Type	Very High	High	Medium	Low
Residential	35	43	59	53
Educational	1	1	3	2
Institutional	0	1	1	3
Assembly	0	3	3	5
Business	8	5	3	3
Mercantile	39	34	30	27
Industrial	12	11	1	3
Storage	5	2	1	3

3.1.5. Predictive Probable Fire Risk Evaluation Results

IDW interpolation techniques are extensively performed for the generation of hotspot analysis outputs on spatial dimensions [30,42]. Figure 9 represents the predictive continuous smooth surfaces classified into five different classes based on the quantile classification method at the extreme limits considering the raising population due to rapid urbanization with the predictive decadal population growth rate for 2031 as 21.2% and 2041 as 20.7% [55]. The very high fire risk zone is represented in red color with high values. The results represented the central part of the city extending towards the lower eastern part under predictive very high fire risk covering 19.20% area of the city as shown in Table 6, and the higher three classes cover 58.69% of the city area, indicating proactive planning by the resource allocation of fire services for the future decades.

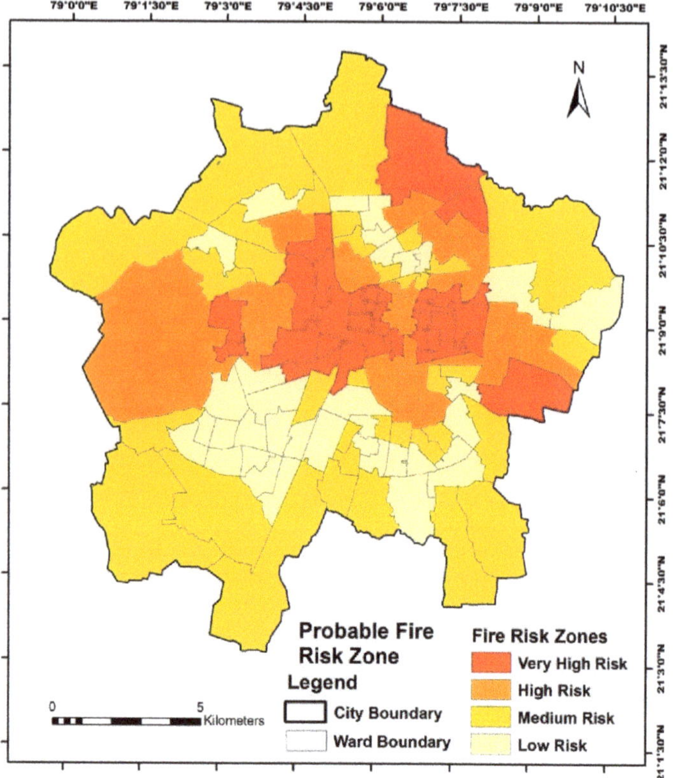

Figure 9. Predictive probable fire risk zones of the city.

Table 6. Predictive probable fire risk zones.

Z-Score	Type	Number of Wards	Percentage Area	Percentage Population 2031	Population Density for 2031 per km^2	Ranking Based on Population Density
>1.96	Very High	34	15	21	22,350	1
1.96 to 0.65	High	34	20	22	18,258	3
0.65 to −0.65	Medium	35	47	30	10,138	4
−0.65 to −1.96	Low	35	18	26	23,726	2

3.2. Temporal Analysis

Fire incidences and the population growth across the study period for the hottest month are evaluated statistically and the output is listed in Table 7 revealing the highest fire incidence frequency in 2019 with the highest Fire Index, z-score, and probability values, while least is in the years 2013 and 2020. The urban activities were restricted in 2020 due to the COVID-19 pandemic lockdown, resulting in a reduced fire incidence frequency. The population growth has the rising trend as per the UN population projection in lacks [57] and the ratio of fire incidences is revealed to be the highest in 2019 and lowest in 2020, as shown in Figure 10. The output reveals the impact of urban and human activities on fire incidences in a rising pattern during the normal conditions, but during the pandemic, the ratio lowers down to the least throughout the decadal period, and furthermore, the trendline of ratio is in a declining pattern with R square value 0.0024.

Table 7. Fire incidences and population matrices.

Year	Fire Incidence Matrices			Population Matrices			Ratio
	Fire Index	Z-Score	Probability	PopulationIndex	Z-Score	Probability	
2011	1.14	0.68	0.75	0.93	−1.54	0.06	2.59
2012	0.93	−0.34	0.37	0.95	−1.21	0.11	2.08
2013	0.70	−1.44	0.08	0.96	−0.88	0.19	1.54
2014	0.88	−0.59	0.28	0.98	−0.54	0.29	1.90
2015	1.04	0.17	0.57	0.99	−0.19	0.42	2.20
2016	0.93	−0.34	0.37	1.01	0.16	0.56	1.95
2017	1.23	1.10	0.86	1.02	0.52	0.70	2.53
2018	1.07	0.34	0.63	1.04	0.88	0.81	2.17
2019	1.39	1.86	0.97	1.06	1.22	0.89	2.77
2020	0.70	−1.44	0.08	1.07	1.58	0.94	1.38

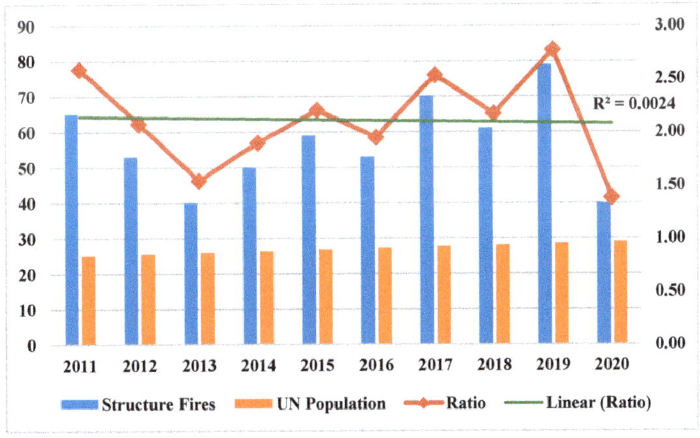

Figure 10. Yearly fire incidences, UN population, and ratio.

Figure 11a depicts the waterfall chart of the evaluated z-score of the fire incidence pattern, as listed in Table 3, revealing the maximum increase in 2019 with a 1.86 z-score and decreasing in 2020 for a pandemic reason. Figure 11b depicts the evaluated proportion of the fire indexes reiterating the fact of maximum rise in 2019 with 39% above the mean value with the trend line of R square value of 0.00292 indicating a very slow but rising pattern.

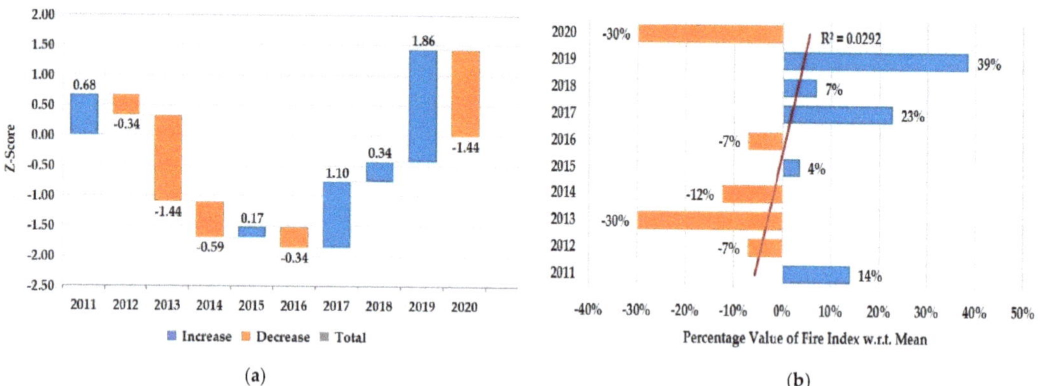

Figure 11. Temporal fire incidence pattern: (**a**) Z-score waterfall chart; (**b**) proportional fire index.

Figure 12 depicts the temporal data of weekdays and hourly fire incidences. The weekdays dataset (as in Figure 12a) has the maximum fire frequencies on Thursday at 17% and a minimum on Monday at 12%, then the weekends. The hourly dataset (as in Figure 12b) has the maximum fire events during the hot afternoon hours from 12:00 to 17:00 with a maximum of 7% indicating the impact of climatic conditions and the human urban activities hours. The late hours from 19:00 to 21:00 correspond to dinner time, with the rise in fires indicating human activities.

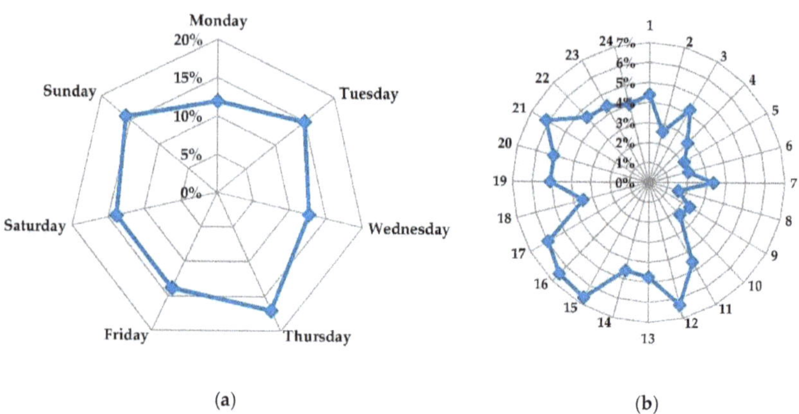

Figure 12. Temporal fire incidence pattern: (**a**) Weekdays; (**b**) Hourly.

The Pearson's correlation between the hourly fire frequencies has the coefficient value of 0.54, indicating a moderate correlation. The sensitivity analysis through autocorrelation test by Durbin–Watson (DW) statistics with the null hypothesis states that fire frequencies have no relation with the hour of the day at which they occur. The DW value ranges from 0 to 4 and the 0 to 2 value indicates a positive correlation and from 2 to 4 indicates negative relation. The DW value nearer to zero indicates strong positive autocorrelation and nearer to 2 indicates no autocorrelation. The DW statistic test evaluated the value of d = 0.35 (nearer to zero) at *p*-value < 0.05, which rejects the null hypothesis indicating the positive autocorrelation between the hour of the day and fire frequencies.

3.3. Cause-Wise Analysis to Understand Urban and Human Activities

Table 8 represents the causes of the fire in the fire risk zones across the study period, classified in five types as gas cylinder leakage (14%), electric short circuits (31%), adjacent garbage fire (2%), other causes (6%) including lamps, cooking, mechanical failure, and electric press, and highest cause being unknown with 48% of total structure fires. All fire risk zones have a maximum of unknown causes followed by electric short circuits and gas cylinder leakages. The electric short circuit cause is generally due to increased load on electricity causing voltage fluctuation due to the operation of electrical cooling devices for comforting the confined spaces during the scorching hot May month, indicating the impact of climatic conditions.

Table 8. Cause-wise percetage of structure fires in the fire risk zones.

Type	Gas Cylinder Leakage	Electric Short Circuit	Garbage Fire	Unknown	Other
Very High	13	33	2	41	11
High	10	30	2	54	3
Medium	16	28	2	51	4
Low	17	31	2	42	8
Total	14	31	2	48	6

Figure 13 depicts the yearly cause-wise fires with the trendlines of the unknown and electric short circuit causes showing a rise with an R^2 value of 0.083 and 0.0333, respectively, indicating a gradual rise, and the gas leakage cause has a declining trendline with an R^2 value of 0.0332. The unknown cause has to be studied in depth for identifying the actual causes for intervention and implications.

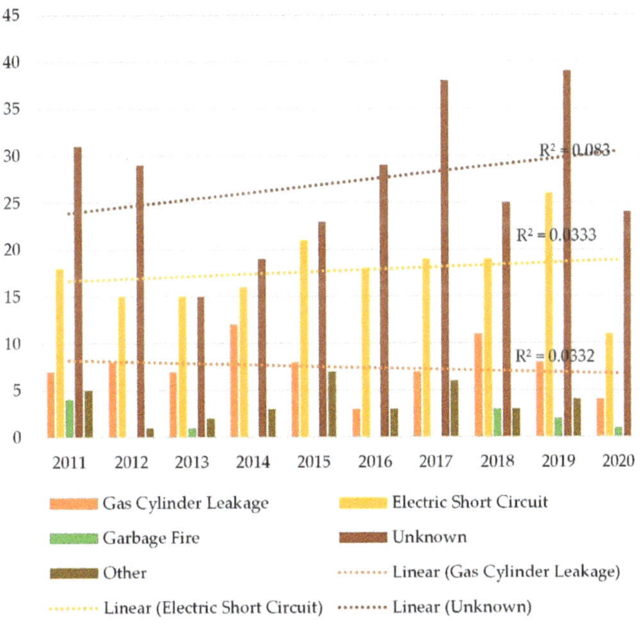

Figure 13. Cause-wise yearly structure fires.

4. Discussion

The geospatial analyses approach connects the data to a spatial visualization map, integrating location data with all types of statistical and descriptive information, providing a foundation for mapping, and an analysis for urban planning, particularly for fire service provision. The quantification of the built-up areas and populations under risk is essential for efficient and effective mitigation. The historical fire incidence data of a decadal period for the hottest month of May with maximum fire frequencies are analyzed with the spatial analyst and statistics tools to identify the spatial variation of fire risk zones across the study area with an increasing population under persistent threat. The fire risk zone hotspot analysis through the KDE and HA(GOG*) is performed, compared, and estimated statistically.

4.1. Result Overview

The detection by KDE identifies one hotspot area through the entire city area, in the central core part of the city under a very high fire risk zone revealing the highest population under persistent threat. The high population size indicates the rapid growth of urbanization due to the prominent commercial and administrative activities associated with increased urban and human activities, resulting in the development of high rises and compact development that challenges the management of fire services. The KDE is a nonparametric technique identifying the probability density estimate of event hotspots, revealing the concentration of the fire events across the study period to identify the influenced areas.

The HA(GOG*) technique evaluates the fire risk zones statistically for the same data set and helps to understand results with statistically significant z-scores in the study. The very high fire risk class is the hotspot area above the critical value of 1.96 at a 95% confidence level, identifying three hotspot areas within the single identified hotspot area of the KDE analysis. The hotspot area comprises a total of 17 wards out of 34 wards of the KDE over an area of 5% out of 9% of the KDE, with a population density of 25,526 persons per km^2 out of 29,860 persons per km^2. The high and medium fire risk classes have wide variations influencing the varied population densities under respective threats. The results from both techniques have wide variation, and the results of HA(GOG*) are more accurate with statistical analysis than the KDE technique, helping decision makers to optimally and sustainably utilize resources.

The built-up areas under the significant fire risk zones are estimated to quantify the actual area and population size under persistent fire threats. The built-up density reveals the development pattern with the influenced population density. The very high fire risk zone has the highest population density at 27,994 persons per km^2, whereas, at the ward level, the population density is 25,526 persons per km^2, indicating the compact development. Meanwhile, the low fire risk zone has the second-highest built-up density, revealing the compactness but comparatively with less population density of the third rank.

Structure fires are most frequent in mercantile occupancy structures followed by residential and industrial within the identified hotspot areas, while residential structure fires are prominent in other fire risk areas followed with the mercantile class. The population density indicates that human activities contribute to the fire incidences, integrating the urbanization trend from the cause-wise analysis with the most common cause being the unknown cause, followed by electric short circuits and gas cylinder leakages. The temporal data analysis reveals that the maximum fire incidences occur on Thursday afternoons during the hot hours rather than the early morning and sleeping night hours. The result indicates the impact of human activities of daytime and hot climatic conditions with increased electrical load for cooling the confined spaces of structures.

4.2. Planning Implications

The geospatial analysis approaches are significantly useful for enhancing the efficiency of the fundamental fire services. The outcomes are helpful to the decision makers for fire disaster management purposes, both in terms of risk estimation [16,80,81] and impacts on the environmental services. The outcomes of the hotspot analysis provide evidence to the decision makers for improving the provision of fire services. The hotspot maps are a useful visualization tool for the policy makers implicating proactive planning with immediate resource allocations in the potential fire risk zones. Proactive planning has the potential to significantly minimize the losses and reduce the economic impact of fire hazards and is a powerful tool for planners attempting to enhance the efficiency of fire service and increase sustainability [82].

The temporal analysis reveals the maximum fire rates in the hot afternoon hours, implicating the resource allocative actions. Awareness programs for fire safety can be implemented regarding human activities to reduce fire incidence rates as assessed from the cause analysis. The fire frequency with an unknown cause has to be controlled by alerting the occupants about towards fire safety through public education. Fire safety behavior for gas cylinder operation is to be conducted repeatedly. Continuous awareness for public participation in prevention and suppression techniques to handle incipient fires through fire safety practices is to be encouraged [83]. Fire prevention and preparedness programs help to reduce the damages, promote the role of understanding in the community, and reduce the adverse impacts of fire in the ecosystem [17]. The policy has to be effectively planned and adopted at regular intervals with maximum participation considering the fire risk zones.

The results of the study can help to enhance the effectiveness of the emergency fire services over the potential spatial dimension with strategic proactive planning and interventions to build community resilience continuing the urban functionality fulfilling all the other fire safety objectives.

4.3. Limitations and Future Scope

The approach adopted for the study is with the data constraint limitations acquired from the administrative authority. The impact of socioeconomic factors on spatial fire incidence patterns can be researched with various interventions. -564. Secondly, the open-source Landsat images used are of 30-m resolution in most bands (4.5 pixels per acre) at a 16-day revisit cycle. The high-resolution images of 5-m have the potential to differentiate the built-up areas comprising buildings and varied sizes of opens spaces which are used as parks, gardens and open grounds enhancing to assess actual structural areas improving the accuracy of the built-up areas and thus is a limitation. In addition, visualizations in immersive virtual reality (VR) provide information in real-time and from a first-person perspective, which can be adopted for conducting future studies in detail at a smaller scale of the zone, ward, or neighborhood level, escalating the potential of fire safety objectives with sustainability. Furthermore, an "environmental approach" for the evaluation of the city is suggested due to the interdependence of various parameters to supplement our analysis and fulfill the fire safety objective of environment protection.

4.4. Linking with Urban Development Schemes and City's Vision

Urban agglomeration integrates the physical, institutional, and socioeconomic infrastructure challenging fire service management [35]. Fire Services are the responsibilities of local bodies and play a significant role in the success of all governmental schemes of India such as AMRUT [46], PMAY [47], HRIDAY [48] carried out for the betterment of citizen's life and infrastructure in the cities for sustainable urbanization. Nagpur City is achieving urban transformation under the Smart City Mission [51], and the second Smart City Proposal has additional convergence of the Solar City Programme, the Safe City Project Crime and Criminal Tracking Network and System (CCTNS) Project, and the National River Conservation Project [84]. The potential Transit-Oriented Development (TOD) due to the Metro Rail project across the city encourages high density mixed development through FSI of four or more along the Metro alignment, resulting in high rise development challenging fire service management. Fire services' responsibility safeguards the functionality of urban centres, enhancing the quality of life and infrastructure. The study has the potential to inform policy makers and ULBs to reassess and restructure fire service targeting resource allocation for vulnerable built environments integrating the Smart Cities Mission [51], and is a significant preparatory tool for updating fire service in the country. This study output acts as a foundational step to achieving the research goals in detail overlapping the Smart City Goals [51], fulfilling the Sustainable Development Goals (SDG-11) [52], and helping urban planners and policy makers to develop a sustainable city, state, and nation.

5. Conclusions

The study investigated the historical structure fire events of the typical urban agglomerating mid-sized Indian city for a decade from 2011 to 2020. The analysis for the peak period of the year in the hottest month of May with the highest fire incidence frequency is evaluated for extreme consideration, revealing the direct impact of climatic conditions on fire incidence frequency affecting urban sustainable development.

The hotspot analysis delineates the areas with the highest vulnerability, along with other significant fire risk areas and revealing the random distribution of fire events, thus alerting urban planners to the need for the provision of appropriate fire services as fire incidences are unplanned events resulting in irrecoverable heavy losses. Fire service is the fundamental civic service to provide fire safety, fulfilling the objectives of protection to citizen's lives, property loss, continuity of operations, environment protection, and heritage conservation, which also overlaps the goals of SDG 11, making cities resilient towards fire hazards developing a sustainable city, state, and nation.

The geoinformation techniques implementing KDE provide the visual concentration of the hotspot area, which is nonparametric, while the HA(GOG*) reveals the statistically significant hotspots delineating vulnerability of fire risk zones, which are within the KDE hotspot. The IDW interpolation of the hotspot analysis reveals the administrative zonal area at the local ward level for proactive fire service planning enhancing the efficiency, reducing the losses of life and property with clear visualization of risk areas. Remote sensing dataset and GIS tool quantified the actual built environment under the significant fire risk zones. The estimation of built-up area density and the population density reveals the urbanization impact with varied human activities in the urban built environment threatening sustainability concentrating in the central core part of the city. The cartographic maps developed through the spatial analysis helps in clear visualization to sense the severity of the structure fires in the city. Thursday of the weekdays had the maximum fire frequencies, and the hourly analysis revealed that the highest fire frequencies occurred during the hot afternoon hours of the day from noon to 5 pm and the hours of the waking human activities up to 10 pm, while night hours had the minimum frequencies. The major cause of electric short circuits, gas cylinder leakages, and unknown causes directly reflect the heavy load on electricity in hot hours for cooling, and gas cylinder leakages in hot hours emphasize the need for users' awareness for minimizing the fire frequencies and hence, the losses.

The spatial statistical technique of GIS is an effective and powerful tool for detecting the significant fire risk zones from the historical dataset in urban planning contexts for proactive resource allocations, and strategic planning for mitigation programs minimizes the losses with the better use of finite resources, reducing the budgetary load on the government as fire service involves high budget investment and management. The outputs can be adopted for enhancing the potentiality of the fire services building community resilience. In addition, the methodology can be standardized for evaluation of similar-sized urban agglomeration of India quantifying the built environment upgrading the NDMA requirement to build urban resilient infrastructure and in accord with Smart City Mission.

Author Contributions: Conceptualization, P.P.S.; methodology, P.P.S.; formal analysis, P.P.S.; validation, P.P.S.; software, P.P.S.; writing—original draft preparation, P.P.S.; writing—review and editing, P.P.S.; supervision, C.S.S. and V.S.K. All authors have read and agreed to the published version of the manuscript.

Funding: This research received no external funding.

Acknowledgments: This research is funded by the scholarship awarded to Priya Singh from the Ministry of Human Resource Development (MHRD), Government of India. The authors are thankful to the Fire Service Department of Nagpur Municipal Corporation for providing the data and information helping to carry out the study. The authors also are thankful to the staff of other departments of Nagpur Municipal Corporation for sharing the data and knowledge regarding the study. Furthermore, the authors are thankful to the anonymous reviewers and editors for providing insightful comments that helped to improve the manuscript.

Conflicts of Interest: The authors declare no conflict of interest.

Appendix A

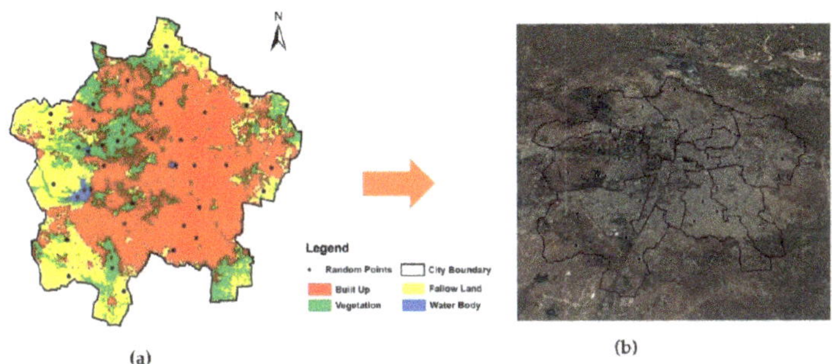

Figure A1. LULC 2011 accuracy assessment random points image in (**a**) GIS (**b**) Google Earth.

Table A1. Accuracy assessment of LULC-2011.

LULC 2011	Built-Up	Vegetation	Fallow Land	Water Bodies	Row Total	User's Accuracy
Built-Up	12	1	1	0	14	86
Vegetation	0	7	1	0	8	88
Fallow Land	0	0	7	0	7	100
Water Bodies	0	0	0	5	5	100
Total	12	8	9	5	34	
Producer's Accuracy	100	88	78	100		
Overall Accuracy	91					
Kappa Coefficient	85					

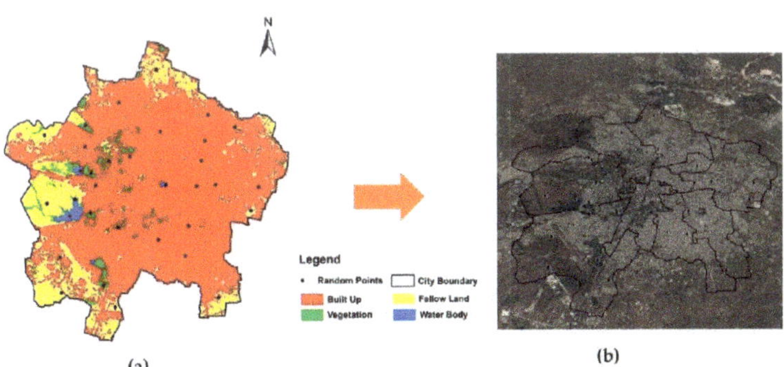

Figure A2. LULC-2020 accuracy assessment random points image in (**a**) GIS (**b**) Google Earth.

Table A2. Accuracy assessment of LULC-2020.

LULC 2011	Built-Up	Vegetation	Fallow Land	Water Bodies	Row Total	User's Accuracy
Built-Up	12	0	1	0	13	92
Vegetation	0	7	1	0	8	88
Fallow Land	1	0	6	0	7	100
Water Bodies	0	0	0	5	5	100
Total	13	7	8	5	33	
Producer's Accuracy	92	100	75	100		
Overall Accuracy	91					
Kappa Coefficient	86					

References

1. Brushlinsky, N.N.; Ahrens, M.; Sokolov, S.V.; Wagner, P. World Fire Statistics, Centre of Fire Statistics 25. 2020. Available online: http://www.ctif.org/sites/default/files/ctif_report23_world_fire_statistics_2018.pdf (accessed on 11 September 2020).
2. Pinkerton and FICCI, India Risk Survey 2019. Available online: https://ficci.in/Sedocument/20487/India-Risk-Survey-2019-ficci.pdf (accessed on 23 April 2019).
3. National Crime Records Bureau 2019. Available online: https://ncrb.gov.in/en/crime-india-2019-0 (accessed on 23 August 2020).
4. Rahardjo, H.A.; Prihanton, M. The most critical issues and challenges of fire safety for building sustainability in Jakarta. *J. Build. Eng.* **2020**, *29*, 101133. [CrossRef]
5. Lee, Y.H.; Kim, M.S.; Lee, J.S. Firefighting in vulnerable areas based on the connection between fire hydrants and fire brigade. *Sustainability* **2021**, *13*, 98. [CrossRef]
6. UNDESA. Department of Economics and Social Affairs. 2017. Available online: https://www.un.org/development/desa/en/news/population/2018-revision-of-world-urbanization-prospects.html (accessed on 11 May 2019).
7. UN-Habitat. World Cities Report 2016, Urbanization and Development: Emerging Futures. Available online: www.unhabitat.org (accessed on 11 May 2019).
8. Mckinsey. India's Urban Awakening: Building Inclusive Cities, Sustaining Economic Growth. 2010. Available online: https://www.mckinsey.com//media/Mckinsey/GlobalThemes/Urbanization/UrbanawakeninginIndia/MGI_Indias_urban_awakening_executive_summary.ashx (accessed on 18 November 2018).
9. Kantakumar, L.N.; Kumar, S.; Schneider, K. SUSM: A scenario-based urban growth simulation model using remote sensing data. *Eur. J. Remote Sens.* **2019**, *52*, 26–41. [CrossRef]
10. Perez, J.; Fusco, G.; Moriconi-Ebrard, F. Identification and quantification of urban space in India: Defining urban macro-structures. *Urban Stud.* **2019**, *56*, 1988–2004. [CrossRef]
11. Zhang, X.; Yao, J.; Sila-Nowicka, K.; Jin, Y. Urban fire dynamics and its association with urban growth: Evidence from Nanjing, China. *ISPRS Int. J. Geo Inf.* **2020**, *9*, 218. [CrossRef]
12. Kiran, K.C.; Corcoran, J. Modelling residential fire incident response times: A spatial analytic approach. *Appl. Geogr.* **2017**, *84*, 64–74. [CrossRef]
13. Yao, J.; Zhang, X. Location optimization of fire stations: Trade-off between accessibility and service coverage. In Proceedings of the 9th International Conference on GIScience Short Paper Proceedings, Montreal, QC, Canada, 27–30 September 2016. [CrossRef]
14. Yao, J.; Zhang, X.; Murray, A.T. Location optimization of urban fire stations: Access and service coverage. *Comput. Environ. Urban Syst.* **2019**, *73*, 184–190. [CrossRef]
15. Špatenková, O.; Virrantaus, K. Discovering Spatio-temporal relationships in the distribution of building fires. *Fire Saf. J.* **2013**, *62*, 49–63. [CrossRef]
16. Masoumi, Z.; van L Genderen, J.; Maleki, J. Fire risk assessment in dense urban areas using information fusion techniques. *ISPRS Int. J. Geo-Inf.* **2019**, *8*, 579. [CrossRef]
17. Raškauskaitė, R.; Grigonis, V. An Approach for the analysis of the accessibility of fire hydrants in urban territories. *ISPRS Int. J. Geo-Inf.* **2019**, *8*, 587. [CrossRef]
18. Corcoran, J.; Higgs, G.; Brunsdon, C.; Ware, A.; Norman, P. The use of spatial analytical techniques to explore patterns of fire incidence: A South Wales case study. *Comput. Environ. Urban Syst.* **2007**, *31*, 623–647. [CrossRef]
19. Gul Guldåker, N.; Hallin, P.O. Spatio-temporal patterns of intentional fires, social stress and socio-economic determinants: A case study of Malmö, Sweden. *Fire Saf. J.* **2014**, *70*, 71–80. [CrossRef]
20. Corcoran, J.; Higgs, G.; Higginson, A. Fire incidence in metropolitan areas: A comparative study of Brisbane (Australia) and Cardiff (United Kingdom). *Appl. Geogr.* **2011**, *31*, 65–75. [CrossRef]
21. Ardianto, R.; Chhetri, P. Modeling spatial-temporal dynamics of urban residential fire risk using a Markov chain technique. *Int. J. Disaster Risk Sci.* **2019**, *10*, 57–73. [CrossRef]

22. Jennings, C.R. Socioeconomic characteristics and their relationship to fire incidence: A review of the literature. *Fire Technol.* **1999**, *35*, 7–34. [CrossRef]
23. Wang, Z.; Zhang, X.; Xu, B. Spatio-temporal features of China's urban fires: An investigation with reference to gross domestic product and humidity. *Sustainability* **2015**, *7*, 9734–9752. [CrossRef]
24. Jennings, C.R. Social and economic characteristics as determinants of residential fire risk in urban neighborhoods: A review of the literature. *Fire Saf. J.* **2013**, *62*, 13–19. [CrossRef]
25. Duncanson, M.; Woodward, A.; Reid, P. Socioeconomic deprivation and fatal unintentional domestic fire incidents in New Zealand 1993–1998. *Fire Saf. J.* **2002**, *37*, 165–179. [CrossRef]
26. Chhetri, P.; Corcoran, J.; Stimson, R.J.; Inbakaran, R. Modelling potential Socio-economic determinants of building fires in southeast Queensland. *Geogr. Res.* **2010**, *48*, 75–85. [CrossRef]
27. Wuschke, K.; Clare, J.; Garis, L. Temporal and geographic clustering of residential structure fires: A theoretical platform for targeted fire prevention. *Fire Saf. J.* **2013**, *62*, 3–12. [CrossRef]
28. Corcoran, J.; Higgs, G.; Brunsdon, C.; Ware, A. The use of comaps to explore the spatial and temporal dynamics of fire incidents: A case study in South Wales, United Kingdom. *Prof. Geogr.* **2007**, *59*, 521–536. [CrossRef]
29. LeBlanc, J.C.; Pless, I.B.; King, W.J.; Bawden, H.; Bernard-Bonnin, A.C.; Klassen, T.; Tenenbein, M. Home safety measures and the risk of unintentional injury among young children: A multicentre case-control study. *CMAJ* **2006**, *175*, 883–887. [CrossRef] [PubMed]
30. Besag, J.; Newell, J. The detection of clusters in rare diseases. *J. R. Stat. Soc. Ser. A Stat. Soc.* **1991**, *154*, 143–155. [CrossRef]
31. Schaefer, A.J.; Magi, B.I. Land-cover dependent relationships between fire and soil moisture. *Fire* **2019**, *2*, 55. [CrossRef]
32. Sequeira, C.R.; Rego, F.C.; Montiel-Molina, C.; Morgan, P. Half-century changes in LULC and fire in two iberian inner mountain areas. *Fire* **2019**, *2*, 45. [CrossRef]
33. Caggiano, M.D.; Hawbaker, T.J.; Gannon, B.M.; Hoffman, C.M. Building loss in WUI disasters: Evaluating the core components of the wildland–Urban interface definition. *Fire* **2020**, *3*, 73. [CrossRef]
34. Dhyani, S.; Lahoti, S.; Khare, S.; Pujari, P.; Verma, P. Ecosystem based disaster risk reduction approaches (EbDRR) as a prerequisite for inclusive urban transformation of Nagpur City, India. *Int. J. Disaster Risk Reduct.* **2018**, *32*, 95–105. [CrossRef]
35. Singh, P.P.; Sabnani, C.S.; Kapse, V.S. Integrating benchmark assessment of emergency fire service using geoinformation technology. *Int. J. Disaster Risk Reduct.* **2021**, *63*, 102432. [CrossRef]
36. Asgary, A.; Ghaffari, A.; Levy, J. Spatial and temporal analyses of structural fire incidents and their causes: A case of Toronto, Canada. *Fire Saf. J.* **2010**, *45*, 44–57. [CrossRef]
37. Han, A.; Qing, S.; Bao, Y.; Na, L.; Bao, Y.; Liu, X.; Zhang, J.; Wang, C. Short-term effects of fire severity on vegetation based on sentinel-2 satellite data. *Sustainability* **2021**, *13*, 432. [CrossRef]
38. Lin, Y.S. Estimations of the probability of fire occurrences in buildings. *Fire Saf. J.* **2005**, *40*, 728–735. [CrossRef]
39. Hanea, D.; Ale, B. Risk of human fatality in building fires: A decision tool using Bayesian networks. *Fire Saf. J.* **2009**, *44*, 704–710. [CrossRef]
40. Wu, J.; Hu, Z.; Chen, J.; Li, Z. Risk assessment of underground subway stations to fire disasters using Bayesian network. *Sustainability* **2018**, *10*, 3810. [CrossRef]
41. ArcGIS Desktop. Arcmap 10.3; An overview of the Mapping Clusters toolset. Available online: https://desktop.arcgis.com/en/arcmap/10.3/tools/spatial-statistics-toolbox/an-overview-of-the-mapping-clusters-toolset.htm (accessed on 11 March 2020).
42. Majumder, R.; Bhunia, G.S.; Patra, P.; Mandal, A.C.; Ghosh, D.; Shit, P.K. Assessment of flood hotspot at a village level using GIS-based spatial statistical techniques. *Arab. J. Geosci.* **2019**, *12*, 1–12. [CrossRef]
43. ZahranEl-Said, M.M.; Jiann, T.S.; Mohamad'Asri, N.A.A.B.; Tan, E.H.A.; Yap, Y.H.; Rahman, E.K.A. Evaluation of various GIS-based methods for the analysis of road traffic accident hotspot. In *MATEC Web of Conferences*; EDP Sciences: Les Ulis, France, 2019; Volume 258, p. 03008. [CrossRef]
44. Andersen, M.A.; Wuschke, K.; Kinney, J.B.; Brantingham, P.; Brantingham, P.J. Cartograms, crime and location quotients. *Crime Patterns Anal.* **2009**, *2*, 31–46.
45. Constitution of India. Twelth Schedule-Arcticle 243 W. 1950. Available online: https://www.constitutionofindia.net/constitution_of_india/article_243w/articles (accessed on 14 August 2020).
46. Atal Mission for Rejuvenation and Urban Transformation. Mission Statement and Guidelines, Ministry of Urban Development, Government of India. Available online: http://amrut.gov.in/upload/uploadfiles/files/AMRUT%20Guidelines%20(1).pdf (accessed on 25 January 2019).
47. Pradhan Mantri Awas Yojna, Housing for All (Urban), Scheme Guidelines, Ministry of Housing and Urban Poverty Alleviation, Government of India, 2015. Available online: https://pmaymis.gov.in/PDF/HFA_Guidelines/hfa_Guidelines.pdf. (accessed on 27 January 2019).
48. Heritage City Development & Augmentation Yojna. Ministry of Urban Development, Government of India. Available online: http://mohua.gov.in/upload/uploadfiles/files/Guidelines%20HRIDAY.pdf (accessed on 25 January 2019).
49. National Disaster Management Authority, Government of India. Available online: https://ndma.gov.in/Response/Fire-Service (accessed on 16 September 2019).
50. Kiran, K.C.; Corcoran, J.; Chhetri, P. Measuring the spatial accessibility to fire stations using enhanced floating catchment method. *Socio Econ. Plan. Sci.* **2020**, *69*, 100673. [CrossRef]

51. Smart City Mission. Transform-nation, Mission Statement and Guidelines, Ministry of Urban Development, Government of India. 2015. Available online: https://smartcities.gov.in (accessed on 11 September 2019).
52. Sustainable Development Goals, SDG-11. Available online: https://www.un.org/sustainabledevelopment/cities/ (accessed on 22 September 2019).
53. Census of India. Census Digital Library. Registrar General & Census Commissioner, India, Ministry of Home affairs, Government of India. 2011. Available online: https://www.censusindia.gov.in/2011-Common/CensusData2011.html (accessed on 22 September 2018).
54. India Meteorological Department (IMD); Ministry of Earth Sciences; Climate Application and User Interface; Climatology of Smart Cities, Climate Smart City-Nagpur (Sonegaon). Available online: https://imdpune.gov.in/caui/smartcities.html (accessed on 23 March 2021).
55. Nagpur Municipal Corporation. Available online: https://www.nmcnagpur.gov.in/assets/250/2018/10/.../Final_CDP_Nagpur_Mar_15.pdf (accessed on 11 August 2019).
56. Zhang, X.; Yao, J.; Sila-Nowicka, K. Exploring spatiotemporal dynamics of urban fires: A case of Nanjing, China. *ISPRS Int. J. Geo Inf.* **2018**, *7*, 7. [CrossRef]
57. World Urbanization Prospects 2018-United Nations Population Estimates and Projections of Major Urban Agglomerations. Available online: https://worldpopulationreview.com/world-cities/nagpur-population. (accessed on 22 March 2019).
58. *National Building Code of India (NBCI) 2016, Bureau of Indian Standards, The National Standards Body of India*; Government of India: New Delhi, India, 2016.
59. Hart, T.; Zandbergen, P. Kernel density estimation and hotspot mapping. *Polic. Int. J. Police Strategy Manag.* **2014**, *37*, 305–323. [CrossRef]
60. Barbosa, M.L.F.; Delgado, R.C.; Teodoro, P.E.; Pereira, M.G.; Correia, T.P.; de Mendonça, B.A.F.; de Ávila Rodrigues, R. Occurrence of fire foci under different land uses in the State of Amazonas during the 2005 drought. *Environ. Dev. Sustain.* **2019**, *21*, 2707–2720. [CrossRef]
61. Colak, H.E.; Memisoglu, T.; Erbas, Y.S.; Bediroglu, S. Hot spot analysis based on network spatial weights to determine spatial statistics of traffic accidents in Rize, Turkey. *Arab. J. Geosci.* **2018**, *11*, 1–11. [CrossRef]
62. ESRI. Available online: https://desktop.arcgis.com/en/arcmap/10.3/tools/3d-analyst-toolbox/how-idw-works.htm#:~{}:text=Inverse%20distance%20weighted%20(IDW)%20interpolation,of%20a%20locationally%20dependent%20variable (accessed on 30 April 2021).
63. Silverman, B.W. *Density Estimation for Statistics and Data Analysis*; Routledge: London, UK, 2018. [CrossRef]
64. Ord, J.K.; Getis, A. Local spatial autocorrelation statistics: Distributional issues and an application. *Geogr. Anal.* **1995**, *27*, 286–306. [CrossRef]
65. Srikanth, L.; Srikanth, I. A case study on kernel density estimation and hotspot analysis methods in traffic safety management. In Proceedings of the 2020 International Conference on COMmunication Systems & NETworkS (COMSNETS), IEEE, Bangalore, India, 7–11 January 2020; pp. 99–104. [CrossRef]
66. Lippi, C.A.; Stewart-Ibarra, A.M.; Romero, M.; Lowe, R.; Mahon, R.; Van Meerbeeck, C.J.; Rollock, L.; Hilaire, M.G.S.; Trotman, A.R.; Holligan, D.; et al. Spatiotemporal tools for emerging and endemic disease hotspots in small areas: An analysis of dengue and chikungunya in Barbados, 2013–2016. *Am. J. Trop. Med. Hyg.* **2020**, *103*, 149–156. [CrossRef]
67. Smith, M.; Goodchild, M.F.; Longley, P.A. Geoespatial Analisys. A Comprehensive Guide to Principles Techniques and Software Tools. Available online: https://www.spatialanalysisonline.com/extractv6.pdf (accessed on 12 May 2021).
68. ESRI. How Hotspot Analysis (Getis-Ord Gi*) Works. Available online: https://pro.arcgis.com/en/pro-app/latest/tool-reference/spatial-statistics/h-how-hot-spot-analysis-getis-ord-gi-spatial-stati.htm (accessed on 11 February 2020).
69. Simpson, S. Resource allocation by measures of relative social need in geographical areas: The relevance of the signed $\chi2$, the percentage, and the raw count. *Environ. Plan. A* **1996**, *28*, 537–554. [CrossRef]
70. Prasannakumar, V.; Vijith, H.; Charutha, R.; Geetha, N. Spatio-temporal clustering of road accidents: GIS-based analysis and assessment. *Procedia Soc. Behav. Sci.* **2011**, *21*, 317–325. [CrossRef]
71. Montiel-Molina, C.; Vilar, L.; Romão-Sequeira, C.; Karlsson, O.; Galiana-Martín, L.; Madrazo-García de Lomana, G.; Palacios-Estremera, M.T. Have historical land use/land cover changes triggered a fire regime shift in central Spain? *Fire* **2019**, *2*, 44. [CrossRef]
72. Chettry, V.; Surawar, M. Assessment of urban sprawl characteristics in Indian cities using remote sensing: Case studies of Patna, Ranchi, and Srinagar. *Environ. Dev. Sustain.* **2021**, *23*, 11913–11935. [CrossRef]
73. Patel, N.; Mukherjee, R. Extraction of impervious features from spectral indices using artificial neural network. *Arab. J. Geosci.* **2015**, *8*, 3729–3741. [CrossRef]
74. Minta, M.; Kibret, K.; Thorne, P.; Nigussie, T.; Nigatu, L. Land use and land cover dynamics in Dendi-Jeldu hilly-mountainous areas in the central Ethiopian highlands. *Geoderma* **2018**, *314*, 27–36. [CrossRef]
75. Korhonen, P.; Moskowitz, H.; Wallenius, J. Multiple criteria decision support-A review. *Eur. J. Oper. Res.* **1992**, *63*, 361–375. [CrossRef]
76. Krackhardt, D. Predicting with networks: Nonparametric multiple regression analysis of dyadic data. *Soc. Netw.* **1988**, *10*, 359–381. [CrossRef]
77. White, K.J. The Durbin-Watson test for autocorrelation in nonlinear models. *Rev. Econ. Stat.* **1992**, 370–373. [CrossRef]

78. Lee, J.H.; Chun, W.Y.; Choi, J.H. Weighting the attributes of human-related activities for fire safety measures in historic villages. *Sustainability* **2021**, *13*, 3236. [CrossRef]
79. Ricotta, C.; Bajocco, S.; Guglietta, D.; Conedera, M. Assessing the influence of roads on fire ignition: Does land cover matter? *Fire* **2018**, *1*, 24. [CrossRef]
80. Nyimbili, P.H.; Erden, T. A hybrid approach integrating entropy-AHP and GIS for suitability assessment of urban emergency facilities. *ISPRS Int. J. Geo Inf.* **2020**, *9*, 419. [CrossRef]
81. Vani, M.; Prasad, P.R.C. Assessment of spatio-temporal changes in land use and land cover, urban sprawl, and land surface temperature in and around Vijayawada city, India. *Environ. Dev. Sustain.* **2020**, *22*, 3079–3095. [CrossRef]
82. Syphard, A.D.; Rustigian-Romsos, H.; Keeley, J.E. Multiple-scale relationships between vegetation, the wildland–urban interface, and structure loss to wildfire in California. *Fire* **2021**, *4*, 12. [CrossRef]
83. Ebenehi, I.Y.; Mohamed, S.; Sarpin, N.; Masrom, M.A.N.; Zainal, R.; Azmi, M.M. The management of building fire safety towards the sustainability of Malaysian public universities. In Proceedings of the IOP Conference Series: Materials Science and Engineering, Birmingham, UK, 13–15 October 2017; Volume 271, p. 012034. [CrossRef]
84. The Smart City Challenge, Stage 2, Smart City Proposal, Nagpur. Available online: http://smartcities.gov.in/upload/uploadfiles/files/Nagpur_SCP.pdf (accessed on 25 January 2020).

 *fire*

Article

Simulation of Evacuation from Stadiums and Entertainment Arenas of Different Epochs on the Example of the Roman Colosseum and the Gazprom Arena

Marina Gravit [1], Ekaterina Kirik [2], Egor Savchenko [3], Tatiana Vitova [2] and Daria Shabunina [1,*]

[1] Peter the Great St. Petersburg Polytechnic University, 195251 St. Petersburg, Russia; marina.gravit@mail.ru
[2] Institute of Computational Modelling of the Siberian Branch of the Russian Academy of Sciences, 660036 Krasnoyarsk, Russia; kirik@icm.krasn.ru (E.K.); vitova@icm.krasn.ru (T.V.)
[3] Design and Research Institute of Air Transport Lenaeroproekt, 198095 St. Petersburg, Russia; e.savchenko@fakt-group.ru
* Correspondence: shabunina.de@edu.spbstu.ru

Abstract: Space-planning decisions of two sports and entertainment arenas with large crowds—the Roman Colosseum (Italy) and the modern Gazprom Arena stadium (St. Petersburg, Russia)— were analyzed to compare the flow of people during evacuation by simulation. It was shown that the space-planning decisions of the Colosseum seem more advantageous compared with the Gazprom Arena in calculation of evacuation time and evacuation organization process: the capacity of the stairs of the Colosseum with a width of 2.8 m is comparable with the capacity of the Gazprom Arena's stairs (4 m). In the Colosseum the average specific flow is $q_{average} = 1.14$ person/s/m, while in the Gazprom Arena the average specific flow is $q_{average} = 0.65$ (with a march width of 2.6 m) and $q_{average} = 0.8$ person/s/m (with a march width of 4 m). It was found that the Colosseum complies with current standards for on-time evacuation; while modern sports and entertainment arenas are currently designed with additional services, infrastructure, comfort and, in general, high commercialization. The antique arenas are currently being reborn and used for concerts and other public events, so the obtained results have practical significance.

Keywords: design; stadiums and arenas; evacuation time; safety; Colosseum; organizing evacuation; computer simulation

Citation: Gravit, M.; Kirik, E.; Savchenko, E.; Vitova, T.; Shabunina, D. Simulation of Evacuation from Stadiums and Entertainment Arenas of Different Epochs on the Example of the Roman Colosseum and the Gazprom Arena. *Fire* **2022**, *5*, 20. https://doi.org/10.3390/fire5010020

Academic Editor: Alistair M. S. Smith

Received: 27 December 2021
Accepted: 29 January 2022
Published: 1 February 2022

Publisher's Note: MDPI stays neutral with regard to jurisdictional claims in published maps and institutional affiliations.

1. Introduction

Sports and entertainment stadiums with a large number of people are high-risk facilities. A source of hazard is the simultaneous presence of thousands of people in them. The greatest danger is posed by the operating conditions with the simultaneous targeted pedestrian movement, including the stadium outflow after events and the emergency evacuation, e.g., during a fire case. An important role belongs to the space-planning decisions of the structure: the size, configuration, and number of evacuation routes to leave the stands and the building in relation to the arena's capacity.

Computer simulation is widely used to analyze the infrastructure during public events and the operation of space-planning decisions [1,2]. For example, in Ronchi et al. [3], three scenarios of the evacuation of music festival locations with a capacity of 65 thousand people were explored. Simulations of pedestrian movement in the stands are considered in Was et al. [4], Wagner et al. [5], and Zhang et al. [6]. Simulation of the evacuation from the Wuhan Sports Center Stadium (one of the largest gymnasiums in China) was considered in Zong et al. [7]. In Wei et al. [8], the simulation technology of fire spread and evacuation in a large stadium was studied. In Kirik et al. [9,10], the authors presented the effects of different stadium features on evacuation times and densities, which were found using simulation.

Computer simulation provide numerical results for analyzing the object, verifying various hypotheses and obtaining reliable conclusions based on the simulation. Many works have been published aimed at the accurate reproduction of cultural heritage objects using digital technology. For example, [11] describes a digital 3D reconstruction of Sinhaya, a X–XIIth century Muslim suburb in the city of Zaragoza, as a result of which its exact models are obtained. The visualization is based on archaeological evidence from excavations and accurate historical documents. Digital reconstruction has helped to preserve some of the city's cultural heritage. In Papagiannakis et al. [12], a digital visualization of the 16th century Mosque of Hagia Sophia is presented in order to introduce virtual cultural heritage objects into an educational and recreational program. In Heigeas et al. [13], a modeling process is presented to produce a realistic crowd simulation in the ancient Greek agora of Argos. This paper considers the movement of crowds submitting to a common flow in a constrained environment. In Cain et al. [14], a study aimed to create a real-time interactive scenario in the ancient Roman Odeon in Aphrodisias based on historical sources is described. The results of the work present the development of the main scenarios of crowd movement.

Buildings with mass gatherings are not only the heritage of the contemporary world but similar arenas were also built in ancient times. The Roman Colosseum, which is the most famous structure of antiquity and was commissioned in 80 A.D., was built for gladiatorial games, mock naval battles, animal hunts, and the execution of criminals. The Colosseum is the largest amphitheater ever built, with an estimated capacity of 40,000 to 50,000 people [15,16]. The Colosseum was built of travertine stone, tuff, and brick, with marble as a facing material [17]. These materials are not combustible, but there was a fuel load in the building: on the upper tier, there were wooden masts and yards with sunshades on them; at the bottom (basement, under-stand galleries), there were wooden cages for animals, hay, fabrics, stretchers, baskets, etc. An open fire was used for lighting.

In Tan et al. [18] and Hernández [19], a goal was to reconstruct the Colosseum building using a computer model, and in Napolitano et al. [20], a model was created. A computer simulation of masonry in the stone structures of the Colosseum was used. In Croci [21], the weakness of the building concerning earthquakes is outlined. The influence of the space-planning decisions of the Colosseum on the evacuation time is partially considered in Gravit et al. [22]. According to [23], the Colosseum has such space-planning decisions that it is possible to fill and leave the amphitheater within a few minutes. It is estimated that due to the efficiency of the stairs, a full audience is able to leave the Colosseum in three minutes, which is disputed by the authors in [24]. This paper presents a digital reconstruction of the Colosseum to simulate crowd movement, which results in the identification of potential bottlenecks preventing rapid (timely) evacuation. As an effective evacuation scenario for the Colosseum, in [25], a comparison was made with one of the stadiums of modern times, the Beijing National Stadium ("Bird's Nest") built for the 2008 Olympic Games, on the TV show Time Scanners (on the National Geographic Channel). The experiment focuses on the ability of both stadiums to evacuate visitors in the shortest possible time: 1/8th of the Colosseum and the Bird's Nest Stadium were created to reproduce the stadium bowls, corridors, and stairs within seating. The experiment was conducted with two control measurements: full evacuation of people from the stands and full evacuation from the stadiums. According to the results of the first part of the experiment, it took 4 min for spectators to leave the stands in Beijing Stadium, while in the Colosseum during this time people were still in the stands, which means that the design of the exits and stands in the Bird's Nest Stadium is better in evacuation compared to the Colosseum. In the second part of the experiment, as the flow continues, the crowd density in the ancient amphitheater begins to decrease over time due to the configuration and width of the stairs, whereas in the modern stadium the flow begins to slow down and accumulate due to the integrated infrastructure. As a result, the last person left the Colosseum in 12 min 44 s and the last person left the Bird's Nest Stadium in 12 min 57 s. Thus, studying ancient objects and comparing them with modern objects is an actual task.

The purpose of this study was to simulate the space-planning decisions of two sports and entertainment arenas of different epochs: the Roman Colosseum (Italy) and Gazprom Arena (Russia) for a comparative analysis of the organization of pedestrian evacuation, with regard to the geometric characteristics of the stairs affecting the carrying capacity. The following tasks are set to achieve this purpose: to analyze the space-planning decisions of the considered arenas and on their basis to develop 3D models; to calculate and compare the movement of people on the stairs; to determine evacuation time, fields intensity of movement, and density of people.

2. Materials and Methods

2.1. Evacuation Modelling

In case of fire, the facility's smoke protection system plays a decisive role in ensuring safe evacuation conditions. The safe conditions are currently defined by the inequality (1):

$$t_{evac} < \alpha \, t_{block} \tag{1}$$

where t_{evac} is the time of the end of evacuation from the building area, t_{block} is the time of reaching the critical value by any dangerous fire factors, and $0 < a < 1$ is a safety factor (for example, it equals 0.8 in Russia) [26].

The quantitative characteristics were obtained using the computer simulation of the movement of people (evacuation) in the Sigma FS (Russia) software package for the advanced fire and evacuation simulation [27,28]. The software was used to check the designs and organize pedestrian areas for the 2018 FIFA World Cup and the 29th Winter Universiade athletics facilities [9,29].

An individual flow model was built to simulate the evacuation. The model suggests the calculation of each person's position, including the positions of other people and obstacles on the plane, and allows one to specify individual characteristics, including the free movement velocity, projected area, path, and movement start time. The individual flow model is best suited for simulating the pedestrian traffic on facilities with stands.

At each time instant t, the position of each person is determined by the previous coordinate by the formula (2):

$$\vec{x}_i(t) = \vec{x}_i(t - \Delta t) + \vec{v}_i(t)\Delta t, \; i = \overline{1,N}, \tag{2}$$

where $\vec{x}_i(t - \Delta t)$ denotes the particle's position at the previous time step; $\vec{v}_i(t), i = \overline{1,N}$ is the particle's current speed measured in [m/s]; and Δt is a time shift equal to 0.25 s.

A person's speed depends on density [30–32]. It is assumed that only conditions in front of the person influence on speed. It is motivated by the front-line effect (that is well pronounced while flow moves in open boundary conditions) in a dense mass of people, which results in the diffusion of the flow.

Thus, only density $F_i(\hat{\alpha})$ in the direction chosen is required to determine the speed. According to [30,33] the current velocity of the particle may be calculated, for example, by formula (3):

$$v_i(t) = \begin{cases} v_i^0 \left(1 - a_l \ln \frac{F_i(\hat{\alpha})}{F^0}\right), & F_i(\hat{\alpha}) > F^0; \\ v_i^0, & F_i(\hat{\alpha}) \le F^0, \end{cases} \tag{3}$$

where F^0 is the limit people density until which free people movement is possible (density does not influence on the speed of people movement); a_l is the factor of people adaptation to current density while moving on l^{th} kind way ($a_1 = 0.295$ is for horizontal way; $a_2 = 0.4$, for downstairs; $a_3 = 0.305$, for upstairs).

An individual flow model was built using the Sigma FS software to simulate the evacuation. The following individual characteristics of people were used in the calculation:

1. The average maximum velocity of a person's free movement was taken to be 1.66 m/s [33];

2. The fundamental diagram of the relation between the velocity and the current flow density was borrowed from [33] (the assumption that this diagram is fully justified for the Colosseum is based on the analysed data in terms of the limiting flow rate and dynamics);

3. The person horizontal projection area used was 0.1 m^2 [33]. The differences in gender, age, health status, and other indicators were ignored.

Simulation of the movement of each individual and the phenomena peculiar to the flow of people: merger, reshaping (spreading, compaction), the non-simultaneous merging of flows, formation and deformation of congestions, flow around turns, and movement in rooms with a developed internal layout, counter-flows, and intersecting flows are performed.

2.2. Description of the Arena Designs—Gazprom Arena Stadium

Gazprom Arena (Russia) is the most visited stadium in Eastern Europe, commissioned at the end of 2016 and hosting the 2018 FIFA World Cup and the 2020 UEFA European Football Championship [34].

According to the technical specifications of the building, a stadium bowl is designed for 68,000 seats, including temporary stands, which can be installed on the third- and sixth-floor stylobates. When the field is involved, the stadium capacity in the concert regime is increased to 80,000 people. The bowl consists of two (lower and upper) tiers. The height difference between the lower tier rows is almost 12 m. There are exits (safety hatches) to the second floor and to the inner stylobate located on the third floor (the attitude is +14.550). The lower bowl is almost symmetrical relative to the minor axis of the field. The height difference between the upper tier rows is almost 20 m. There are exits (safety hatches) to the fifth (+25,200) and sixth (+32,850) floors. The upper bowl can be considered symmetrical with respect to both axes. Figure 1 shows a north-eastern view of the Gazprom Arena and a 3D model of the Gazprom Arena (north-eastern view), built with Sigma FS software.

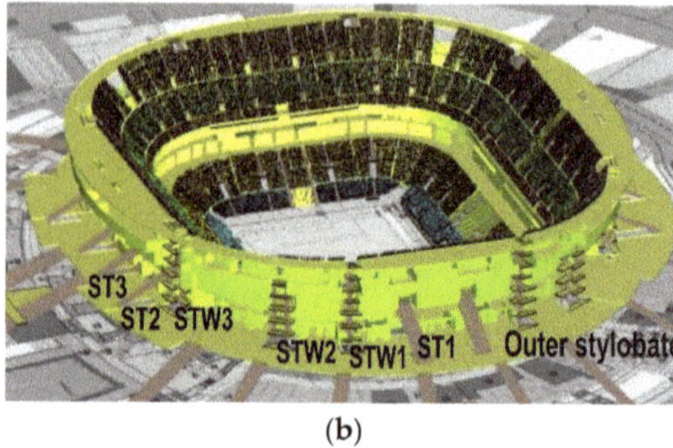

(a)   (b)

Figure 1. (**a**) Northeast side view of the Gazprom Arena stadium; (**b**) 3D model of the Gazprom Arena stadium (north-eastern view) built in the Sigma FS software.

The emergency exits from the building for the lower bowl audience are located mainly on the third floor (only the eastern-sector audience can exit outside directly from the second floor below the third-floor outer stylobate). The exit outside from the upper bowl is also located at the third-floor level. For this purpose, there are stairs accessed from the fifth and sixth floors. The audience members go out to the third floor outer stylobate from the stairs outside. There are 12 such access stairs along the stadium perimeter. In Figure 1, there are marching staircases STW with a number corresponding to the north-eastern quarter of the arena and running from the sixth floor. In addition, straight (no marches) stairs ST with a number are available to descend from the fifth floor directly to the third floor of the stylobate. The audience members descend from the third-floor stylobate to the grade.

In this study, the evacuation of the Gazprom Arena was considered from the upper bowl of the investigated quadrant. We assumed that the exit from the upper bowl would be the exit to the stylobate, located on the third floor, due to the space-planning similarity and comparable capacity of the Gazprom Arena and the Colosseum. Figure 2a shows the plan of the upper bowl of the north-eastern part of the Gazprom Arena, specifying the number of people in the stands. The numbers of people going to the fifth and sixth floors are shown, and the stairs that can be used to descend are indicated (STW1 and STW3). Figure 2b shows a plan of the fifth-floor under-stand space. Stairs accessible from the fifth floor to the third floor (STW1, STW2, and STW3) and straight descents directly to the third floor outer stylobate (ST1, ST2, and ST3) are marked. The arrows show the directions of movement from the hatches to the nearest exits from the floor; the numbers of people for whom the corresponding exit is the nearest one are indicated (the total number of people is 4454). The stairs are distributed around the fifth floor fairly uniformly. In this case, the loads on the adjacent stairs differ by a factor of up to 2. The stairs-to-sector ratio is 6/9. Figure 2c presents a plan of the sixth-floor under-stand space. The stairs accessible for descending from the sixth to third floor (STW1, STW2, and STW3) are shown. The arrows show the directions of movement from the hatches to the nearest exits from the floor; the numbers of people for whom the corresponding exit is the nearest one are indicated (the total number of people is 4692). The analysis of the sixth-floor plan shows that the number of stairs in it is twice as small as on the fifth floor, while the number of audience members on the former is greater. The stairs-to-sector ratio is 3/9. The stairs are nonuniformly distributed relative to the hatches, the loads on the stairs differ by a factor of more than 2, and the minimum load is twice as high as that on the fifth floor.

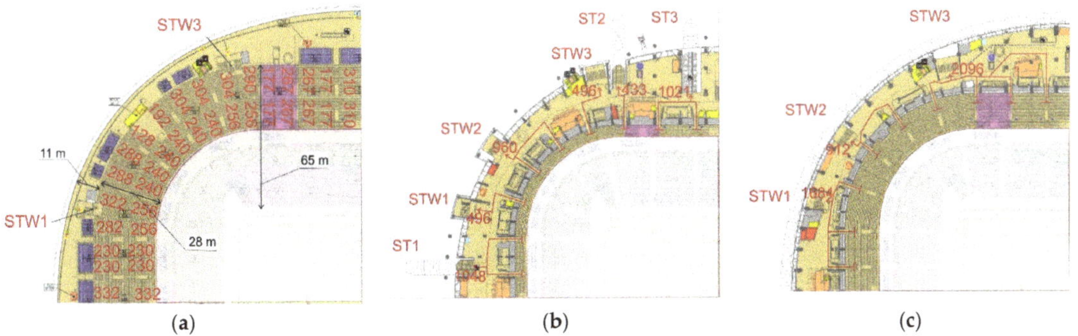

Figure 2. (**a**) Plan of the upper bowl of the north-eastern part of the Gazprom Arena and the number of people in the stands; (**b**) plan of the north-eastern part of the Gazprom Arena fifth floor; (**c**) plan of the north-eastern part of the Gazprom Arena sixth floor.

For further analysis, only stairs STW1, STW2, and STW3 are considered, since they are used by people descending from two (fifth and sixth) floors. In addition, the design of stairs STW1 is significantly different from that of stairs STW2 and STW3 (Figure 3). The quantitative data are given in Table 1.

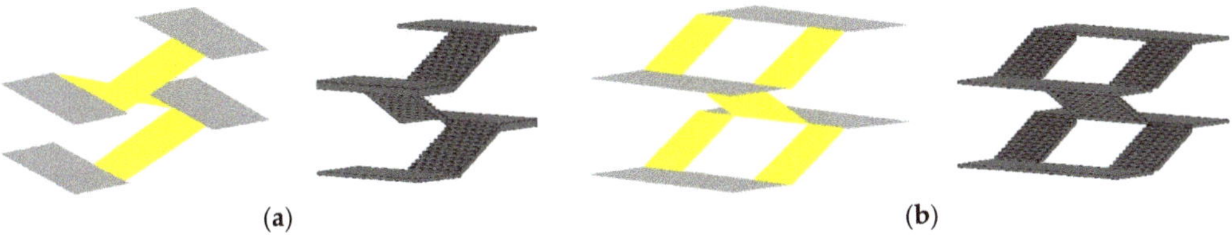

Figure 3. (**a**) Fragment of the stairs STW1 with a stair flight of 2.6 m; (**b**) fragment of the stairs STW2 and STW3 with a stair flight of 4 m (2 + 2 m).

Table 1. Loads on stairs STW1, STW2, and STW3 and their geometric dimensions.

Name of Stairs	Fifth Floor, Persons	Sixth Floor, Persons	Total, Persons	Minimum Width, m	I, Person/m of Width	Stair Length along the Axis of Movement, m
STW1	496	1684	2180	2.6	838.5	63
STW2	960	912	1872	4.0	468	63
STW3	496	2096	2592	4.0	648	63
Total, persons	1952	4692	6644			

The minimum path width for stairs STW1 is 1.5 times less than for the stairs STW2 and STW3, although the number of people evacuating on the stairs STW1 (2180) is comparable to the number evacuating on the stairs STW2 (1872) and STW3 (2592). The ratio between the discharge values for these stairs is the same. Calculating the stairs loading according to the nearest stairs principle, it is clear that the staircase with the lowest discharge value (STW1) on the sixth floor has an almost maximum load: the stairs take half of the northern part of the sixth floor. At the same time, the adjacent stairs STW2 with a discharge value greater by a factor of 1.5 are only accessed for two sectors located directly on the corner. The load on stairs STW1 on the fifth floor is reduced by the presence of exit ST1.

2.3. Description of the Arena Designs—Colosseum

There has still been no consensus among historians and architects about an antique amphitheater's design features and appearance. The characteristics that are important for the study and included in the three-dimensional computer model of the building to simulate evacuation and analyze the results obtained are considered. The computer model of the Colosseum is based on Durm's structural scheme [16]. During the simulation, the main attention is paid to the under-stand space, and the stairs for descending from the upper tiers since this part of the building affects the evacuation time the most.

The Colosseum central part is an oval stage surrounded by a flat strip of seats; the ratio between the major and minor axes of the entire building is 1.22. An oval cone with seats is around the arena. It is based on 80 parting walls directed radially inward and interconnected by ring walls and arched rows. Between them, there are a corresponding number of radially directed crossings and staircases; ring galleries stretching along the entire amphitheater between the ring walls and arcades connect walkways and stairs. The exterior galleries of the second and third floors serve as lounges. The gallery height on the floors is 10–11 m.

There are 80 arches along the outer perimeter that form 80 amphitheater entryways (Figure 4). The entrances/exits are located at the ground level (the so-called datum). Therefore, the evacuation can only occur top-down.

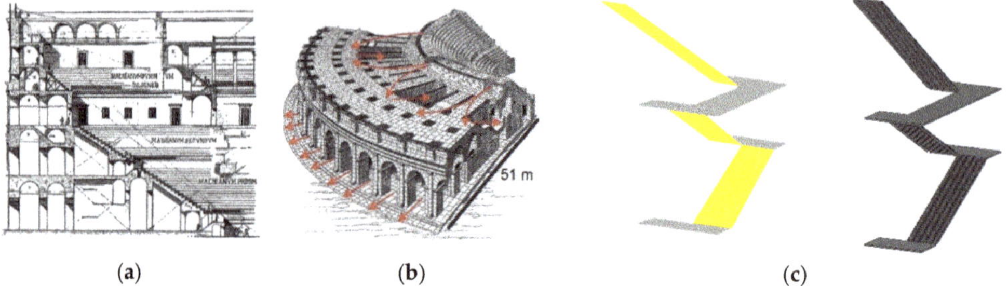

(a) (b) (c)

Figure 4. (a) Schematic of the Colosseum architectural design according to Durm's representations [16]; (b) schematic view of the Colosseum second floor (Gyuade) [16]; (c) fragment of the stairs.

The amphitheater can be conventionally divided into three tiers, each containing under-stand galleries, stands, and walkways to the seats (Figure 5).

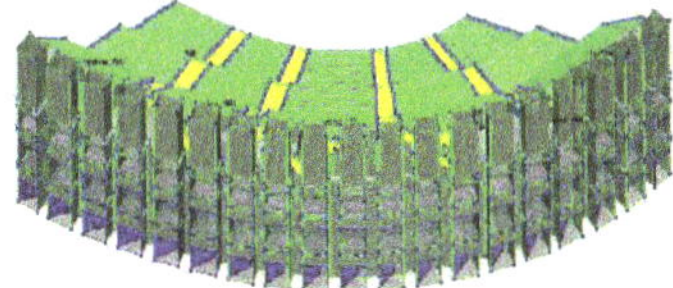  

Figure 5. Sigma FS software 3D models of the Colosseum.

The model was built assuming that the access to the ground-tier stands was mainly through the second floor; to the second-tier stands, through the third floor; and to the third-tier stands, through the fourth floor (the attitude of the latter is about +40,000). The first two tiers represent sequences of 20 stand rows, and the upper tier contains 16 rows. The data on the maximum arena capacity reported by different authors are inconsistent and vary between 40 and 50 thousand audience members simultaneously [16], so the conventional number of people is 48,000.

The Colosseum has the line-of-sight downstairs on both sides of each exit to the under-stand gallery (Figure 4b). The simulation considered 1/4 of the Colosseum (calculation sector), where 4 stairs are taken to evacuate people, which are located in this sector. The extreme stairs on two opposite sides of the calculation sector take the remainder of the flow for each subsequent sector. The stairs are uniformly distributed along the floor perimeter. The number of stairs is consistent with the number of exits to the under-stand space, i.e., it is equal to the number of tier sectors. The stairs path width along the axis of movement ranges from 2 m for descending from the upper tier of stands to the third floor to 4.5 m in the lower part.

2.4. Initial Data for the Evacuation Simulation

To compare the two arenas, a quarter of the Colosseum and a quarter of the Gazprom Arena's upper bowl are considered. This is justified by the symmetry of the Colosseum and the Gazprom Arena upper bowl with respect to both axes; in addition, the buildings have comparable capacities (12,000 and 9500 people, respectively), and the only way to evacuate is down the stairs.

Figures 6 and 7 show the 3D models built for the arenas. The Gazprom Arena computer model was built using modern drawings. The entire stadium was modeled and used not only within the limits of this study.

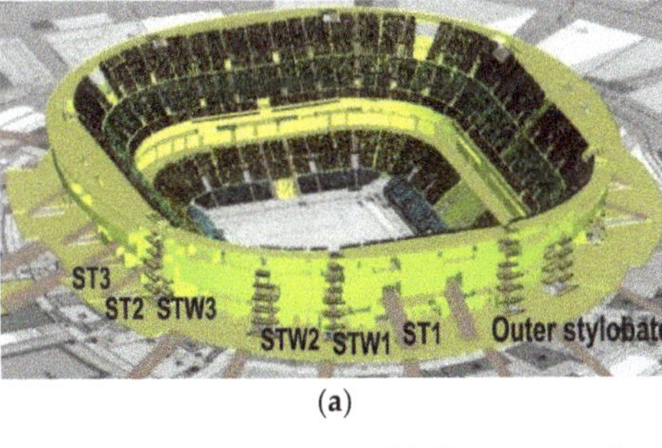

(a)

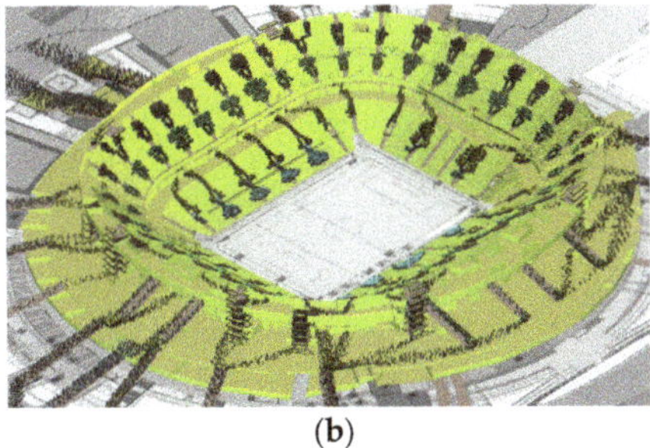

(b)

Figure 6. (**a**) The position of people in the stands of the Gazprom Arena before the evacuation; (**b**) the position of people during evacuation from the Gazprom Arena at the hundredth second from its beginning.

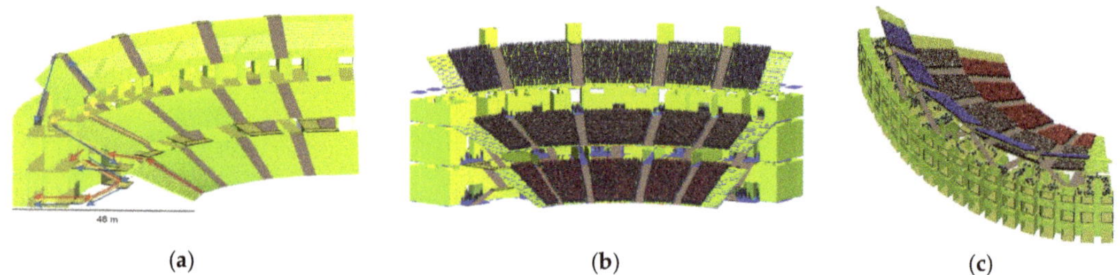

(a)   (b)   (c)

Figure 7. (a) View from the side of the Colosseum; (b) view from the center of the arena; (c) view from the front of the building.

The Colosseum computer model is based on Durm's structural scheme [16]. The attention was mainly paid to the under-stand space, and the stairs for descending from the upper tiers since this part of the building affects the evacuation time the most. The arrangement of the stairs for descending from the upper tiers is approximately the same around the perimeter of the arena, so, when building the computer model, the approximate length and width of each unit path down from the upper floors and the number of paths (stairs) are provided.

Many geometrical dimensions of the interior space of the Colosseum were taken at a scale relative to the known dimensions given in the drawings. The descriptions provide limited data on the configuration of the stairs used to descend from the upper tier to the third floor. However, it is known that people from the upper tier merged into the streams of people from the corresponding sectors of the lower tier. Therefore, the stairs for descending from the upper-tier were conditionally restored to ensure the descent of a number of persons significant for further consideration in the general flow to the third floor. Each sector of the stands on each tier has a staircase for descending from the sector to the underlying floor, where people use the nearest stairs to descend further. The model includes 5 sectors. They are secured by 5 access staircases. In order to exclude boundary effects, the dynamics of human movement in the central part was analyzed, i.e., in the three central sectors and the four central staircases. For the same reason, the extreme sectors in the model are only half-filled (Figure 7b).

The computational domain involved the stands, under-stand galleries, and stairs. At the initial instant of time, people were in the stands or in the under-stand space. The evacuation of people from the building was simulated before exiting outside at the first-floor level for the Colosseum and before exiting beyond the exterior perimeter to the stylobate for the Gazprom Arena.

3. Results and Discussion

3.1. Comparative Analysis of the Arenas Using the Numerical Simulation of Human Movement

Figure 8 shows a fragment of the Colosseum evacuation at the hundredth second from its beginning and mass gathering intensity field on the Colosseum third floor. Figure 9 shows the fields intensity of movement and crowding.

There were 7 calculations (scenarios) for the Gazprom Arena with different staircase loads STW1 and STW2-3. The last two scenarios (6 and 7) are proposed in the absence of flow control on the fifth and sixth floors with an uneven distribution of stairs, which is explained by the use of certain sectors for the needs of different client groups. The data on number N of the persons who passed the stairs, spent time t, and flow rate Q, determined by formula (4), are given in Table 2.

$$Q = N/t \tag{4}$$

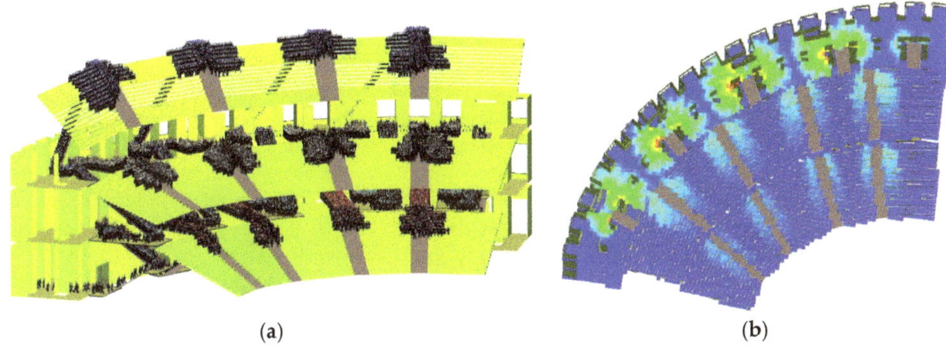

(a)   **(b)**

Figure 8. (**a**) Fragment of the Colosseum evacuation at the hundredth second from its beginning; (**b**) mass gathering intensity field on the Colosseum third floor.

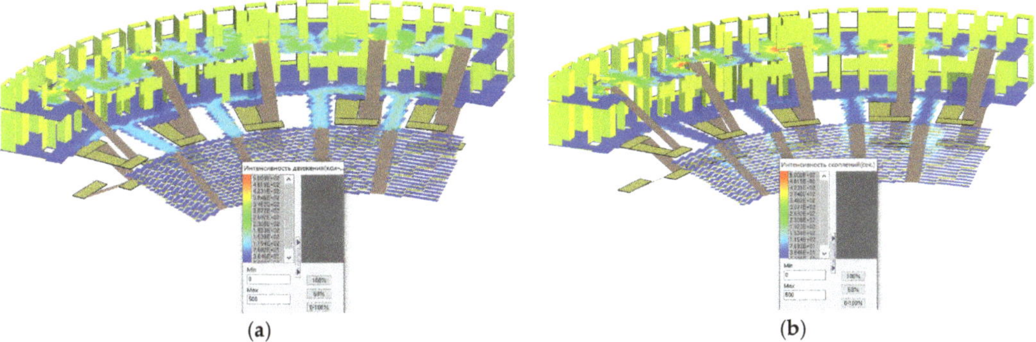

(a)   **(b)**

Figure 9. (**a**) Field of total traffic intensity in seconds on the second and third floors of the Colosseum; (**b**) intensity field of crowding in seconds on the second and third floors of the Colosseum.

Table 2. Numerical characteristics of the Colosseum and Gazprom Arena.

	Gazprom Arena. the Height Difference Is 18.3 m						Colosseum. the Height Difference Is 22 m		
	STW1, the Width Is 2.6 m			STW2-3, the Width Is 4 m			the Width Is 2.8 m		
	N	t, s	Q, Person/s	N	t, s	Q, Person/s	N	t, s	Q, Person/s
1	2	3	4	5	6	7	8	9	10
1	1680	990	1.7	1810	520	3.5	2150	705	3.1
2	1800	1075	1.7	1970	600	3.3	2405	760	3.2
3	2030	1175	1.7	1980	640	3.1	2480	740	3.4
4	2380	1400	1.7	2010	660	3.0	2720	840	3.2
5	2850	1570	1.8	2150	640	3.4			
6	3580	2080	1.7	2380	725	3.3			
7	3670	2025	1.8	4410	1290	3.4			
8	Mean		1.7	Mean		3.3	Mean		3.2

According to Table 2, rows 1–4 do not account the remaining number of evacuees in the Colosseum, which are on the extreme staircases on two opposite sides of the calculation sector.

The data are given in columns 4 and 7 confirm the expected difference (by a factor of about 2) between the flow intensity estimates for stairs STW1 and STW2-3 because of the similar difference between the path widths. At similar numbers of persons, the evacuation time for stairs STW1 is twice as long as for stairs STW2-3.

It is worth noting that the capacity of the Colosseum stairs is comparable with that of stairs STW2-3 in the Gazprom Arena. Meanwhile, the staircase width in the Colosseum is smaller by a factor of $\sim$1.5. The construction of the stairs causes this effect. In the Colosseum, the height difference between the third and first floors is 22 m; in the Gazprom Arena, the height difference between the investigated sixth and third floors is 18.3 m. These values can be considered similar. The structure of the Gazprom Arena stairs was accurately reconstructed in the computer model. The main important features are that all the stairs connecting the upper floors are outside the bowl. There are eight 180° turns between the sixth and third floors (a stair flight has an average height difference of 2.1 m and an average slope of 30°; the flight widths are given in Table 1).

The evacuation time for the considered part of the Gazprom Arena ranges from 520 to 2080 seconds and depends on the load of the stairs and can be regulated by the organisation of the human flow. The evacuation time from the Colosseum is 14.5 min, taken as the sum of the maximum time to leave the stairs of the sector (840 s) and the additional time to exit from the structure (30 s).

In order to assess the results obtained for the Colosseum, it should be noted that the interior space (in particular the staircases) has been reconstructed approximately. However, the space-planning decisions of the Colosseum floors, which is still accessible for research, and the data on the under-stand space structure and the axes lengths in the plan allow to consider the geometry of the Colosseum vertical connections used in the model to be sufficient for this study. In particular, the descent from the third to second floor was reconstructed as straight (without turns, its length is 21.5 m); it occupies the under-stand space of the second tier. The stair flights going down from the floors are codirected; to reach the next flight, one needs to make two 180° turns. There is one 180° turn between the second and first floors, and there are three turns to make in total when descending from the third and first floors; the average flight slope is 30°.

Table 3 generalizes the numerical characteristics of the investigated stairs for the two arenas. The Colosseum stairs are characterized by the highest specific flow (column 5). With conditionally the same length, slope, and height difference parameters, this fact is ensured by the layout of the Colosseum stairs, specifically, by the number of turns (column 6), which is twice as small as in the Gazprom Arena stadium. The result obtained is consistent with the data of a full-scale experiment [35], in which the movement downstairs in a nine-storied building was examined; there were 180° turns on the stairs, and the specific flow decreased with a decrease in the floor (and an increase in the number of turns).

Table 3. Summary table with the numerical characteristics of the Colosseum and Gazprom Arena stairs.

	Stairs	Width, m	$Q_{av\prime}$ Person/s	$q_{av\prime}$ Person/s/m	Number of 180° Turns	Height Difference, m	Length, m	Slope, Deg
1	2	3	4	5	6	7	8	9
1	Gazprom Arena, STW1	2.6	1.7	0.65	8	18.3	63	30
2	Gazprom Arena, STW2-3	4	3.3	0.8	8	18.3	63	30
3	Colosseum	2.8	3.2	1.14	3	22	63	30

Thus, under other conditions, which can be assumed to be the same or slightly different for the investigated arenas, a key characteristic that determines the building evacuation rate was found to be the geometric characteristic of the stairs determining the number of 180° turns. The relation between the specific flow and the number of turns is nonlinear. In addition, as can be seen from rows 1 and 2 of column 5, the configuration of the stairs (Figure 3) also affects the flow rate. Table 4 shows the main geometric characteristics of the Colosseum and Gazprom Arena stairs.

Table 4. Summary table with the numerical characteristics of the Colosseum and Gazprom Arena stairs.

	Characteristic	Colosseum	Gazprom Arena
1	Minimum downstairs flow rate, person/m/s	1.14	0.65
2	Number of 180° turns per stairs	3	8
3	Average mass gathering time, s	360	900
4	Evacuation control (routing) to balance the load on the stairs and reduce the time of mass gathering in front of the stairs	not required	required
5	Stage-by-stage evacuation	not required	required
6	Fencing the escape routes from the main space	no	yes
7	Protection against the dangerous fire factors	Stairs configuration ensuring the high velocity of movement	Fenced-off staircases protected from the spread of the dangerous fire factors
8	Free path to the adjacent hatch along the stand	yes	no
9	Availability of a staircase for each stand (stairs/stand)	1/1	2/3 (fifth floor); 1/3 (sixth floor)

3.2. Discussion

The most reliable smoke protection methods are the use of optimal space-planning decisions of buildings and structures.

The Colosseum is an open structure, where, in case of fire, there are almost no obstacles for spreading the dangerous fire factors, including, first of all, smoke, in the under-stand space; therefore, the speed of evacuation from the building is a decisive factor. The high velocity of movement of people from the upper tiers is ensured by the escape routes maximally straightened using the optimal configuration of the stairs and providing each stand with its own downstairs and own exit from the building. In the Colosseum, the people gathering places with a density of 6 [person/m^2] and higher are the exits to the downstairs on the third floor (Figure 8b), since the capacity of these stairs is lower than the intensity of flows from the second and third tiers. Therefore, the time of mass gathering on the third floor can be minimized by the phased evacuation.

In the Gazprom Arena, the under-stand space is fenced off from the environment (in contrast to the bowl, which, in general, can be considered open). The smoke protection by design is implemented via walling off the staircases and making them smoke-free. The availability of downstairs in the Gazprom Arena upper bowl ranges within 1/3–2/3 on different floors. This leads to the discrepancy between the intensities of the suitable flow and the discharge values of the doors on the staircase and causes the long-term (up to 900 s) mass gathering (Table 4). The problem can be solved by organizing the phased evacuation. To enhance the efficiency of using the vertical lines, it is necessary to control the human flows on the fifth floor in order to relieve stairs STW1-likewise, which take a significant load in the south and north sectors of the sixth floor.

4. Conclusions

Currently, ancient arenas are being reborn: They are used for concerts and other public events, so research on the calculation of evacuation times from such structures is relevant and meaningful. In addition, the Colosseum is the prototype of most modern sports facilities in the present (Fisht Stadium (Sochi, Russia) and the Bird's Nest Stadium (Beijing, China)).

The evacuation process from Colosseum (Italy, Rome) and the Gazprom Arena (Russia, St. Petersburg) is investigated using pedestrian dynamics simulation. The effect of the design of evacuation paths on evacuation time(s) is studied, and the need to optimally organize evacuation (assist in loading stairs) is found.

According to results of investigation, the Colosseum design seems advantageous over the Gazprom Arena. The most significant difference is the higher stability and weak need of the evacuation process in the control factors. The key issue is the uniform distribution of vertical communication ways around the perimeter of the arena, the balance of the capacity of the escape routes and the intensity of the flow, which is also achieved due to the geometric features of the escape routes—the straighter the path, the higher the speed of movement.

The greatest intensity of human flows in the Colosseum is recorded on the third floor, because spectators are flocking here from the two tiers (second and third). There are also the longest crowds (the average duration is 200–250 s).

In the Colosseum, the high speed of movement of people from the tiers is realized by maximally straightened evacuation routes (staircase configuration and provision of each tribune with its own staircase). The stairwells at the Gazprom Arena are walled off and separated from the general volume of the stadium bowl, in particular, from the under-stands premises. The availability of staircases for the upper tier stands at the Gazprom Arena varies between 1/3 and 2/3 of the floors. The key characteristic determining the building's evacuation rate is the number of 180° turns.

According to the simulation results, the evacuation from the upper bowl of the Gazprom Arena to the stylobate of the 3rd floor ranges from 9 to 35 minutes. The evacuation depends on the location and load of the stairs, which is uneven and can be regulated by organising the flow of people. The evacuation from the Colosseum is 14.5 minutes, as the stairs are designed to be evenly loaded and symmetrically arranged. When the flow is organised appropriately in the Gazprom Arena, the structures have similar evacuation times. With an average march width of 2.8 m, the average specific flow $q_{average} = 1.14$ person/s/m (in the Colosseum), and 0.65 and 0.8 person/s/m (in the Gazprom Arena on the STW1 and STW2-3 types of stairs, respectively).

The Colosseum is designed with large, long staircases using the principle of Vomitoria, which means eruption. This study proved the effectiveness of the stairs used in the Colosseum. In the construction of a structure in order to ensure the shortest possible evacuation time, this solution is the most effective. According to this study, the Colosseum complies with current standards for timely evacuation and can be operated as a modern sports and entertainment facility and host public events.

The main difference between modern sports and entertainment arenas is that they are designed with additional services, infrastructure, comfort and, in general, high commercialization, which has an impact on evacuation times and requires additional resources for the application of organizational management of the flow of the people.

Author Contributions: Conceptualization, M.G.; software, E.K.; investigation, T.V.; formal analysis, E.S.; data curation, D.S. All authors have read and agreed to the published version of the manuscript.

Funding: The research is partially funded by the Ministry of Science and Higher Education of the Russian Federation under the strategic academic leadership program "Priority 2030" (Agreement 075-15-2021-1333 dated 30 September 2021).

Institutional Review Board Statement: Not applicable.

Informed Consent Statement: Not applicable.

Acknowledgments: The authors would like to thank Nikolai Ivanovich Vatin, Peter the Great St. Petersburg Polytechnic University, St. Petersburg, Russia, for valuable and profound comments.

Conflicts of Interest: The authors declare no conflict of interest.

References

1. Kuligowski, E.D. Computer Evacuation Models for Buildings. In *SFPE Handbook of Fire Protection Engineering*; Springer: New York, NY, USA, 2016; pp. 2152–2180.
2. Schadschneider, A.; Klingsch, W.; Klüpfel, H.; Kretz, T.; Rogsch, C.; Seyfried, A. Evacuation Dynamics: Empirical Results, Modeling and Applications. In *Extreme Environmental Events*; Springer: New York, NY, USA, 2011; pp. 517–550.
3. Ronchi, E.; Uriz, F.N.; Criel, X.; Reilly, P. Modelling Large-Scale Evacuation of Music Festivals. *Case Stud. Fire Saf.* **2016**, *5*, 11–19. [CrossRef]

4. Was, J.; Lubaś, R. Towards Realistic and Effective Agent-Based Models of Crowd Dynamics. *Neurocomputing* **2014**, *146*, 199–209. [CrossRef]
5. Wagner, N.; Agrawal, V. An Agent-Based Simulation System for Concert Venue Crowd Evacuation Modeling in the Presence of a Fire Disaster. *Expert Syst. Appl.* **2014**, *41*, 2807–2815. [CrossRef]
6. Zhang, L.; Wang, J.; Shi, Q. Multi-Agent Based Modeling and Simulating for Evacuation Process in Stadium. *J. Syst. Sci. Complex.* **2014**, *27*, 430–444. [CrossRef]
7. Zong, X.; Wang, C.; Du, J.; Jiang, Y. Tree Hierarchical Directed Evacuation Network Model Based on Artificial Fish Swarm Algorithm. *Int. J. Mod. Phys. C* **2019**, *30*, 1950097. [CrossRef]
8. Wei, Z.; Hui, D.; Tong, W. The Application of Fire Spread and Evacuation Simulation Technology in Large Stadium. *Stoch. Environ. Res. Risk Assess.* **2009**, *23*, 433. [CrossRef]
9. Kirik, E.; Malyshev, A.; Vitova, T.; Popel, E.; Kharlamov, E. Pedestrian Movement Simulation for Stadiums Design. *IOP Conf. Ser. Mater. Sci. Eng.* **2018**, *456*, 012074. [CrossRef]
10. Kirik, E.; Vitova, T.; Malyshev, A.; Popel, E.; Kharlamov, E.; Moiseichenko, V.; Kalinin, E.; Smirnov, N. Computer Simulation of Pedestrian Dynamics in the Design and Operation of Stadiums. *Constr. Unique Build. Struct.* **2021**, *94*, 9401. [CrossRef]
11. Gutierrez, D.; Seron, F.J.; Magallon, J.A.; Sobreviela, E.J.; Latorre, P. Archaeological and Cultural Heritage: Bringing Life to an Unearthed Muslim Suburb in an Immersive Environment. *J. Cult. Herit.* **2004**, *5*, 63–74. [CrossRef]
12. Papagiannakis, G.; Foni, A.; Magnenat-Thalmann, N. Real-Time Recreated Ceremonies in VR Restituted Cultural Heritage Sites. In Proceedings of the CIPA XIXth International Symposium, Antalya, Turkey, 30 October 2003; pp. 1–6.
13. Heigeas, L.; Luciani, A.; Thollot, J.; Castagné, N. A physically-based particle model of emergent crowd behaviors. In Proceedings of the International Conference Graphicon, Moscow, Russia, 5–10 September 2003; pp. 1–9.
14. Cain, K.; Chrysanthou, Y.; Niccolucci, F.; Silberman, N. A Case Study of a Virtual Audience in a Reconstruction of an Ancient Roman Odeon in Aphrodisias. In Proceedings of the ACM SIGGRAPH 2005 Courses, Los Angeles, CA, USA, 4 August 2005. [CrossRef]
15. Di Salvo, C.; Mancini, M.; Cavinato, G.P.; Moscatelli, M.; Simionato, M.; Stigliano, F.; Rea, R.; Rodi, A. A 3d Geological Model as a Base for the Development of a Conceptual Groundwater Scheme in the Area of the Colosseum (Rome, Italy). *Geosciences* **2020**, *10*, 266. [CrossRef]
16. Tsires, A.G. *Architecture of the Colosseum*; Publishing House of the Academy of Architecture of the USSR: Moscow, Russia, 1940.
17. Devoti, L. *Circhi e Stadi di Roma Antica*; Tascabili Economici Newton: Rome, Italy, 1997.
18. Tan, A.H.; Croft, F.M.; Tan, F.H. Computer Graphic Modeling for the Reconstruction of the Roman Colosseum. *J. Geom. Graph.* **2015**, *19*, 65.
19. Hernández, A.J. Roman Amphitheatres in Baetica: Considerations on Geometry, Design and Drawing. *Arch. Español Arqueol.* **2015**, *88*, 127–148. [CrossRef]
20. Napolitano, R.; Glisic, B. Understanding the Function of Bonding Courses in Masonry Construction: An Investigation with Mixed Numerical Methods. *J. Cult. Herit.* **2019**, *39*, 120–129. [CrossRef]
21. Croci, G. The Colosseum: Safety evaluation and preliminary criteria of intervention. *Struct. Anal. Hist. Constr.* **1996**, *18*, 154–165.
22. Gravit, M.; Kirik, E.; Savchenko, E. Effect of Design on the Evacuation Time for the Colosseum of Rome. *Constr. Unique Build. Struct.* **2021**, *95*, 9504. [CrossRef]
23. Vellek, G.F.; Pearson, J. Arena. The Story of the Colosseum. Class. *World* **1976**, *69*, 350. [CrossRef]
24. Gutierrez, D.; Frischer, B.; Cerezo, E.; Gomez, A.; Seron, F. AI and Virtual Crowds: Populating the Colosseum. *J. Cult. Herit.* **2007**, *8*, 176–185. [CrossRef]
25. Roman Engineering: Crowd Control and Evacuation of the Roman Colosseum vs. the Modern Beijing Bird's Nest Stadium (Computer Simulation). Available online: http://thehomesteadingboards.com/forums/the-compost-bin-1/roman-engineering-crowd-control-and-evacuation-of-the-roman-colosseum-vs-the-modern-beijing-birds-nest-stadium-computer-simulation/ (accessed on 16 January 2022).
26. Russian Federal Law No. 123-FZ. Technical Regulation of Fire Safety. Available online: http://www.consultant.ru/document/cons_doc_LAW_78699/ (accessed on 25 November 2021).
27. Kirik, E.; Vitova, T.; Malyshev, A.; Popel, E. A Conjunction of the Discrete-Continuous Pedestrian Dynamics Model SigmaEva with Fundamental Diagrams. In Proceedings of the International Conference on Parallel Processing and Applied Mathematics, Bialystok, Poland, 8–11 September 2019; pp. 457–466. [CrossRef]
28. Kirik, E.; Litvintsev, K.; Dekterev, A.; Khasanov, I.; Gavrilov, A.; Malyshev, A.; Harlamov, E.; Popel, E. Simulations of fire evacuations in "Sigma FS" software as a fire safety training instrument. In Proceedings of the Ninth International Seminar on Fire and Explosion Hazards, Saint Petersburg, Russia, 21–26 April 2019; pp. 1281–1291. [CrossRef]
29. Kirik, E.; Malyshev, A. Computer Simulation of Pedestrian Flows for Universiade 2019 Sport Facilities Versus Hand Calculations. *Proc. Pedestr. Evacuation Dyn.* **2016**, *A11*, 446–454.
30. Kholshevnikov, V.V.; Shields, T.J.; Boyce, K.E.; Samoshin, D.A. Recent Developments in Pedestrian Flow Theory and Research in Russia. *Fire Saf. J.* **2008**, *43*, 108–118. [CrossRef]
31. Gwynne, S.M.V.; Rosenbaum, E.R. *Employing the Hydraulic Model in Assessing Emergency Movement. SFPE Handbook of Fire Protection Engineering*; Springer: New York, NY, USA, 2016; pp. 2115–2151.

32. Weidmann, U. Trasporttechnik der Fussgänger. Institut für Verkehrsplanung, Transporttechnik, Strassen-und Eisenbahnbau. *Schr. IVT* **1993**, *90*, 115. [CrossRef]
33. Kholshevnikov, V. Forecast of human behavior during fire evacuation. In Proceedings of the International Conference Emergency Evacuation of People from Buildings—EMEVAC, Belstudio, Warsaw, Poland, 30 March–2 April 2011; pp. 139–153.
34. Gazprom Arena Stadium. Available online: https://gazprom-arena.com/ (accessed on 17 January 2022).
35. Huo, F.; Song, W.; Chen, L.; Liu, C.; Liew, K.M. Experimental Study on Characteristics of Pedestrian Evacuation on Stairs in a High-Rise Building. *Saf. Sci.* **2016**, *86*, 165–173. [CrossRef]

Review

Influence of Natural Fire Development on Concrete Compressive Strength

Robert Kuehnen, Maged A. Youssef * and Salah F. El-Fitiany

Department of Civil and Environmental Engineering, Faculty of Engineering, Western University, London, ON N6A 5B9, Canada; rkuehnen@uwo.ca (R.K.); selfitia@uwo.ca (S.F.E.-F.)
* Correspondence: youssef@uwo.ca

Abstract: With increasing acceptance of performance-based design principles in the field of fire safety, it is imperative to accurately define the behaviour of materials during fire exposure. Real-world fire events, otherwise referred to as natural fires, are defined by four characteristics: heating rate, maximum temperature, exposure duration, and cooling rate. Each of these four characteristics influences concrete's behaviour in a different manner. In this paper, the available experimental work for concrete, tested at elevated temperatures, is examined to identify the influence of the four natural fire characteristics on concrete compressive strength. This review focuses on normal strength concrete tests only, omitting parameters such as unique additives and confinement. The intent is to provide a fundamental understanding of normal strength concrete. The findings show that maximum temperature and cooling rates have a significant influence on concrete strength. Exposure duration has a moderate impact, particularly at shorter durations. Variable rates of heating have minimal influence on strength. Detailed conclusions are provided along with review limitations, practical considerations for designers, and future research needs.

Keywords: natural fire; concrete strength; exposure duration; maximum temperature; heating rate; cooling rate

Citation: Kuehnen, R.; Youssef, M.A.; El-Fitiany, S.F. Influence of Natural Fire Development on Concrete Compressive Strength. *Fire* **2022**, *5*, 34. https://doi.org/10.3390/fire5020034

Academic Editor: Wojciech Węgrzyński

Received: 19 January 2022
Accepted: 24 February 2022
Published: 28 February 2022

Publisher's Note: MDPI stays neutral with regard to jurisdictional claims in published maps and institutional affiliations.

1. Introduction

In contrast with timber and steel construction, one of the primary advantages of using concrete as a building material is that it can withstand fire events without burning, melting, or needing additional protective materials. Concrete, however, is not completely unaffected by fire exposure. Studies on normal strength concrete (NSC) have shown that an exposure temperature of 600 °C can reduce concrete compressive strength by up to 55% [1].

There are a number of material properties that are known to affect concrete strength at elevated temperatures, such as ambient strength, aggregate type, water-cement ratio, additives, and prestress level. The influence of these properties is well investigated in the existing experimental work and detailed in numerous textbooks and literature-review publications [2,3]. Concrete strength is also affected by fire characteristics, such as: rate of heating, maximum temperature level, exposure duration, and rate of cooling [4]. The influence of these four fire characteristics is less thoroughly addressed in the existing literature.

It is imperative for designers to understand the behaviour of fires and its influence on concrete compressive strength. Existing performance-based models have shown that understanding these four fire characteristics is a necessary step to accurately modeling reinforced concrete beam and column behaviour [5]. By implementing the findings of this review into existing models, designers can produce performance-based solutions with increased confidence in safety, reliability, and efficiency.

2. Natural Fire Definition

Fire events are typically represented by temperature-time relationships, as shown in Figure 1. The term natural fire is used to define a fire event as it would occur in the real world. No two natural fires will ever be identical, as these fires are influenced by a wide range of compartment and environmental properties. Three examples of potential fire profiles are shown in Figure 1. Fire events can have high temperature over short duration, low temperature over long duration, or anywhere in between.

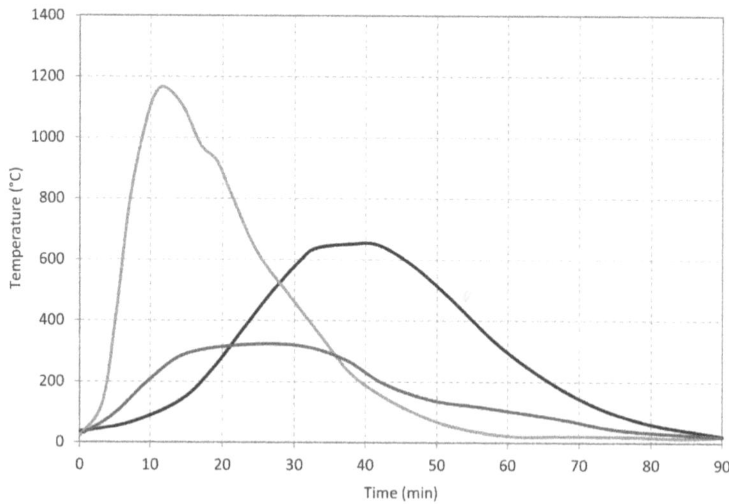

Figure 1. Examples of natural fire temperature–time curves.

To define a natural fire, four fire characteristics can be calculated: heating rate, maximum temperature, overall exposure duration, and cooling rate [4]. During the growth of a fire, variable rates of heating can occur, ranging from slow heating to almost instantaneous flashover. The rate of heating is greatly dependent on available oxygen and the presence of highly combustible materials. At the peak of a fire event, the value of the maximum temperature as well as its exposure duration vary based on reliability of fuel and oxygen supply. Once a fire begins to decay, variable rates of cooling can be present, ranging from slow air cooling in a smoldering compartment to rapid water cooling achieved by firefighting efforts. Each of these four fire characteristics plays a notable and different role in the deterioration of concrete strength. It is the influence of these characteristics on concrete strength that is evaluated in this paper.

3. Available Experimental Work

The available experimental work features a wide range of testing parameters. To narrow the scope of this review, several concrete and testing parameters are controlled. Only tests with the following attributes have been reviewed in this paper: (a) unstressed tests, (b) unconfined tests, (c) unsealed tests, (d) ordinary Portland cement (no additives such as fly ash, silica, fibers, etc.), and (e) NSC with ambient strength less than or equal to 50 MPa. The intent of the control parameters is to focus the evaluation on basic NSC. Doing so highlights the influence of fire characteristics on behaviour and allows future researchers to identify when newly introduced parameters present unusual responses. Additionally, the majority of existing work is based on NSC, allowing for a wide range of sample data points.

In addition to the controlled parameters, there are other parameters that are known to affect compressive strength during fire exposure. These parameters include water-cement ratio, aggregate-cementitious material ratio, aggregate type, size and content, geometric dimensions, and testing procedure [3]. Because there is so much variation in the existing experimental work, it is difficult to control all these parameters. Furthermore, the variation of these parameters is acceptable within the definition of NSC. To provide meaningful review, specific sections of this paper control the various parameters when possible, and when not possible, a selection of similar tests are averaged and presented for evaluation.

Table 1 shows a summary of the experimental tests investigated during this review. Full details are available in the referenced work. The required fire characteristics of each test, shown in Figure 2, are identified in the table. The concrete ambient strength ("$f'_{c,20}$") and reported "aggregate" type are also recorded for additional context. "Testing time" refers to when the compressive strength of the sample was taken. A "residual" testing time indicates that testing occurred after the specimen cooled back to ambient temperature. A "hot" testing time indicates testing occurred while the sample was still at the maximum applied temperature.

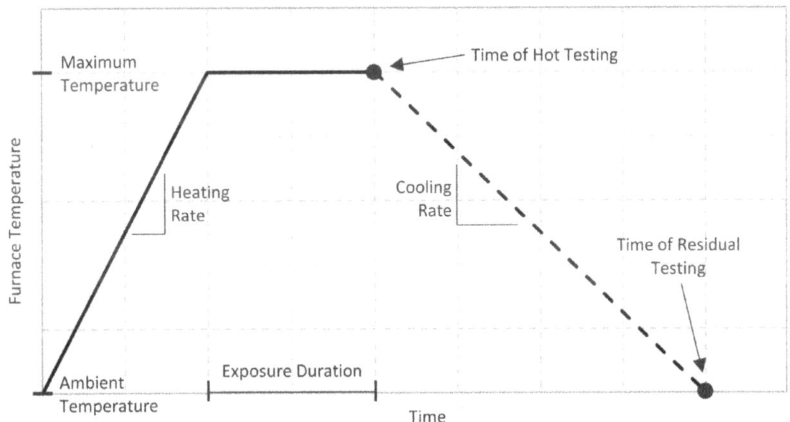

Figure 2. Furnace heating profile during specimen testing.

For experimental work with a variable "heating rate", the average rate is provided in the table. A heating rate of "instant" indicates that the specimen was placed in a preheated furnace. A rate of "standard" indicates that the standard fire curve was applied for the heating profile. The term "measured" is used for tests where the heating rate was controlled based on measuring the internal temperature of the specimen and maintaining some maximum difference from the furnace temperature. Maximum temperature ("max temp") is recorded as the maximum temperature of the furnace. Exposure "duration" is recorded in hours from the time when heating ends to the time when hot testing or residual cooling begins. An exposure duration of "uniform" indicates the specimen's internal temperature was measured and that heat was applied for a continuous duration until the specimen's internal temperatures uniformly reached the furnace temperature. "Cooling rate" is stated as either "slow" or "rapid". Comprehensive definitions of the two cooling rates are provided in Section 4.4. It should be noted that similar to maximum temperature, the heating and cooling rates refer to the temperature change in the furnace, not the specimen itself. Although the furnace temperature is not necessarily an ideal way to represent these values, it is easier to record and is widely reported in the literature as such.

Table 1. List of Evaluated Experimental Work with Test Parameters.

Label	Ref.	$f'_{c,20}$	Agg Type	Duration	Testing Time	Heating Rate	Cooling Rate
		(MPa)		(hr)		(°C/min)	
T-6A	[6]	27	Siliceous	uniform	residual	2.4	slow
T-6B	[6]	27	Siliceous	uniform	residual	2.4	rapid
T-7A	[7]	27	Calcareous	uniform	hot	measured	—
T-7B	[7]	27	Calcareous	uniform	residual	measured	slow
T-7C	[7]	27	Siliceous	uniform	hot	measured	—
T-7D	[7]	27	Siliceous	uniform	residual	measured	slow
T-8	[8]	50	Quartzite	2.00	hot	5.0	—
T-9A	[9]	35	Calcareous	3.00	residual	16.0	slow
T-9B	[9]	35	Calcareous	3.00	residual	16.0	rapid
T-10	[10]	45	Siliceous	15.00	residual	1.0	rapid
T-11	[11]	47	Limestone	varied	residual	0.3	slow
T-12	[12]	31	Limestone	0.17	hot	7.5	—
T-13A	[13]	27	Siliceous	2.00	residual	2.8	slow
T-13B	[13]	40	Siliceous	2.00	residual	2.8	slow
T-14	[14]	33	Siliceous	2.00	hot	2.0	—
T-15	[15]	34	Granite	1.00	hot	2.0	—
T-16A	[16]	21	Siliceous	2.00	hot	1.0	—
T-16B	[16]	42	Siliceous	2.00	hot	1.0	—
T-17A	[17]	25	Sandstone	1.00	residual	1.5	slow
T-17B	[17]	23	Gravel	1.50	hot	7.5	—
T-18	[18]	50	Limestone	2.00	hot	2.0	—
T-19A	[19]	20	Granite	4.00	residual	2.0	slow
T-19B	[19]	20	Granite	4.00	residual	2.0	rapid
T-20	[20]	44	Basalt	none	residual	standard	slow
T-21A	[21]	24	Gravel	1.00	hot	measured	—
T-21B	[21]	24	Gravel	1.00	residual	measured	slow
T-22	[22]	32	Basalt	varied	residual	5.0	slow
T-23	[23]	39	Siliceous	1.00	residual	1.0	slow
T-24A	[24]	43	Gravel	uniform	residual	standard	slow
T-24B	[24]	43	Gravel	uniform	residual	standard	rapid
T-25A	[25]	47	Gravel	1.50	residual	instant	slow
T-25B	[25]	46	Dolomite	1.50	residual	instant	slow
T-26	[26]	37	Limestone	1.00	residual	1.0	slow
T-27A	[27]	50	Limestone	uniform	hot	5.0	—
T-27B	[27]	50	Limestone	uniform	residual	5.0	slow
T-28	[28]	38	Granite	1.00	residual	2.5	slow
T-29	[29]	35	Gravel	0.25	hot	2.7	—
T-30	[30]	49	Limestone	2.00	residual	2.5	slow
T-31	[31]	28	Siliceous	0.50	residual	8.0	rapid
T-32	[32]	28	Siliceous	0.50	residual	2.0	slow
T-33	[33]	40	Siliceous	0.50	hot	5.0	—

4. Influence of Fire Characteristics

In this section, the influence of each fire characteristic is evaluated. Contrary to the chronological order of a natural fire event, the influence of maximum temperature is discussed first as it is the most well-documented characteristic in the literature. It is intended that by recognizing the effects of maximum temperature first, the less-documented fire characteristics can be subsequently evaluated with greater clarity.

4.1. Influence of Maximum Temperature

Figure 3 presents the averaged relative strength of hot and residual tests for a range of maximum temperature exposures. The averaged values consist of findings from 37 different studies. To provide an understanding of the variation in existing data, upper and lower limits of the evaluated test data are given (dotted line). Eurocode prescribed strength reductions for siliceous aggregate (dashed line) are also given [34,35].

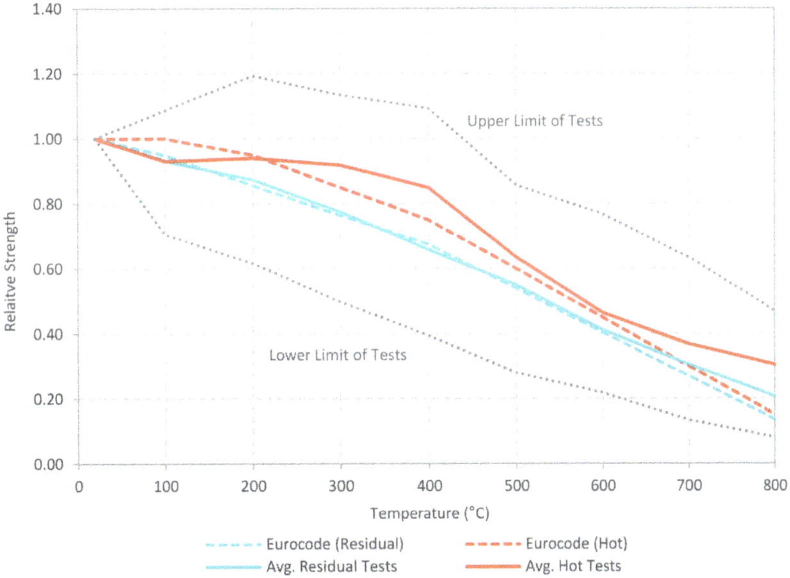

Figure 3. Relative strength of concrete for hot and residual conditions. [Includes Tests from T-7A to T-10, T-12 to T-20, T-21A to T-28, T-30 to T-34].

The averaged experimental work shows that increasing maximum temperature has a significant influence on concrete strength. Concrete tested after cooling exhibits lesser strength at every temperature compared with hot tested concrete. This relationship is largely due to the influence of cooling, which is examined in Section 4.4. To address the influence of maximum temperature specifically, discussion focuses on the response of the hot tested profile.

Concrete strength exhibits three trends when exposed to elevated temperature. At lower temperatures below 350 °C, strength loss is relatively minor. Some of the experimental work, such as by Diederichs et al. [14] and Fu et al. [15], even observed moderate strength gains in the low temperature ranges. The extent of these gains can be seen in the steep rise of the dotted upper limit line. Castillo and Duranni [12] proposed that this strength gain results from stiffening of the cement gel due to the evaporation of concrete moisture. As such, changing concrete properties, such as porosity and moisture content, can have a notable impact on delaying strength loss at low maximum temperatures.

In the mid-range temperatures, 350–600 °C, strength drops sharply. By 600 °C, relative strength levels of 45% and 41% can be expected for hot and residual test averages, respectively. In this temperature range, the concrete becomes substantially dehydrated, such that the full influence of micro-cracking, cement and aggregate decomposition, and thermal expansion stresses is realized [36].

Above 600 °C, severe degradation can be expected, with as much as 90% strength loss by 800 °C. This reduction illustrates the substantial influence that maximum temperature has on the strength of concrete. At these higher temperatures, specimens can often be broken up into gravel by hand [37]. The rate of strength loss above 600 °C, however, is slightly less severe than in the mid-range temperatures. This lessening rate may be attributed to the calcination or crystallization of aggregates [12].

4.2. Influence of Heating Rate

During concrete heating, a thermal gradient develops between a section's outer layers and inner core. This gradient induces thermal stresses between the different constituents of the concrete, which in turn produces micro-cracking and compressive strength loss. It is by this mechanism that variable rates of heating can influence concrete strength.

For evaluation, experimental work is divided into low and high heating rates. A low heating rate is defined as a rate less than 3 °C/min, with high heating being that greater than 3 °C/min. This definition of low and high rates is based off the median heating rate of the available experimental work. For comparison, the standard fire has an average heating rate of 33 °C/min (between 0 °C and 800 °C) and the Cardington fire tests give an average rate of 18 °C/min for a typical compartment fire [38]. Although 3 °C/min is a comparably much lower rate of heating, the experimental work has focused on this level due to the relative simplicity of its application. These low heating rate tests are also not without merit, as they are still valid for potentially smaller natural fire events.

To control for the effects of the other fire characteristics, only tests with a similar exposure duration have been included. Hot and residual tests have been separated for comparison. All residual tests feature a similar cooling regime.

Figures 4 and 5 present the relative concrete strength of hot tested specimens for high and low rates of heating. The average profile for the plotted tests is indicated by the dashed line. The experimental work is found to be in good agreement, with only a few outliers from the test average. Figures 6 and 7 similarly present the relative concrete strength of residually tested specimens. The low heating rate tests show very good agreement, but greater fluctuation is observed for high heating. This may be due to the wider selection of heating rates presented on the plot, ranging from 5 °C/min to instantaneous heating.

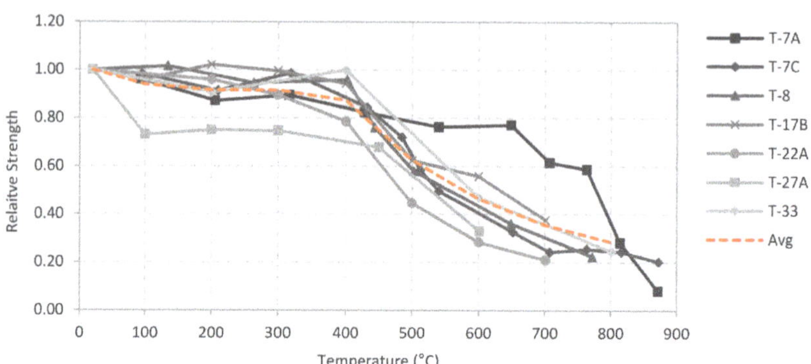

Figure 4. Relative strength of hot tested concrete with high heating rates.

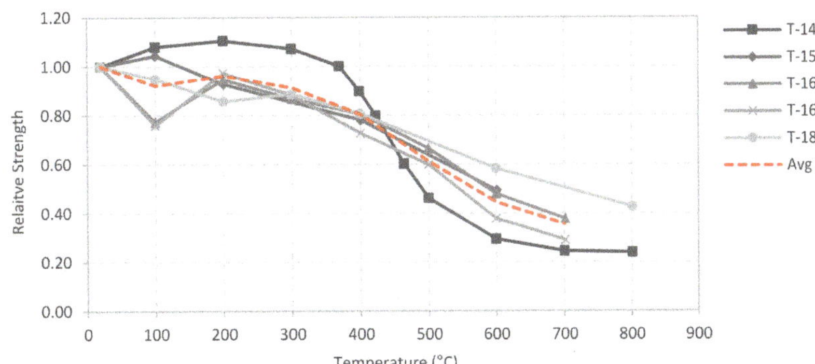

Figure 5. Relative strength of hot tested concrete with low heating rates.

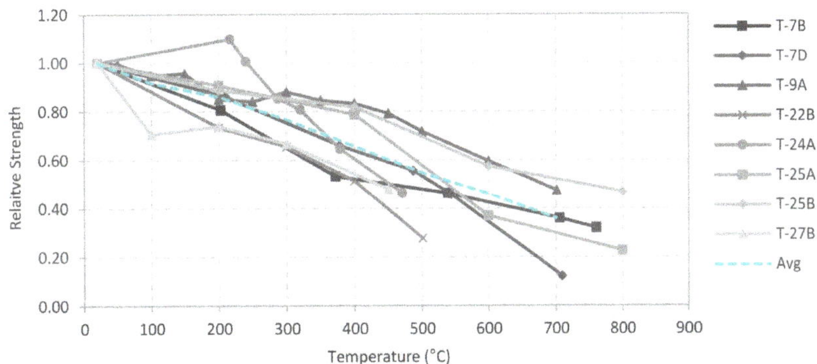

Figure 6. Relative strength of residually tested concrete with high heating rates.

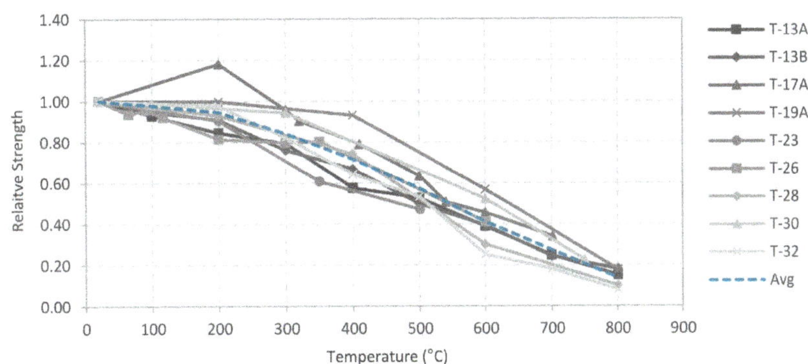

Figure 7. Relative strength of residually tested concrete with low heating rates.

Figure 8 records the average strengths of the experimental work for direct comparison. The average profiles have been truncated at 700 °C due to a shortage of available tests beyond this temperature. The Eurocode prescribed profiles for hot and residual siliceous concrete are also given as a baseline [34,35].

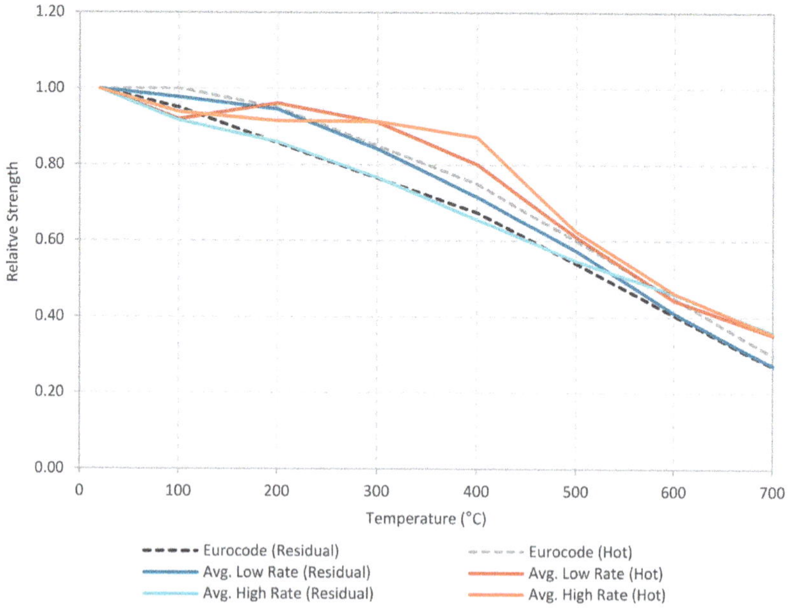

Figure 8. Relative strength of concrete due to low or high rates of heating.

Considering the average profiles, no clear trend emerges. In general, high rates appear to result in slightly greater strength reduction. This is most notable for the residually tested concrete at lower temperatures around 200 °C. However, at any given temperature, the effect of heating can produce higher, lower, or identical strengths. In particular, beyond 500 °C, all four heating regimes converge and result in comparable strength levels.

A justification for this minor and fluctuating influence may be due to the conflicting nature of heating mechanisms. At higher heating rates, large thermal gradients develop, causing greater strength reduction due to extensive micro-cracking. However, at the same time, the rapid expulsion of moisture from the concrete strengthens the adhesive action of the cement gel. These two mechanisms act in contrast resulting in similar concrete strengths for low and high heating rates. Mohamidbhai [22] proposed that at temperatures above 600 °C, the majority of the moisture is removed, and the micro-cracking occurs regardless of the heating rate. Therefore, low and high rates can be expected to result in similar strength losses at high temperatures, which is reflected in Figure 8.

It should be noted that although heating rate does not have a large impact on concrete strength, it is often cited as having a significant impact on explosive spalling [39]. Explosive spalling is a phenomenon in which exterior portions of a concrete specimen violently spall off during heating. This effect significantly reduces the elements cross-section and potentially exposes internal reinforcement, greatly reducing sectional strength. Castillo and Durrani [12], Noumowe et al. [26], and Phan and Carino [27] all reported major spalling in their high-strength concrete (HSC) samples but none in their NSC. Noumowe et al. [39] observed explosive spalling in HSC specimens at heating rates as low as 1 °C/min. It is well documented that NSC is often unaffected by spalling compared with HSC. However, in view of the potential severity of explosive spalling, heating rate is a factor that should be given due consideration.

4.3. Influence of Exposure Duration

Exposure duration refers to the time for which concrete is subjected to elevated temperatures. For a natural fire, exposure duration would intuitively be taken from the time when the fire starts to when it is fully extinguished. This overall duration, however, is not often reported in the literature. Instead, exposure duration is typically reported as the time from when heating ends to the time when hot testing or residual cooling begins. During this period, the concrete is exposed consistently to the maximum temperature. Defining exposure duration in this way makes temperature control easier during testing. It also has the added benefit of allowing its influence on concrete strength to be separated from that of variable heating and cooling rate.

To evaluate the influence of exposure duration, this section focuses on the work of Carette et al. [11] and Mohamidbhai [22]. Both studies specifically investigated variable exposure durations, ranging from hours to months. For comparison, complimentary experimental work has been selected with similar heating, residual cooling, calcareous aggregates, and specimen sizes.

Figure 9 presents the relative strength reductions for concrete when exposed to a maximum temperature of 400 °C for various durations. Figure 10 provides the same for a 600 °C temperature. An exposure duration of "uniform" indicates continuous exposure was applied until the specimen's internal temperatures were measured to match the furnace temperature. An exposure duration of "0-hr" indicates the specimen began cooling immediately after maximum furnace temperature was reached.

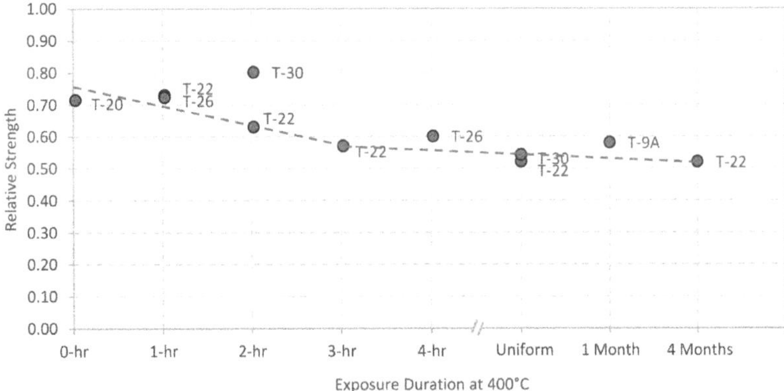

Figure 9. Relative strength of concrete at 400 °C with various exposure durations.

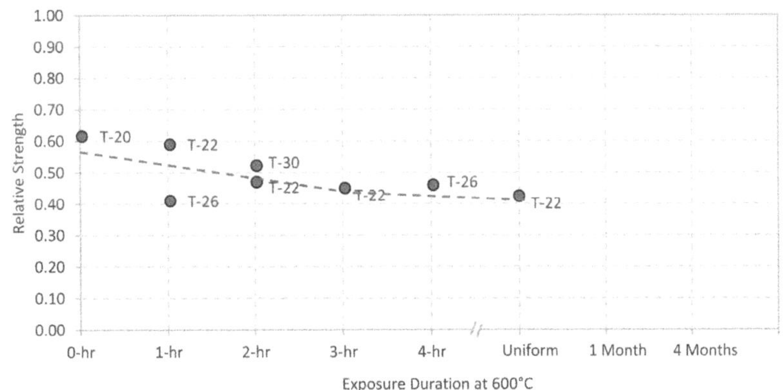

Figure 10. Relative strength of concrete at 600 °C with various exposure durations.

The results show that the majority of strength loss occurs early in the exposure process. Test T-20 with an exposure duration of 0 h, exhibited relative strength loss of 29% at 400 °C and 38% at 600 °C.

As exposure duration increases, strength reduction follows two trends. Up until 3 h, moderate strength reduction continues to occur. Beyond 3 h, insignificant further strength reduction is observed. Even at extreme durations of one and four months, strength levels are comparable to the 3 h and 4 h exposure durations. Those two trends are presented in Figures 9 and 10 by the dashed lines.

The rationale behind the relationship can be attributed to the internal temperatures within the concrete. At shorter durations, there is a temperature lag between the outside surfaces of the concrete and the inside. During this period, continued cracking and strength degradation occurs as the internal temperature increases. Once a uniform internal temperature is reached, the mechanisms of strength loss become minimal.

Based on the reviewed experiments, a uniform internal temperature can be expected in typical laboratory test specimens after 3 h of constant exposure. For larger concrete cross-sections, the time it takes to reach a uniform internal temperature varies greatly.

4.4. Influence of Cooling Rate

As previously observed in Figure 8, the residual strength of concrete after cooling is notably lesser when compared with its hot strength. The cause of this additional strength loss is due to the development of internal temperature gradients, similar to the heating process. Because these gradients form in the opposite direction of heating, they generate new stresses and new cracks that further reduce concrete strength [26].

Considering the effects of a natural environment, variable rates of cooling can be present, ranging from slow cooling in a smoldering compartment to rapid cooling from firefighting efforts. To evaluate the effect of cooling, the reviewed experimental testing is divided into two rates: slow and rapid cooling. In this paper, cooling rate is taken from the time furnace temperature begins to decline until the furnace reaches ambient temperature.

Slow cooling occurs when a test specimen is either cooled within the test furnace or taken outside into the ambient environment. Internal specimen temperature by Lee et al. [19] showed that these two different cooling methods produce very comparable cooling rates. Savaa et al. [30] and Morita et al. [23] indicated that slow cooling results in a rate of 0.4 °C/min to 1.0 °C/min. Slow cooling can subsequently be defined as having a rate less than or equal to 1.0 °C/min.

Rapid cooling is achieved in experimental work by exposing the specimen to water during the cooling stage. Water quenching or spraying techniques are typically applied by submerging or spraying the specimen with ambient temperature water for a prolonged duration. In the specific case of 150 mm cubed specimens, Botte and Caspeele [10] identified that from an elevated temperature of 600 °C, quenching is equivalent to a cooling rate of 30–40 °C/min. The results of this experiment demonstrate the magnitude of possible cooling rates that can occur during natural fire scenarios.

Figures 11 and 12 display the relative concrete strength of specimens exposed to slow and rapid cooling. Only tests of similar heating rate and exposure duration are presented. All the rapid cooling studies were conducted immediately after cooling was complete, avoiding the influence of potential strength recovery. The overall profile of the experiments for both cooling regimes were found to be in good agreement with one another. Due to the additional inconvenience of conducting rapid cooling tests, their number in the literature is very small.

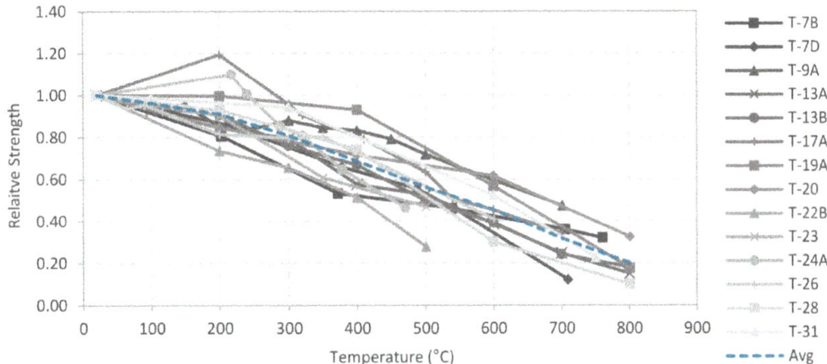

Figure 11. Relative strength of concrete with slow cooling.

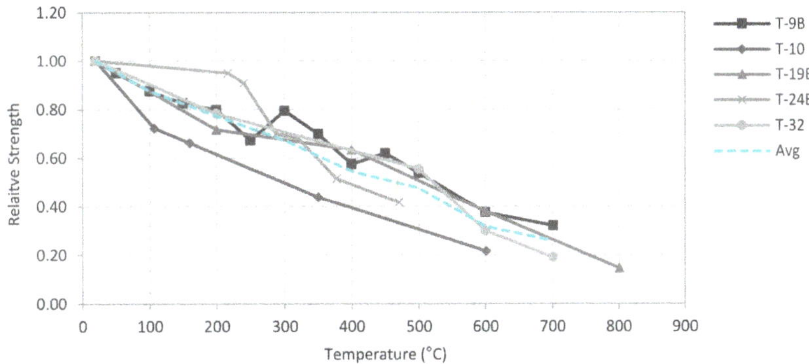

Figure 12. Relative strength of concrete with rapid cooling.

Figure 13 presents the averaged cooling profiles along with the Eurocode siliceous baseline and the averaged profile of the hot tested concrete from Figure 7. It can be seen that an increased cooling rate results in greater strength reduction. Even at a slow rate of cooling, lower residual strengths can be expected when compared with the hot tested specimens. On average, 6% greater strength loss is observed between hot to slow and an additional 10% is observed between slow to rapid.

The strength loss due to cooling is not constant with temperature. At low temperatures circa 100 °C, the residual concrete exhibits only minorly lesser strength levels as compared with hot tested concrete. However, in the mid-temperature range of 200 °C to 500 °C, the influence of cooling becomes significant. The maximum difference between hot and rapid cooling is 29% when at 300 °C. This trend indicates the extreme importance of considering the influence of cooling rates in moderate fire events. At the higher temperatures above 600 °C, the three profiles display some convergence. Due to a shortage of test data, the rapid cooling profile is discontinued early. Specific testing by Lee et al. [19] indicates that at temperatures of 800 °C, slow and rapid cooling continue to converge and reach comparable strength levels.

The lack of agreement between the rapid cooling and Eurocode profile should also be noted from Figure 13 [34]. This is the only fire characteristic for which a significant and unconservative relationship is observed between the code and test results. When assessing the residual strength of concrete, this potential limitation in the code prescribed values should be considered.

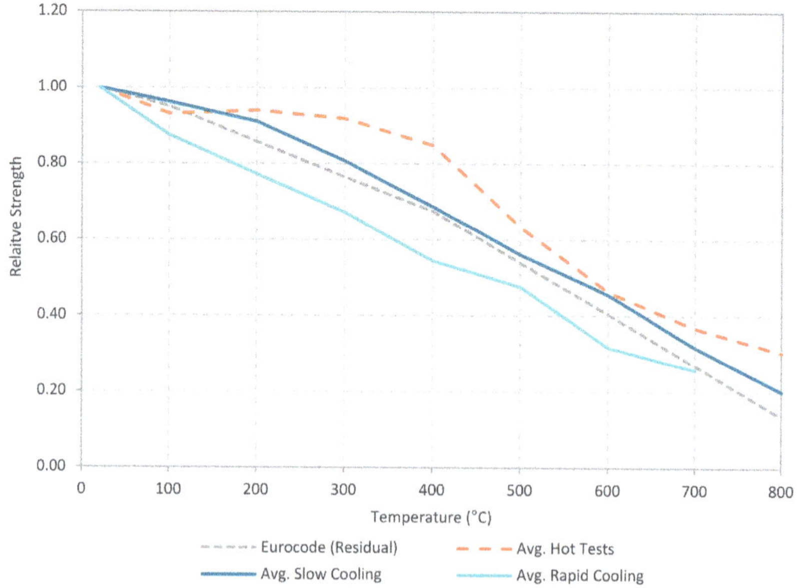

Figure 13. Relative strength of concrete due to slow or rapid cooling.

Post-Fire Strength Recovery Due to Cooling Rate

Post-fire strength recovery is a process by which fire damaged concrete can significantly regain strength when cooled with water. This recovery is attributed to the rehydration of the cement [40]. Maximizing water exposure and allowing time for recuring are important factors in facilitating recovery.

The concept of strength recovery has been well investigated in the literature since it was first observed in 1970 by Crook and Murray [41]. Experimental work and reviews often focus on the influence of long-term recuring techniques, such as soaking specimens for weeklong durations [40]. From the perspective of a natural fire event, this duration of water exposure is unlikely. The following experimental work has been reviewed to demonstrate the influence of short-term recuring.

Poon et al. [40] performed experimentation involving a continuation of the test data presented in Table 1 for T-28. After slow air cooling from 600 °C, NSC specimens were recured by water spraying for 2 hrs and then tested after 7, 28, and 56 days. The results show that after 7 days, the concrete recovered 14% of its strength, and after 56 days recovered 19%. This represents a significant recovery. The researchers identified that the 2 hr spraying duration was selected after many trials to be the minimum soaking time for optimized results.

Abramowicz and Kowalski [6] explored the concept of very short duration water cooling. Specimens were either slow cooled in ambient air or rapid cooled by quenching for 10 s, followed by further slow cooling in ambient air. Strength testing was completed the next day. This very short duration immersion and quick testing time produced no significant effect on the specimen's strength compared with the baseline slow cooled specimens.

Based on these findings, rapid water cooling is not sufficient to induce notable strength recovery that is reliably useful for design purposes. This is due to two reasons. Firstly, recuring requires time. When considering the strength of concrete during the natural fire event and the safety of occupants and first responders, insufficient time will have been provided for recuring regardless of water exposure. Secondly, it is important to also consider the geometry of the concrete involved. Far larger amounts of water would be required for a building, versus 100 mm specimens. To reliably recreate the findings of Poon et al. [40], an extended and intentional recuring effort would be required.

5. Conclusions

Based on the reviewed literature, the following conclusions can be made regarding the influence of each of the four natural fire characteristics on concrete compressive strength:

1. Maximum temperature causes the most significant strength reduction to concrete. At temperatures below 350 °C, strength losses are relatively minor. Strength increase is even possible in the low temperature range depending on concrete mix properties. Beyond 350 °C, strength drops rapidly. By 600 °C, hot and residual tested concrete can be expected to have lost 55% of their ambient strength. In the high temperature ranges, maximum temperature dominates the other fire characteristics and is the principal source of strength reduction.
2. Heating rates have minimal influence on the strength of concrete. At lower temperatures, higher heating rates were found to result in marginally lower strengths. However, the findings fluctuated greatly such that no decisive conclusion can be made. At higher temperatures above 500 °C, both low and high heating rates produced comparable strength losses. The impact of explosive spalling is likely the primary concern when considering the influence of heating rate on the strength of a specific concrete section.
3. Exposure duration was found to have a major but diminishing impact on concrete strength. The majority of strength loss happens very rapidly, within the first minutes and hours of exposure. After several hours of constant exposure, strength loss is comparable to that of concrete exposed for month-long durations. This finding demonstrates the importance of understanding and defining a section's internal temperature gradient. Once a section's internal temperature becomes uniform, negligible further degradation is expected regardless of extended exposure. One item for future consideration that was not found in the literature is the influence of very short duration high-temperature heating.
4. Cooling rate was found to have an important influence on strength. On average, residual tested concrete exhibited 10% greater strength loss compared with hot tested concrete. Comparing slow and rapid cooled specimens demonstrated that higher cooling rates result in even further strength loss. The greatest impact on cooling rate was in the mid temperature range of 200 °C to 500 °C. Above 500 °C, hot tested, slow cooled, and rapid cooled profiles begin to converge and reach comparable strength levels. The possibility of strength recovery due to rapid water cooling was found to be unlikely using typical water-cooling techniques alone.

The intent of this paper is to provide a general understanding of NSC behaviour during natural fire exposure. This allows designers to focus on the parameters that have the largest impact on concrete behaviour and researchers to identify when newly introduced parameters present unusual responses. To achieve this goal, several assumptions were made which limit the validity of the conclusions of this paper. The reviewed experimental work was limited to: unstressed tests, unconfined tests, unsealed tests, ordinary Portland cement, and NSC. Additionally, other NSC parameters were left uncontrolled and broadly accepted within this review. These parameters include water-cement ratio, aggregate-cementitious material ratio, aggregate type, size and content, and geometric dimensions. Future research is needed to address these limitations.

Author Contributions: Conceptualization, R.K., M.A.Y. and S.F.E.-F.; data curation, R.K.; investigation, R.K.; writing—original draft preparation, R.K.; writing—review and editing, M.A.Y. and S.F.E.-F.; acquisition, M.A.Y. All authors have read and agreed to the published version of the manuscript.

Funding: This research was funded by the Natural Sciences and Engineering Research Council of Canada (NSERC), the India–Canada Center for Innovative Multidisciplinary Partnerships to Accelerate Community Transformation and Sustainability (IC-IMPACTS), and Western University.

Data Availability Statement: The data presented in this literature review is available in the referenced papers.

Conflicts of Interest: The authors declare that they have no known competing financial interests or personal relation-ships that could have appeared to influence the work reported in this paper.

References

1. Phan, L.T.; Carino, N.J. Code provisions for high strength concrete strength-temperature relationship at elevated temperatures. *Mater. Struct.* **2003**, *36*, 91–98. [CrossRef]
2. Purkiss, J.A. *Fire Safety Engineering, Design of Structures*; Butterworth-Heinemann: Burlington, MA, USA, 2007.
3. Khoury, G.A. Effect of fire on concrete and concrete structures. *Prog. Struct. Eng. Mater.* **2000**, *2*, 429–442. [CrossRef]
4. Zhang, B.; Bicanic, N.; Pearce, C.J.; Balabanic, G.; Purkiss, J.A. Discussion on: Residual fracture properties of normal and high strength concrete subject to elevated temperature. *Mag. Concr. Res.* **2001**, *53*, 221–224. [CrossRef]
5. Kuehnen, R.; Youssef, M.A.; El-Fitiany, S.F. Performance-based design of RC beams using an equivalent standard fire. *J. Struct. Fire Eng.* **2020**, *12*, 98–109. [CrossRef]
6. Abramowicz, M.; Kowalski, R. The influence of short time water cooling on the mechanical properties of concrete heated up to high temperatures. *J. Civ. Eng. Manag.* **2005**, *6*, 85–90. [CrossRef]
7. Abrams, M.S. *Compressive Strength of Concrete at Temperatures to 1600F*; ACI Special Publication; American Concrete Institute: Farmington Hills, MI, USA, 1971; Volume 25, pp. 33–58.
8. Anderberg, Y.; Thelandersson, S. *Stress and Deformation Characteristics of Concrete at High Temperatures—Experimental Investigation and Material Behavior Model*; Bulletin 34—Division of Structural Mechanics and Concrete Construction, Lund Institute of Technology: Lund, Sweden, 1976.
9. Bingöl, A.F.; Gül, R. Effect of elevated temperatures and cooling regimes on normal strength concrete. *Fire Mater.* **2008**, *33*, 79–88. [CrossRef]
10. Botte, W.; Caspeele, R. Post-cooling properties of concrete exposed to fire. *Fire Saf. J.* **2017**, *92*, 142–150. [CrossRef]
11. Carette, G.G.; Painter, K.E.; Malhotra, V.M. Sustained high temperature effects on concrete made with normal Portland cement, normal Portland cement and slag, or normal Portland cement and fly ash. *Concr. Int.* **1982**, *4*, 41–51.
12. Castillo, C.; Durrani, A.J. Effect of transient high temperature on high-strength concrete. *ACI Mater. J.* **1990**, *87*, 47–53.
13. Chang, Y.F.; Chen, Y.H.; Sheu, M.S.; Yao, G.C. Residual stress-strain relationship for concrete after exposure to high temperatures. *Cem. Concr. Res.* **2006**, *36*, 1999–2005. [CrossRef]
14. Diederichs, U.; Jumppanen, U.; Penttala, V. Material properties of high strength concrete at elevated temperatures. In Proceedings of the 13th Congress of International Association for Bridge and Structural Engineering, Helsinki, Finland, 6–10 June 1988; pp. 489–494.
15. Fu, Y.F.; Wong, Y.L.; Poon, C.S.; Tang, C.A. Stress–strain behaviour of high-strength concrete at elevated temperatures. *Mag. Concr. Res.* **2005**, *57*, 535–544. [CrossRef]
16. Furumura, F.; Abe, T.; Shinohara, Y. Mechanical properties of high strength concrete at high temperatures. In *Fire Performance of High-Strength Concrete: A Report of the State-of-the-Art*; NISTIR 5934; Building and Fire Research Laboratory: Gaithersburg, MD, USA, 1996.
17. Harada, T.; Takeda, J.; Yamane, S.; Furumura, F. *Strength, Elasticity, and Thermal Properties of Concrete Subject to Elevated Temperatures*; ACI Special Publication: Concrete for Nuclear Reactors; American Concrete Institute: Farmington Hills, MI, USA, 1972; Volume 34, pp. 377–406.
18. Khaliq, W. Performance Characterization of High Performance Concretes under Fire Conditions. Ph.D. Thesis, Michigan State University, East Lansing, MI, USA, 2012.
19. Lee, J. Experimental Studies and Theoretical Modelling of Concrete Subjected to High Temperatures. Ph.D. Thesis, University of Colorado, Boulder, CO, USA, 2006.
20. Li, M.; Qian, C.; Sun, W. Mechanical properties of high-strength concrete after fire. *Cem. Concr. Res.* **2004**, *34*, 1001–1005. [CrossRef]
21. Malhotra, H.L. The effect of temperature on the compressive strength of concrete. *Mag. Concr. Res.* **1956**, *8*, 85–94. [CrossRef]
22. Mohamedbhai, G.T.G. Effect of exposure time and rates of heating and cooling on residual strength of heated concrete. *Mag. Concr. Res.* **1986**, *38*, 151–158. [CrossRef]
23. Morita, T.; Saito, H.; Kumagai, H. Residual mechanical properties of high strength members exposed to high temperature. In *Fire Performance of High-Strength Concrete: A Report of the State-of-the-Art*; NISTIR 5934; Building and Fire Research Laboratory: Gaithersburg, MD, USA, 1996.
24. Nassif, A. Postfire full stress–strain response of fire-damaged concrete. *Fire Mater.* **2005**, *30*, 323–332. [CrossRef]
25. Netinger, I.; Kesegic, I.; Guljas, I. The effect of high temperatures on the mechanical properties of concrete made with different types of aggregates. *Fire Saf. J.* **2011**, *46*, 425–430. [CrossRef]
26. Noumowé, A.; Clastres, P.; Debicki, G.; Costaz, J.-L. Transient heating effect on high strength concrete. *Nucl. Eng. Des.* **1996**, *166*, 99–108. [CrossRef]

27. Phan, L.T.; Carino, N.J. *Mechanical Properties of High Strength Concrete at Elevated Temperatures*; NISTIR 6726; Building and Fire Research Laboratory: Gaithersburg, MD, USA, 1996.
28. Poon, C.S.; Azhar, S.; Anson, M.; Wong, Y.-L. Comparison of the strength and durability performance of normal- and high-strength pozzolanic concretes at elevated temperatures. *Cem. Concr. Res.* **2001**, *31*, 1291–1300. [CrossRef]
29. Purkiss, J.A.; Dougill, J.W. Apparatus for compression tests of concrete at elevated temperatures. *Mag. Concr. Res.* **1973**, *25*, 102–108. [CrossRef]
30. Savva, A.; Manita, P.; Sideris, K.K. Influence of elevated temperatures on the mechanical properties of blended cement concretes prepared with limestone and siliceous aggregates. *Cem. Concr. Compos.* **2005**, *27*, 239–248. [CrossRef]
31. Shen, L.M. Study on Concrete at High Temperature in China: An Overview (Fire-resistance analysis on carbonated concrete columns). *Fire Saf. J.* **2004**, *39*, 89–103.
32. Tan, W. A Research on Reinforced Concrete Beams Subjected to High Temperature and Expert Systems. Ph.D. Thesis, Tongji University, Shanghai, China.
33. Yao, Y.X. Research on Fire Response of Reinforced Concrete Frames and Determination of Temperature Reached During a Fire. Master's Thesis, Tongji University, Shanghai, China.
34. *EN 1992-1-2*; Eurocode 2: Design of Concrete Structures—Part 1–2: General Rules–Structural Fire Design; European Committee for Standardization: Brussels, Belgium, 2004.
35. *Implementation of Eurocodes—Handbook 5—Design of Buildings for the Fire Situation*; Leonardo Da Vinci Joint Research Project; European Commission: Brussels, Belgium, 2005.
36. Guo, Z.; Shi, X. *Experiment and Calculation of Reinforced Concrete at Elevated Temperatures*; Elsevier: Amsterdam, The Netherlands, 2011. [CrossRef]
37. Hertz, K.D. Concrete strength for fire safety design. *Mag. Concr. Res.* **2005**, *57*, 445–453. [CrossRef]
38. Lennon, T. *Results and Observations from Full-Scale Fire Test at BRE Cardington, No. 215-741*; Building Research Establishment: Cardington, UK, 2004.
39. Jansson, R. Fire Spalling of Concrete. Ph.D. Thesis, KTH Royal Institute of Technology, Stockholm, Sweden, 2013.
40. Poon, C.S.; Azhar, S.; Anson, M.; Wong, Y.-L. Strength and durability recovery of fire-damaged concrete after post-fire-curing. *Cem. Concr. Res.* **2001**, *31*, 1307–1318. [CrossRef]
41. Crook, D.N.; Murray, M.J. Regain of strength after firing of concrete. *Mag. Concr. Res.* **1970**, *22*, 149–154. [CrossRef]

Article

Residual Stress-Strain Relationship of Scoria Aggregate Concrete with the Addition of PP Fiber after Fire Exposure

Bin Cai [1], Yu Tao [1] and Feng Fu [2,*]

[1] School of Civil Engineering, Jilin Jianzhu University, Changchun 130118, China; caibin@jlju.edu.cn (B.C.); 15842033023@163.com (Y.T.)

[2] School of Mathematics, Computer Science and Engineering, City, University of London, London EC1V 0HB, UK

* Correspondence: feng.fu.1@city.ac.uk

Abstract: Scoria aggregate concrete (SAC) as new green material has been gradually used in some construction projects for its lightweight and high strength, which can reduce the environmental impact of construction materials. In this paper, the residual mechanical properties and intact compressive stress-strain relationships of polypropylene (PP) fiber-reinforced Scoria aggregate concrete after high-temperature exposure at 20, 200, 400, 600, and 800 °C were investigated. The failure modes of PP fiber-reinforced Scoria aggregate concrete specimens and the effect of high temperatures on the peak stress, secant modulus, and peak strain were obtained. The results showed that the residual compressive strength of heated concrete is significantly reduced when the temperature exceeds 400 °C. The residual strength and residual secant modulus of PP fiber-reinforced Scoria aggregate concrete are significantly higher than those of ordinary concrete. The Scoria aggregate concrete specimens with PP fibers exhibited fewer surface cracks and fewer edge bursts under high temperatures. The residual stress-strain equation of the Scoria aggregate concrete was established by regression analysis, which agreed well with the experimental results.

Keywords: Scoria aggregate concrete; PP fiber; high temperature; stress-strain curve

Citation: Cai, B.; Tao, Y.; Fu, F. Residual Stress-Strain Relationship of Scoria Aggregate Concrete with the Addition of PP Fiber after Fire Exposure. *Fire* **2021**, *4*, 91. https://doi.org/10.3390/fire4040091

Academic Editor: Maged A. Youssef

Received: 9 November 2021
Accepted: 2 December 2021
Published: 5 December 2021

1. Introduction

With the development of modern building structures with large spans, high rises, and super high rises, concrete that is lightweight, high-strength, and sustainable is needed to lower the structural weight and improve the thermal insulation. Lightweight aggregate concrete has a high quality, is widely used [1], and has long-term performance in buildings [2]. Although it is possible to meet structural strength requirements by using artificial aggregates to make lightweight aggregate concrete, the consumption of materials and energy is often substantial. Therefore, it is important to find natural aggregates with good material properties. Scoria aggregate, which is abundant in Northeast China [3] and is very clean, is one such material. Scoria aggregate concrete is characterized by its high strength, heat insulation, light weight, fire resistance, good deformation performance and low modulus of elasticity [4], and these characteristics can reduce environmental impacts when this material is used in buildings.

Scoria aggregate concrete has excellent compatibility, strength, and water permeability characteristics [5] and good chemical resistance, which enables it to maintain a stable state under acidic conditions with less mass loss [6]. Blocks made from Scoria aggregate concrete are 30–40% lighter in weight than normal concrete with the same strength [7]. As a result, it has wide application prospects [8]. In recent years, the use of this material has increased owing to the increasing demand for environmentally friendly materials and green buildings, and research on it has further developed. Willy H. Juimo Tchamdjou et al. [9] prepared two sets of natural light-aggregate concrete specimens and investigated the performance of Cameroonian Scoria aggregate concrete compared to ordinary lightweight concrete. The tests showed that the compressive strength of this concrete increased by 27.42–35.36%, proving that Cameroonian Scoria aggregate concrete can be used as structural concrete. Ahmed A. Abouhussien et al. [10] conducted an experimental study on lightweight slag aggregate concrete and showed that concrete beams mixed with lightweight slag aggregates exhibited higher shear strength compared to normal concrete beams. Kozo Onoue et al. [11] investigated the shock absorption capacity of volcanic pumice lightweight concrete through experiments. The results showed that lightweight concrete had a better damping capacity than the control concrete, which used limestone powder as the coarse aggregate, and was on average 28 and 41% more efficient than the two control concretes in decreasing the maximum impact loads at impact velocities of 1.5 and 4.5 m/s, respectively. J. Alexandre Bogas [12] studied the application of nonstructural lightweight concrete produced using Scoria aggregate in building floor slabs and investigated its mechanical properties. Scoria aggregate concrete containing slag exhibited a similar mechanical strength and better high-temperature properties compared to normal concrete. Aref M. al-Swaidani [13] analyzed the effect of parameters such as the cement content, the Scoria aggregate content, and the water content on other properties by building a neural network model.

Accidental fires are a major cause of durability problems in concrete structures. Aggregate replacement and fiber addition are feasible methods to improve the fire resistance of concrete structures. Generally, concrete structures perform well in fires, but concrete without significant damage may also show a decrease in strength because of an increase in temperature [14]. Therefore, it is important to investigate the reduction in the mechanical properties of concrete after the fire to evaluate and repair fire-damaged concrete elements. At present, the main studies include the basic mechanical properties of lightweight aggregate concrete after exposure to high temperatures and the full stress-strain curve of light-aggregate concrete [15]. Chang, Y. F. et al. [16] conducted an experimental study on the complete stress-strain relationship of concrete after high temperature. The temperature effect on the mechanical properties of the material and the full curve model of stress-strain were obtained by regression analysis. Krzysztof Drozdzol [17] studied the feasibility of using perlite concrete blocks for chimneys. The tests proved that although the thermal loading reduced the compressive strength of the chimney blocks, they still showed an adequate average strength of 4.03 MPa. Shoroog Alraddad [18] analyzed volcanic rocks by temperature difference analysis and thermogravimetric analysis and found that volcanic rocks have good thermal stability and are a highly available and low-cost natural material. Khandaker M. Anwar Hossain [19] studied high-strength Scoria aggregate concrete at 800 °C for strength and durability, and compared with a high-strength concrete control, this material showed a better performance in terms of the residual strength, resistance to chloride ion attack, and resistance to high-temperature deterioration. Waqas Latif Baloch [20] studied the effect of incorporation of multi-walled carbon nanotubes on concrete. The results showed that the incorporation of multi-walled carbon nanotubes could improve the strength of concrete both prior to and after exposure to fire. C. Maraveas [21,22] conducted a sensitivity study on the performance of 19th century fireproof flooring systems at high temperatures, and the applicability of the Eurocode expressions to 19th century fireproof flooring systems is satisfactory. Incorporating appropriate amounts of polypropylene (PP) fibers into concrete not only improves the material properties but also enhances its fire resistance and prevents high-temperature bursting [23]. Nicolas Ali Libre et al. [24] tested experimentally the effectiveness of nine mixtures of steel and polypropylene fibers with different volume fractions to improve the ductility of lightweight pumice aggregate concrete. Studies have shown that steel fibers show a very significant improvement in flexural properties, while the improvement in compressive strength is smaller. PP fiber incorporation has little effect on the mechanical properties of concrete. Xi Liu et al. [25] conducted 30 group experiments to investigate whether the incorporation of fibers could positively affect the mechanical properties and axial stress-strain behavior of lightweight confined carbon fiber aggregate concrete. The peak stresses and corresponding strains were modeled and were in good agreement with the experimental results. According to the test results, the optimum dosing of both steel fiber and carbon fiber is 0.6%. Vahid Afroughsabet [26] investigated the effect of incorporation of steel and polypropylene fiber mixture with 1% volume dose on the mechanical properties and some durability of high-strength concrete. The results showed that the incorporation of 1% volume dose of steel and polypropylene fiber mixture significantly improved the mechanical properties of high-strength concrete. Li Jing Jun et al. [27] analyzed the influence of high-performance PP fibers on the mechanical properties of light-aggregate concrete. It was found that the incorporation of high-performance PP fibers significantly improved their mechanical properties, with an increase in the bending strength, splitting tensile strength, bending toughness, and impact resistance but no significant influence on the compressive strength. To achieve more accurate experimental values in prismatic uniaxial compressive experiments, Liu [28] used the digital image correlation (DIC) method to measure the displacement and strain values of the specimens to obtain their strain clouds, which can reveal phenomena such as regions of crack occurrence and stress concentration. Scoria aggregate concrete has been widely studied because of its many advantages, but the mechanical properties of PP fiber-reinforced Scoria aggregate concrete after exposure to high temperatures and its stress-strain relationship have not been reported.

The purpose of this paper is to obtain the mechanical property and residual stress-strain relationship of Scoria aggregate concrete after high temperature. The data obtained are very important for the design and analysis of building structures, but there are few studies on the stress-strain constitutive relationship of Scoria aggregate concrete after fire exposure. Therefore, we conducted some mechanical tests to obtain the compressive strength and splitting tensile strength of Scoria aggregate concrete after high temperature and measured the residual compressive stress-strain relationship after high-temperature exposure by an advanced DIC system to obtain the changes in peak strain, secant modulus, ultimate strain, and deformation capacity of Scoria aggregate concrete at different temperature exposure levels. In addition, the regression analysis of the experimental results led to the establishment of the constitutive relation equation of Scoria aggregate concrete after high temperature, which provides a reference for the fire design of PP fiber-reinforced Scoria aggregate concrete structures and the evaluation and repair after fire.

2. Experimental Program

2.1. Materials and Mix Proportion

The high-strength Scoria aggregate concrete used in this study was designed according to the JGJ51-2002 Technical Specification for Lightweight Aggregate Concrete to produce a design strength grade of C30 and an average density of 1900 kg/m^3. The mixture proportions are shown in Table 1. Scoria aggregate concrete is made of volcanic slag aggregate, normal Portland cement, styrene–acrylic emulsion, grade II fly ash, PP fibers, and water. The raw material is shown in Figure 1. The porous Scoria aggregate with a bulk density of 815 kg/m^3 was produced in Gushanzi, Huinan County, and artificially pulverized into a continuous gradation of 5~40 mm. The stone for normal concrete is ordinary gravel with a grading size of 5–25 mm, and the sand is ordinary river sand with a fineness modulus of 2.8. P O 42.5 ordinary silicate concrete was used in the mix, initial setting time: 85 min, final setting time: 260 min, compressive strength: 43 MPa (28 days). The bundled monofilament PP fibers had a length of 9 mm, 400 MPa tensile strength, 160 °C melting point, and a PP fiber content of 0.22% by volume. Tap water was used for mixing.

Table 1. Mix proportion (kg/m^3).

Type	W/B	SA	Gravel	Sand	Cement	Fly Ash	SAE	WRA	PP Fiber	Water
NC	0.5	-	1192	596	370	54	5.4	0.47	-	185
SAC	0.46	1110	-	-	496	54	5.4	0.47	2	230

W/B = Water/Binder ratio; SA = Scoria aggregate; SAE = Styrene–acrylic emulsion; WRA = Water-reducing admixture.

Figure 1. Materials: (**a**) Scoria aggregate, (**b**) PP fiber.

In this paper, three tests were conducted: cube compressive strength, splitting tensile strength, and the axial compressive strength of prismatic with dimensions of 100 mm × 100 mm × 300 mm. A test procedure similar to [29–33] was used. The test specimens for three groups of tests need to be heated to 20, 200, 400, 600, and 800 °C. Therefore, a total of 45 specimens of Scoria aggregate concrete and 45 specimens of normal concrete were made for comparison. The prepared test block was placed at room temperature for 24 h and then demolded and cured at 20 °C with 95 ± 5% relative humidity for 28 days.

2.2. Test Procedure

All the following tests were performed in the Structures Laboratory of Jilin Jianzhu University, China. The heating equipment was a resistance furnace with a heating rate of 5 °C/min and the temperature in the furnace up to 1000 °C. This replicated the standard fire tests [34,35]. After the specimens were heated to 200, 400, 600, and 800 °C, the furnace temperature was maintained for 3 h to ensure that the internal temperature of the specimens also reached the target value.

The compressive strength of the cube specimens was tested by a YAR-2000 hydraulic testing machine with a loading rate of 0.5 MPa/s. The axial compressive strength was obtained. Axial compressive tests were performed on the prismatic blocks using a Type YAW-5000 electrohydraulic servo universal testing machine. The testing machine applied load through displacement control and set the loading rate to 0.1 mm/min. The stress was controlled by the test machine, and the strain was measured by DIC. The stress-strain curve was then plotted. The test machine is shown in Figure 2.

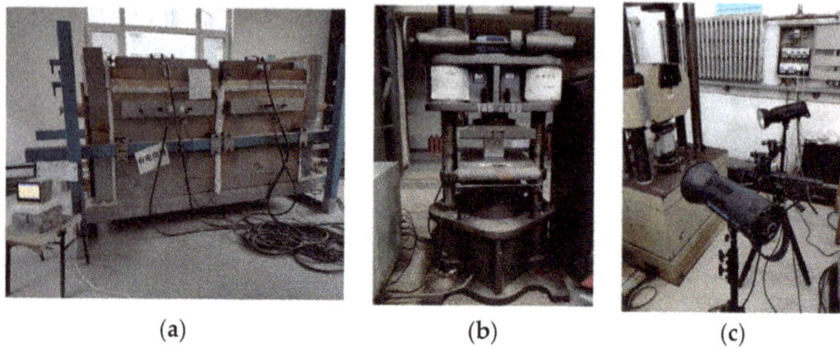

Figure 2. Testing machine: (**a**) Resistance furnace, (**b**) Hydraulic pressure testing machine, (**c**) DIC.

DIC is a noncontact deformation test method in which marking spots are randomly scattered on a specimen surface, and the changes in the relative positions of these scatter spots during the loading process are compared. The displacement field on the specimen surface was calculated to obtain the strain distribution field for further analysis. A high-speed camera with a resolution of 1280 × 800 was used for image acquisition. The power of the dimmable LED lamp was 1000 W. The camera was placed symmetrically with the fill light, and the specimen surface with all the scatter spots was placed in the center of the visible range of the camera. The camera was used to photograph the prismatic axial pressure test at a frequency of one photograph per second. The photographs were acquired and analyzed using DIC software to obtain the vertical displacement and transverse strain clouds of the specimen.

3. Test Results and Discussion

3.1. Failure Mode

3.1.1. Color Changes

Figure 3 shows the changes in the color and the surface cracks of NC and SAC after exposure to different high temperatures. The specimens changed to yellowish gray, brownish-gray, brown, and whitish gray after being subjected to high temperatures of 200, 400, 600 and 800 °C, respectively. Temperatures of 800 and 600 °C produced significantly different color changes in the specimens. The Scoria aggregate concrete specimens with PP fibers exhibited fewer surface cracks and less edge bursting after exposure to 800 °C. This result may have occurred because the PP fibers melted at 160 °C, forming channels through which vapor could diffuse outward and thereby reduce the vapor pressure inside the concrete. These channels were effective in preventing bursting phenomena when the temperature did not exceed 800 °C. PP fibers have a melting point of approximately 160 °C and were therefore visible inside the specimens at room temperature. When the fire temperature reached 400 °C, the PP fibers inside the specimens disappeared.

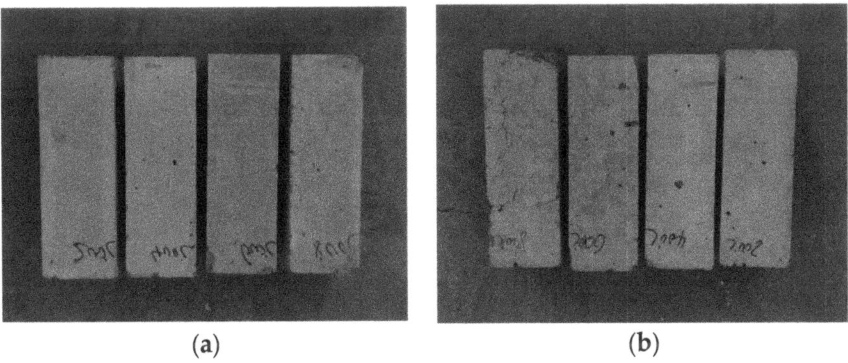

(a)   (b)

Figure 3. The changes in the color and surface cracks of specimens after high temperatures: (**a**) SAC, (**b**) NC.

3.1.2. Cracking Behavior

Figure 4 shows the maximum crack width on the test block surface after high-temperature treatment, as measured by a fracture-width tester. The maximum crack of Scoria aggregate concrete at 800 °C is 0.301 mm, which is 4.43 times higher than that at 200 °C. The maximum crack of normal concrete at 800 °C is 0.562 mm, which is 5.11 times higher than that at 200 °C. The maximum crack width for normal concrete was up to 1.87 times greater than that of the Scoria aggregate concrete.

Figure 5 shows similar patterns of diagonal shear cracking and splitting cracking for the uniaxial compressive damage to the Scoria aggregate concrete specimens treated at different temperatures. The specimens were in the elastic stage at the beginning of loading. The load and displacement increased linearly, and there were no noticeable features on the specimen surfaces. As the load increased, internal microcracks gradually formed, and the specimen stiffness began to decrease. The load increased to a peak, beyond which cracks parallel to the force direction appeared on the surface of the specimens. When the load was reduced to 60–70% of the peak load, the axial deformation of the concrete block continued to increase and eventually resulted in the destruction of the specimens. The Scoria aggregate concrete specimens showed more noticeable brittle damage than the conventional concrete specimens, with fewer cracks on the specimen surface and less transverse deformation.

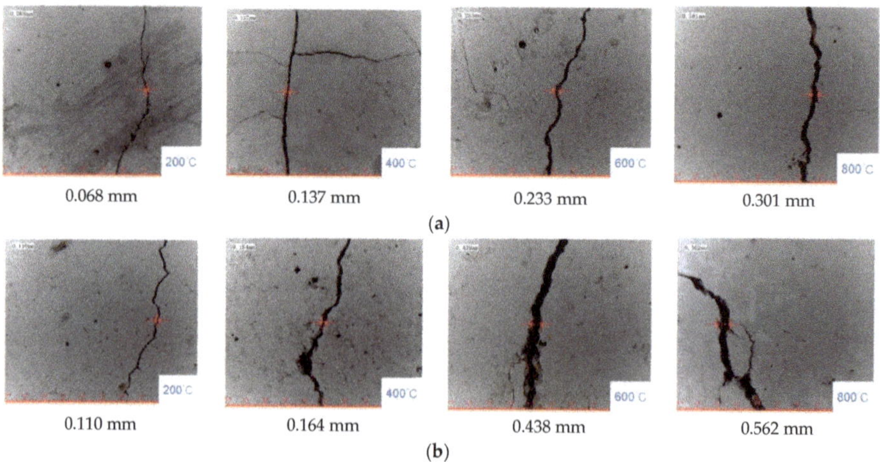

Figure 4. Maximum crack on the surface of specimens after high temperature: (**a**) SAC, (**b**) NC.

Figure 5. Destruction mode of SAC at different temperatures.

3.2. Residual Strength

By testing the compressive strength and splitting tensile strength of concrete specimens treated at different temperatures, it was concluded that the strength of Scoria aggregate concrete decreases with increasing temperature. The effects of temperature on the compressive strength and the strength reduction coefficients of Scoria aggregate concrete are shown in Table 2.

Table 2. Compression strength of specimens after high temperature.

T (°C)	Compression Strength (Mpa)		Strength Reduction	
	SAC	NC	SAC	NC
20	41.66	39.95	1	1
200	37.33	34.25	0.89	0.86
400	34.83	32.67	0.84	0.82
600	20.74	17.66	0.5	0.44
800	11.1	8.09	0.27	0.2

The reduction in compressive strength of Scoria aggregate concrete specimens below 400 °C was relatively low, and the compressive strength reduced to 89 and 84% of that at room temperature. However, the reduction in compressive strength of specimens above 400 °C was relatively large and the compressive strength reduced to 50 and 27% of that at room temperature. This result was obtained because the hydration products in the specimens gradually decomposed after 450 °C, and the thermal mismatch between cement paste and aggregates, resulting in a rapid increase in the number of cracks inside the concrete and a significant decrease in the strength of the specimens. As shown in Figure 6, the compressive strength reduction coefficient of the Scoria aggregate concrete at all temperatures was greater than that of normal concrete, which indicated that the effect of high temperature on the compressive strength of the Scoria aggregate concrete was smaller. The data in Table 2 were fitted to obtain Equations (1) and (2) for the residual axial compressive strengths of SAC and NC after high-temperature treatment.

$$f_{cs,T}/f_{cs} = 0.9872 + 0.005142\left(\tfrac{T}{100}\right) - 0.0169\left(\tfrac{T}{100}\right)^2$$
$$+ 0.0006114\left(\tfrac{T}{100}\right)^3 \quad 20\,°C \leq T \leq 800\,°C \tag{1}$$

$$f_{cn,T}/f_{cn} = 0.9865 - 0.006324\left(\tfrac{T}{100}\right) - 0.01573\left(\tfrac{T}{100}\right)^2$$
$$+ 0.0005038\left(\tfrac{T}{100}\right)^3 \quad 20\,°C \leq T \leq 800\,°C \tag{2}$$

where $f_{cs,T}$ and f_{cs} are the axial compressive strength of SAC at high temperature and at room temperature, respectively (MPa); $f_{cn,T}$ and f_{cn} are the axial compressive strength of NC at high temperature and at room temperature, respectively (MPa); and T is the temperature, °C. The R^2 value in Equation (1) is 0.957. The R^2 value in Equation (2) is 0.946.

The splitting tensile strengths of the Scoria aggregate concrete after high temperature and its strength reduction coefficients are shown in Table 3. The splitting tensile strength of the Scoria aggregate concrete specimens decreased with increasing temperature to 84, 73, 37, and 22% of the splitting tensile strength at room temperature. Similarly, Figure 7 shows that the splitting tensile strength reduction coefficient of the Scoria aggregate concrete was greater than that of normal concrete at all temperatures. Equations (3) and (4) for the residual axial splitting tensile strength of the specimen were similarly obtained by curve fitting.

$$f_{ts,T}/f_{ts} = 0.9902 - 0.009663\left(\tfrac{T}{100}\right) - 0.02479\left(\tfrac{T}{100}\right)^2$$
$$+ 0.001725\left(\tfrac{T}{100}\right)^3 \quad 20\,°C \leq T \leq 800\,°C \tag{3}$$

$$f_{tn,T}/f_{tn} = 1.009 - 0.1383\left(\tfrac{T}{100}\right) + 0.01342\left(\tfrac{T}{100}\right)^2$$
$$- 0.001301\left(\tfrac{T}{100}\right)^3 \quad 20\,°C \leq T \leq 800\,°C \tag{4}$$

where $f_{ts,T}$ and f_{ts} are the splitting tensile strength of SAC at high temperature and at room temperature, respectively (MPa); $f_{tn,T}$ and f_{tn} are the splitting tensile strength of NC at high temperature and at room temperature, respectively (MPa), and the R^2 value in Equation (3) is 0.961. The R^2 value in Equation (4) is 0.931.

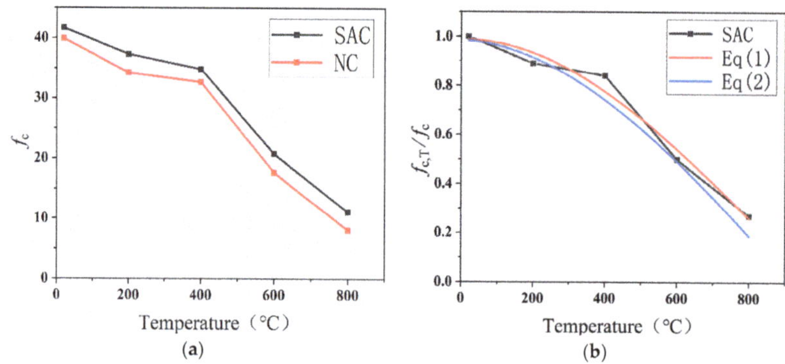

Figure 6. (**a**) Reduction of compressive strength. (**b**) Relative reduction of compressive strength.

Table 3. Splitting tensile strength of specimens after high temperature.

T (°C)	Splitting Tensile Strength (Mpa)		Strength Reduction	
	SAC	NC	SAC	NC
20	2.69	2.02	1	1
200	2.26	1.44	0.84	0.71
400	1.96	1.37	0.73	0.68
600	1	0.65	0.37	0.32
800	0.6	0.22	0.22	0.11

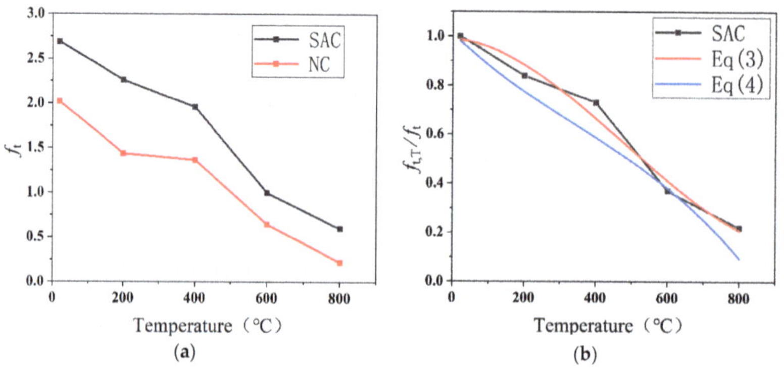

Figure 7. Splitting tensile strength. (**a**) Reduction of splitting tensile strength. (**b**) Relative reduction of splitting tensile strength.

3.3. Stress-Strain Relationship of Scoria Aggregate Concrete

3.3.1. Strain Distribution Analysis of DIC

Figure 8 shows the strain distribution from the DIC image of the peak strain and 80% peak strain of the Scoria aggregate concrete at different temperatures. It can be seen that the strain values gradually increased as the temperature increased. As shown in Figure 8b, for the specimen with the lower temperature, when the peak strain was reached, only the transverse strain concentration zone appeared at the bottom of the specimen, and no cracks appeared in the middle of the specimen. For the specimens subjected to higher temperatures, when the peak strain was reached, transverse strain concentration zones appeared at the top and bottom of the specimens, and these strain concentration zones were connected to form cracks, leading to damage of the specimen. The cracks were consistent with the failure mode of the test block in Figure 5, mainly diagonal shear cracking and splitting cracking, as shown in Figure 8h.

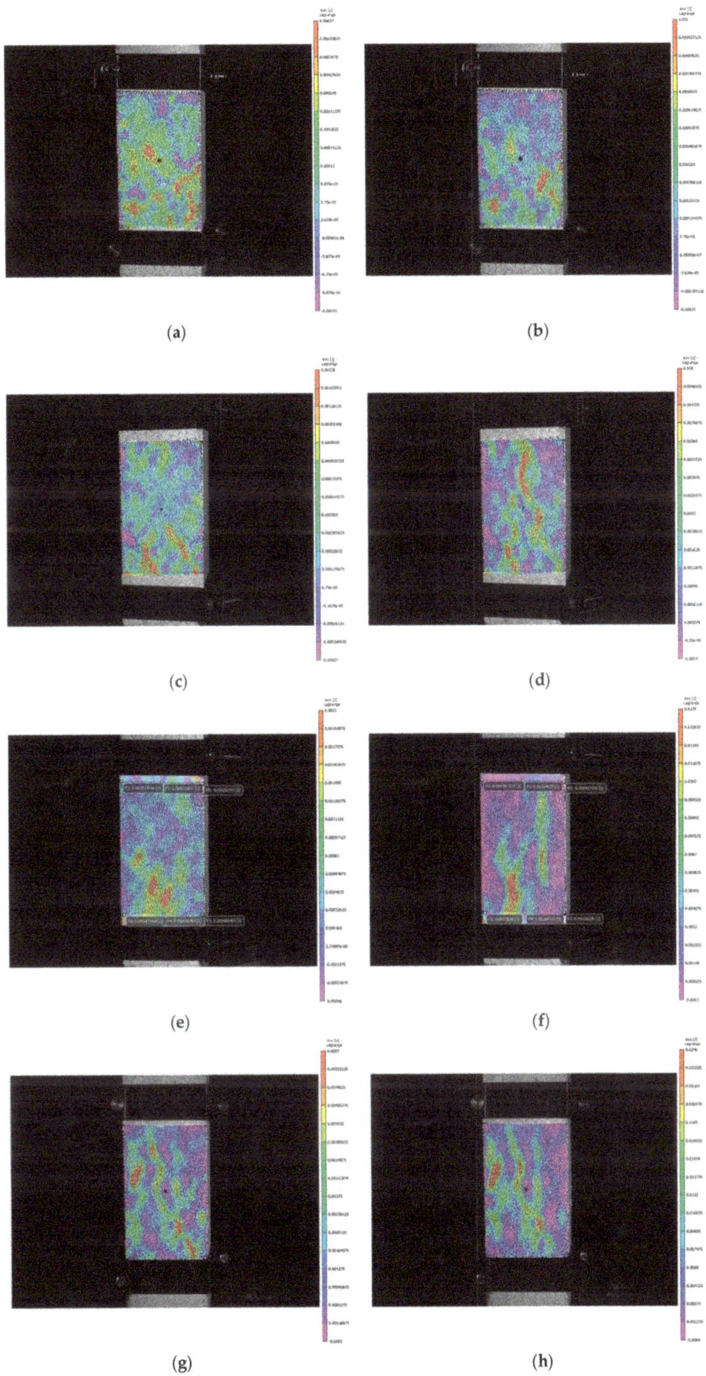

Figure 8. Strain nephogram of SAC: (**a**) 200 °C 80% peak strain; (**b**) 200 °C peak strain; (**c**) 400 °C 80% peak strain; (**d**) 400 °C peak strain; (**e**) 600 °C 80% peak strain; (**f**) 600 °C peak strain; (**g**) 800 °C 80% peak strain; (**h**) 800 °C peak strain.

3.3.2. Peak Strain

As shown in Table 4, the peak strains of both normal concrete and Scoria aggregate concrete increased with increasing temperature. However, the change in the peak strain of the Scoria aggregate concrete after high-temperature treatment was smaller than that of normal concrete. The difference between the two peak strains increased significantly above 400 °C. The peak strains of normal concrete after exposure to high temperatures were 5.42 and 5.97 times greater than those at room temperature, while the peak strains of the Scoria aggregate concrete after high temperatures were 1.9 and 3.11 times greater than those at room temperature. The excellent fire resistance of the Scoria aggregate concrete specimens resulted in fewer cracks and a smaller peak strain compared to the conventional concrete. The effect of temperature on peak strain is shown in Figure 9. Equations (5) and (6) for the relative peak strain of the specimen were similarly obtained by curve fitting.

$$\varepsilon_{cps,T}/\varepsilon_{cps} = 0.9874 - 0.08143\left(\tfrac{T}{100}\right) + 0.03687\left(\tfrac{T}{100}\right)^2 \\ +0.0007655\left(\tfrac{T}{100}\right)^3 \quad 20\ ^{\circ}\mathrm{C} \leq T \leq 800\ ^{\circ}\mathrm{C} \tag{5}$$

$$\varepsilon_{cpn,T}/\varepsilon_{cpn} = 1.4 - 1.599\left(\tfrac{T}{100}\right) + 0.6556\left(\tfrac{T}{100}\right)^2 \\ -0.04786\left(\tfrac{T}{100}\right)^3 \quad 20\ ^{\circ}\mathrm{C} \leq T \leq 800\ ^{\circ}\mathrm{C} \tag{6}$$

where $\varepsilon_{cps,T}$ and ε_{cps} are the peak strain of SAC at high temperature and at room temperature, respectively (MPa); $\varepsilon_{cpn,T}$ and ε_{cpn} are the peak strain of NC at high temperature and at room temperature, respectively (MPa); and the R^2 value in Equation (5) is 0.9763. The R^2 value in Equation (6) is 0.954.

Table 4. Peak strain of specimens after high temperature.

T (°C)	Peak Strain		Relative Peak Strain	
	SAC	NC	SAC	Equation (3)
20	0.00187	0.001887	1	0.9786
200	0.001646	0.001551	0.88	0.9781
400	0.002683	0.003564	1.44	1.301
600	0.003549	0.010222	1.9	1.991
800	0.005809	0.011266	3.11	3.088

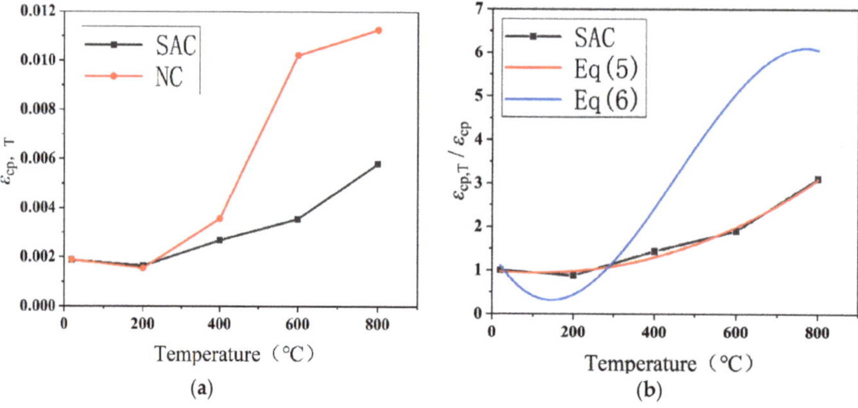

Figure 9. Peak strain of specimens after high temperature: (**a**) Peak strain, (**b**) Relative peak strain.

3.3.3. Secant Modulus

In this paper, the secant modulus corresponding to the rising section of the stress-strain curve of concrete after high temperature from the origin to the 40% peak stress point was taken as the elasticity modulus. Table 5 shows that the secant modulus of the specimens decreased with different degrees of temperature increase, but the secant modulus of Scoria aggregate concrete specimens at different temperatures was greater than that of ordinary concrete. The secant modulus for the Scoria aggregate concrete specimens treated at 200, 400, 600, and 800 °C was approximately 70, 30, 13.6, and 3.5% of the secant modulus of the unheated concrete, respectively. The decreasing trend is shown in Figure 10. Equations (7) and (8) for the relative secant modulus of the specimen were obtained by curve fitting.

$$S_{cs,T}/S_{cs} = 1.045 - 0.1864\left(\frac{T}{100}\right) - 0.002428\left(\frac{T}{100}\right)^2 + 0.001255\left(\frac{T}{100}\right)^3 \quad 20\,°C \leq T \leq 800\,°C \tag{7}$$

$$S_{cn,T}/S_{cn} = 1.04 - 0.1499\left(\frac{T}{100}\right) - 0.02559\left(\frac{T}{100}\right)^2 + 0.003537\left(\frac{T}{100}\right)^3 \quad 20\,°C \leq T \leq 800\,°C \tag{8}$$

where $S_{cs,T}$ and S_{cs} are the secant modulus of SAC at high temperature and at room temperature, respectively (MPa); $S_{cn,T}$ and S_{cn} are the secant modulus of NC at high temperature and at room temperature, respectively (MPa), and the R^2 value in Equation (7) is 0.9908. The R^2 value in Equation (8) is 0.989.

Table 5. Secant modulus of specimens after high temperature.

T (°C)	Residual Secant Modulus		Relative Residual Secant Modulus	
	SAC	NC	SAC	Equation (4)
20	57,851	33,898	1	1.0076
200	40,441	23,691	0.7	0.6725
400	17,456	7100	0.302	0.341
600	7861	499	0.136	0.11
800	2039	229	0.035	0.0212

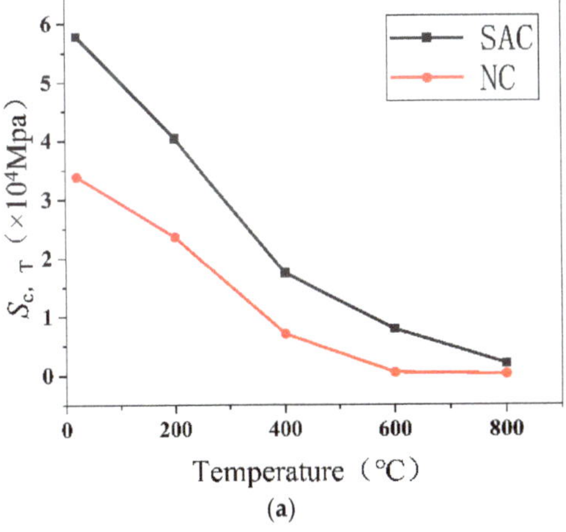

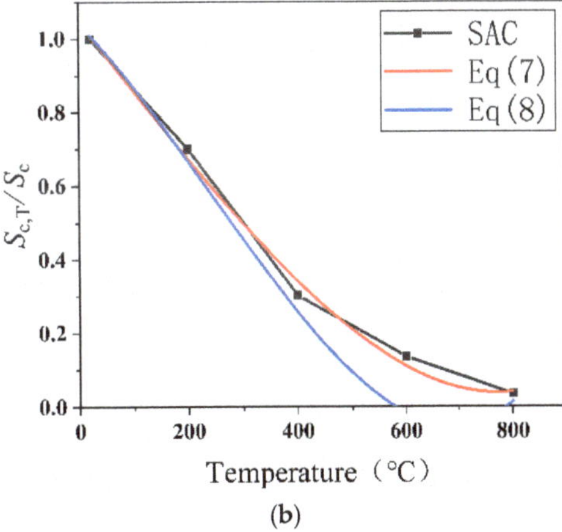

Figure 10. Secant modulus of specimens after high temperature: (**a**) Residual secant modulus. (**b**) Relative residual secant modulus.

3.3.4. Ultimate Strain

The descending section of the stress-strain curve was used to determine the ultimate strain of concrete, corresponding to the strain at the $0.5 f_{cr}$ stress value. The ultimate strain of the specimens after high-temperature treatment is shown in Table 6. The ultimate strain of the concrete increased with temperature, and the amplitude of the variation increased for temperatures above 400 °C. The ultimate strains of the Scoria aggregate concrete specimens subjected to temperatures above 400 °C were 4.24 and 7.4 times greater than those at room temperature, where the relationship between the temperature and the ultimate strain are shown in Figure 11. Although the ultimate strain reduction coefficient of Scoria aggregate concrete was greater than that of normal concrete, the ultimate strain of Scoria aggregate concrete was still smaller than that of normal concrete even when the temperature reached 800 °C because the ultimate strain of Scoria aggregate concrete was very small at room temperature. Equations (9) and (10) for the relative ultimate strain of the specimen were obtained by curve fitting.

$$\varepsilon_{cus,T}/\varepsilon_{cus} = 1.105 - 0.5083\left(\tfrac{T}{100}\right) + 0.1971\left(\tfrac{T}{100}\right)^2 - 0.004385\left(\tfrac{T}{100}\right)^3 \quad 20\,^{\circ}\mathrm{C} \leq T \leq 800\,^{\circ}\mathrm{C} \tag{9}$$

$$\varepsilon_{cun,T}/\varepsilon_{cun} = 1.1 - 0.6517\left(\tfrac{T}{100}\right) + 0.2525\left(\tfrac{T}{100}\right)^2 - 0.01653\left(\tfrac{T}{100}\right)^3 \quad 20\,^{\circ}\mathrm{C} \leq T \leq 800\,^{\circ}\mathrm{C} \tag{10}$$

where $\varepsilon_{cus,T}$ and ε_{cus} are the ultimate strain of SAC at high temperature and at room temperature, respectively (MPa); $\varepsilon_{cun,T}$ and ε_{cun} are the ultimate strain of NC at high temperature and at room temperature, respectively (MPa); and the R^2 value in Equation (9) is 0.9996. The R^2 value in Equation (10) is 0.993.

Table 6. Ultimate strain of specimens after high temperature.

T (°C)	Ultimate Strain		Relative Ultimate Strain	
	SAC	NC	SAC	Equation (5)
20	0.00209	0.005243	1	1.0112
200	0.00184	0.003438	0.88	0.842
400	0.003959	0.008294	1.89	1.945
600	0.008862	0.013817	4.24	4.204
800	0.015472	0.018866	7.4	7.41

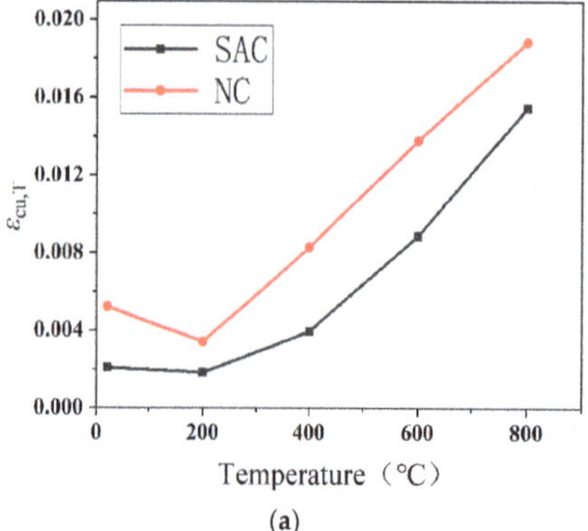

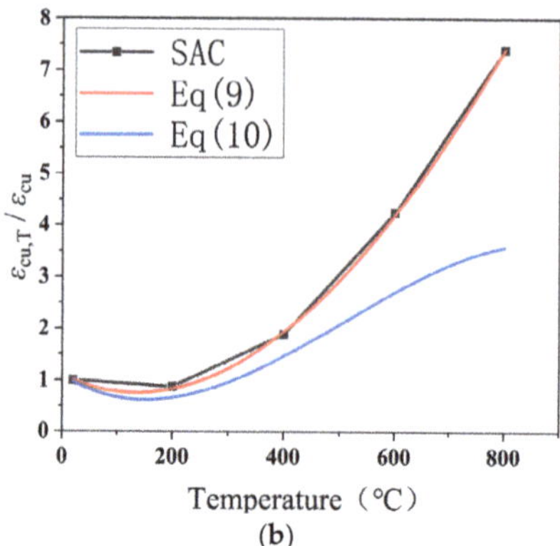

Figure 11. Ultimate strain of specimens after high temperature: (**a**) Ultimate strain. (**b**) Relative ultimate strain.

3.3.5. Deformation Capacity

Different indicators are used in the literature to quantitatively evaluate the deformation capacity of concrete. The ratio of strain at 50% of the peak stress to the peak strain of concrete ($\varepsilon_{cu}/\varepsilon_{cp}$) was used to evaluate the deformation capacity of the concrete specimens under the action of axial pressure in this study. The larger the ratio is, the higher the deformation capacity of the specimen is. Figure 12 shows how $\varepsilon_{cu}/\varepsilon_{cp}$ varies with temperature. Compared to that of the Scoria aggregate concrete, the deformation capacity of the conventional concrete was higher below 400 °C and lower (by approximately half) above 600 °C. Thus, the Scoria aggregate concrete performed better and had a more stable structure than the conventional concrete under high-temperature conditions.

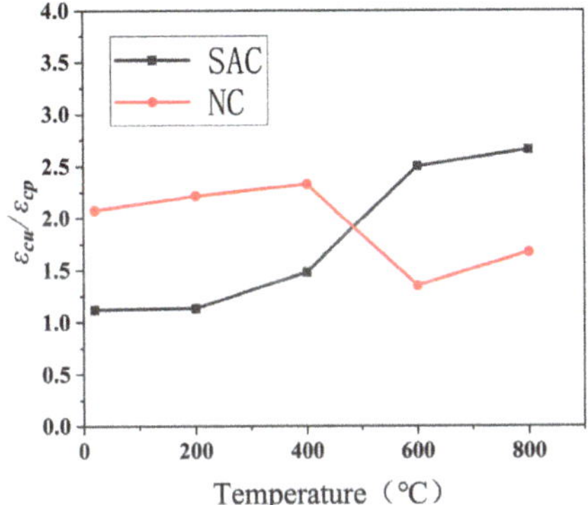

Figure 12. Effect of temperature on $\varepsilon_{cu}/\varepsilon_{cp}$.

4. Development of Constitutive Equations for Scoria Aggregate Concrete

4.1. Stress-Strain Curve

Figure 13 summarizes the axial stress–axial strain curves and axial stress–transverse strain curves of the Scoria aggregate and conventional concrete specimens at different temperatures. Compared to the normal concrete curves, the Scoria aggregate concrete curve has a longer linear ascending section, a steeper descending section (particularly at lower temperatures), and more pronounced brittle damage, which was consistent with the findings of Bing Han [36].

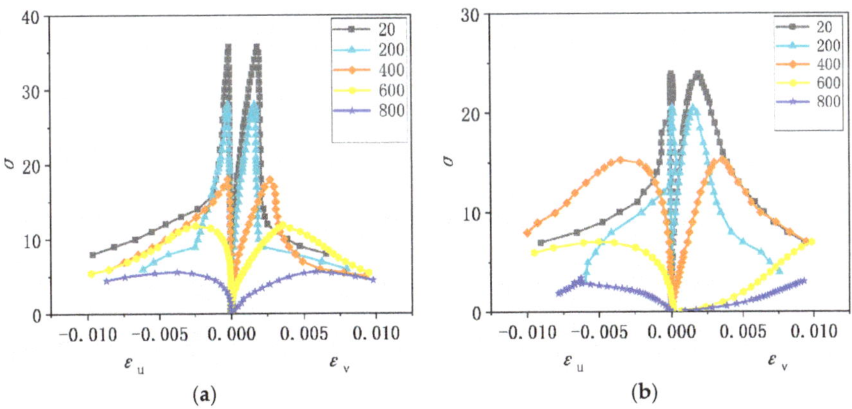

Figure 13. stress-strain curve of specimens after high temperature: (a) SAC, (b) NC.

Where ε_u is the transverse strain of the specimens and ε_v is the vertical strain of the concrete. The detailed axial stress-strain curves for Scoria aggregate concrete at different fire temperatures are shown in Figure 13b. As the temperature increases, the curve area gradually decreases, and the peak strain moves to the right and increases, whereas the secant modulus decreases sharply. The descending section of the curve is very steep at room temperature, and the curve becomes increasingly flat as the temperature increases. This finding showed that when subjected to high temperatures, Scoria aggregate concrete exhibited better mechanical properties than conventional concrete.

4.2. Constitutive Equations

Similar normalized stress-strain curves were obtained for the Scoria aggregate concrete and the conventional concrete, and these curves were divided into four stages, as shown in Figure 14.

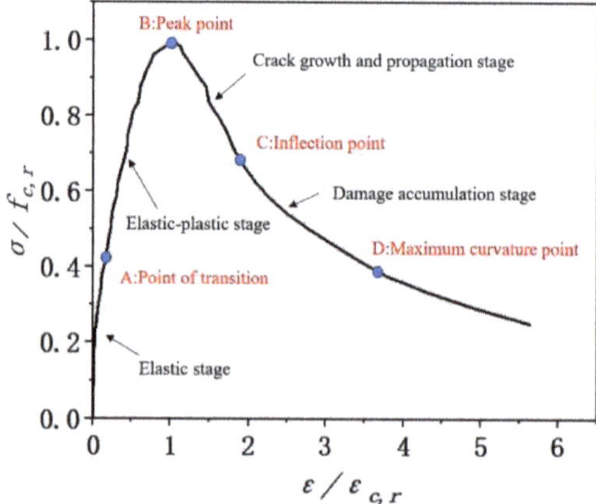

Figure 14. Typical uniaxial compression stress-strain curve of concrete.

Where $\varepsilon_{c,r}$ is the peak strain, and $f_{c,r}$ is the peak stress. In this study, Equation (11) was adopted from the constitutive relation presented in the Code for Design of Concrete Structures (GB50010-2010). The curve consists of ascending and descending sections, the shapes of which are controlled by the independent parameters n and α, respectively. As shown in Equation (12), the parameter n is identified by peak strain, the secant modulus, and peak stress of concrete, and reflects the stress-strain curve characteristics of the rising section of concrete. The parameter α determines the stress-strain curve characteristics of the falling section of concrete. The larger the value of α, the steeper the descending section of the stress-strain curve. The change in the magnitude of α value can reflect the changing characteristics of the concrete stress-strain curve. The concrete plastic deformation properties deteriorate as α increases.

$$\begin{cases} y = nx/n - 1 + x^n & x \leq 1 \\ y = x/\alpha(x-1)^2 + x & x \geq 1 \end{cases} \tag{11}$$

where $x = \varepsilon/\varepsilon_{c,r}, y = \sigma/f_{c,r}$.

$$n = E_c \varepsilon_{c,r} / E_c \varepsilon_{c,r} - f_{c,r}. \tag{12}$$

The experimental data were regressed to obtain n and α for the Scoria aggregate concrete at different fire temperatures. The results are shown in Table 7. Figure 15 is a comparison of Equations (11) and (12) with the experimental curves that were used to validate the proposed constitutive model. In the ascending phase, the normalized stress-strain curves at different temperatures have nearly the same shape with only slight changes in n. The descending phase is very steep at lower temperatures, where the Scoria aggregate concrete was brittle and had poor ductility. The α value was large at lower temperatures and gradually decreased as the temperature increased. The R^2 value reached approximately 0.8. Thus, the constitutive model effectively reproduced the complete test stress-strain curve.

Table 7. Equation parameters of stress-strain curves of SAC after elevated temperatures.

T (°C)	n	R^2	α	R^2
20	1.898	0.805	49.31	0.81
200	2.17	0.886	72.44	0.84
400	1.827	0.78	9.993	0.79
600	2.201	0.79	1.079	0.98
800	1.991	0.976	0.676	0.91

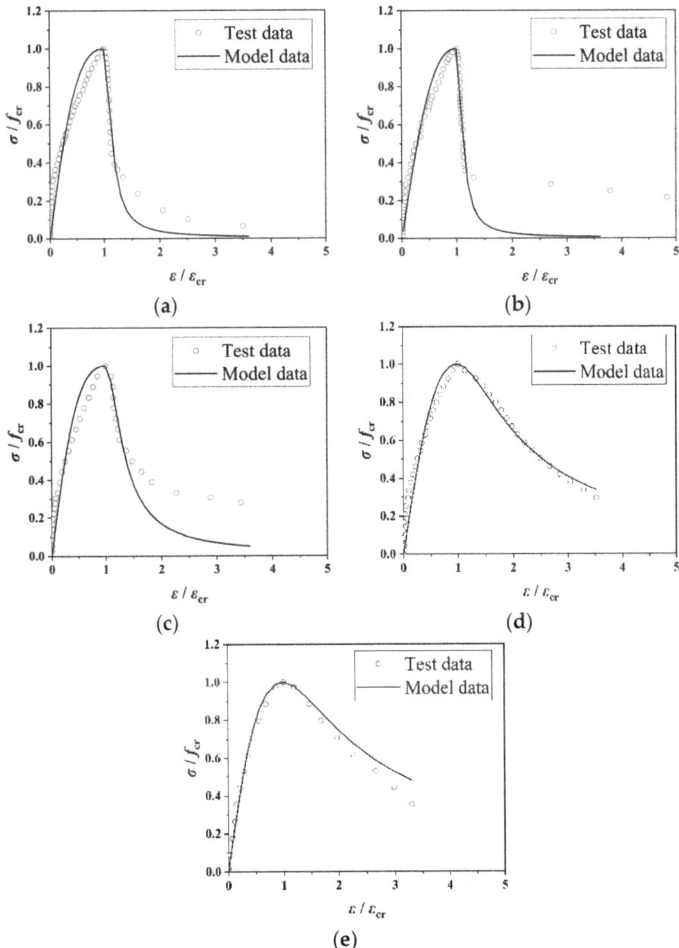

Figure 15. Comparison between calculated and tested curves of HPLWC after different temperatures: (**a**) Ambient temperature, (**b**) 200 °C, (**c**) 400 °C, (**d**) 600 °C, (**e**) 800 °C.

5. Conclusions

After sorting and comparing the data obtained, the following conclusions related to the residual mechanical properties of Scoria aggregate concrete after fire exposure can be made.

(1) The specimens turned yellowish gray, brownish-gray, brown, and whitish gray after being subjected to temperatures of 200, 400, 600, and 800 °C, respectively. The incorporation of the PP fibers effectively prevented bursting phenomena when the temperature did not exceed 800 °C.

(2) The effect of temperature on the maximum crack width was significant, with the maximum crack width of SAC at 800 °C being 4.43 times that at 200 °C. The maximum crack width of the normal concrete was much larger than that of the PP fiber-reinforced Scoria aggregate concrete, reaching a value that was 1.87 times greater than that of Scoria aggregate concrete at 800 °C.

(3) The temperature significantly affected the damage mode of the PP fiber-reinforced Scoria aggregate concrete. Strength reductions began to increase above 400 °C (an important demarcation point). After 200 °C, the PP fiber-reinforced Scoria aggregate concrete retained approximately 89% of its unheated compressive strength, which was further reduced to 50 and 27% after exposure to temperatures of 600 and 800 °C, respectively.

(4) The compressive strength reduction coefficient of Scoria aggregate concrete at 800 °C was 0.27, while the compressive strength reduction coefficient of the normal concrete after 800 °C was 0.2. The degradation of the secant modulus caused by high temperatures was more serious than the degradation of the residual strength, and the secant modulus reduction coefficient of Scoria aggregate concrete at 800 °C was 0.035, while the secant modulus reduction coefficient of the normal concrete after 800 °C was 0.021.

(5) The deformation capacity of the PP fiber-reinforced Scoria aggregate concrete at 600 and 800 °C was generally better than that of the normal concrete, and at 200 and 400 °C, the deformation capacity of the PP fiber-reinforced Scoria aggregate concrete was lower than that of the normal concrete. However, PP fiber-reinforced Scoria aggregate concrete has better mechanical properties compared to normal concrete at high temperatures.

(6) Mathematical expressions for the full stress-strain curve of Scoria aggregate concrete were established based on n and α, and these expressions can be used to numerically simulate the intrinsic structural relationship of Scoria aggregate concrete, as in the present study.

Author Contributions: Conceptualization, B.C. and F.F.; methodology, B.C. and F.F.; test, Y.T.; validation, Y.T.; investigation, Y.T.; resources, B.C.; data curation, B.C.; writing—original draft preparation, Y.T.; writing—review and editing, Y.T.; visualization, B.C.; supervision, B.C.; project administration, B.C.; funding acquisition. All authors have read and agreed to the published version of the manuscript.

Funding: This research was funded by the scientific research projects of the Education Department of Jilin province, grant number JJKH20210279KJ.National Natural Science Foundation of China, 51968013.

Conflicts of Interest: The authors declare that they have no competing interests.

References

1. Ting, T.Z.H.; Rahman, M.E.; Lau, H.H.; Ting, M.Z.Y. Recent development and perspective of lightweight aggregates based self-compacting concrete. *Constr. Build. Mater.* **2019**, *201*, 763–777. [CrossRef]
2. Wang, Z.; Li, X.; Jiang, L.; Wang, M.; Xu, Q.; Harries, K. Long-term performance of lightweight aggregate reinforced concrete beams. *Constr. Build. Mater.* **2020**, *264*, 120231. [CrossRef]
3. Donglin, Y.; Li, B.; Li, W.; Xia, J.; Zhou, Y. Research and Application of Scoria aggregate Lightweight Aggregate Concrete at Home and Abroad. *World Build. Mater.* **2017**, *38*, 26–30.
4. Siddique, R. Properties of concrete made with volcanic ash. *Resour. Conserv. Recycl.* **2012**, *66*, 40–44. [CrossRef]
5. Khandaker, M.A. Blended cement using volcanic ash and pumice. *Cem. Concr. Res.* **2003**, *33*, 1601–1605.
6. Labbaci, Y.; Abdelaziz, Y.; Mekkaouic, A.; Alouanic, A.; Labbacia, B. The use of the volcanic powders as supplementary cementitious materials for environmental-friendly durable concrete. *Constr. Build. Mater.* **2017**, *133*, 468–481. [CrossRef]

7. Demirdag, S.; Gunduz, L. Strength properties of Scoria aggregate aggregate lightweight concrete for high performance masonry units. *Constr. Build. Mater.* **2008**, *22*, 135–142. [CrossRef]
8. Lemougna, P.N.; Wang, K.; Tang, Q.; Nzeukou, A.N.; Billong, N.; Melo, U.C.; Xue-Min, C. Review on the use of volcanic ashes for engineering applications. *Resour. Conserv. Recycl.* **2018**, *137*, 177–190. [CrossRef]
9. Tchamdjou, W.H.J.; Cherradi, T.; Pereira-de-Oliveira, L.A. Mechanical properties of lightweight aggregates concrete made with cameroonian volcanic scoria: Destructive and non-destructive characterization. *J. Build. Eng.* **2018**, *16*, 134–145. [CrossRef]
10. Abouhussien, A.A.; Hassan, A.A.A.; Ismail, M.K. Properties of semi lightweight self consolidating concrete containing lightweight slag aggregate. *Constr. Build. Mater.* **2015**, *75*, 63–73. [CrossRef]
11. Onoue, K.; Tamai, H.; Suseno, H. Shock-absorbing capability of lightweight concrete utilizing volcanic pumice aggregate. *Constr. Build. Mater.* **2015**, *83*, 261–274. [CrossRef]
12. Bogas, J.A.; Cunha, D. Non-structural lightweight concrete with volcanic scoria aggregates for lightweight fill in building's floors. *Constr. Build. Mater.* **2017**, *135*, 151–163. [CrossRef]
13. Al-Swaidani, A.M.; Khwies, W.T. Applicability of Artificial Neural Networks to Predict Mechanical and Permeability Properties of Volcanic Scoria-Based Concrete. *Adv. Civ. Eng.* **2018**, *2018*, 5207962. [CrossRef]
14. Arioz, O. Effects of elevated temperatures on properties of concrete. *Fire Saf. J.* **2007**, *42*, 516–522. [CrossRef]
15. Wei, H.; Wu, T.; Liu, X.; Zhang, R. Investigation of stress-strain relationship for confined lightweight aggregate concrete. *Constr. Build. Mater.* **2020**, *256*, 119432. [CrossRef]
16. Chang, Y.; Chen, Y.; Sheu, M.; Yao, G. Residual stress-strain relationship for concrete after exposure to high temperatures. *Cem. Concr. Res.* **2006**, *36*, 1999–2005. [CrossRef]
17. Drozdzol, K. Thermal and Mechanical Studies of Perlite Concrete Casing for Chimneys in Residential Buildings. *Materials* **2021**, *14*, 2011. [CrossRef] [PubMed]
18. Alraddadi, S. Surface and thermal properties of fine black and white volcanic ash. *Mater. Today Proc.* **2020**, *26 Pt 2*, 1964–1966. [CrossRef]
19. Khandaker, M.; Hossain, A. High strength blended cement concrete incorporating volcanic ash: Performance at high temperatures. *Cem. Concr. Compos.* **2006**, *28*, 535–545.
20. Baloch, W.L.; Khushnood, R.A.; Memon, S.A.; Ahmed, W.; Ahmad, S. Effect of Elevated Temperatures on Mechanical Performance of Normal and Lightweight Concretes Reinforced with Carbon Nanotubes. *Fire Technol.* **2018**, *54*, 1331–1367. [CrossRef]
21. Maraveas, C.; Wang, Y.; Swailes, T. Thermal and mechanical properties of 19th century fireproof flooring systems at elevated temperatures. *Constr. Build. Mater.* **2013**, *48*, 248–264. [CrossRef]
22. Maraveas, C.; Wang, Y.; Swailes, T. Fire resistance of 19th century fireproof flooring systems: A sensitivity analysis. *Constr. Build. Mater.* **2014**, *55*, 69–81. [CrossRef]
23. Chen, M.; Hou, X.; Zheng, W.; Wu, B. Review and analysis of researches on the critical temperature of concrete high-temperature burst and fiber content. *J. Build. Struct.* **2017**, *1*, 161–170.
24. Libre, N.A.; Shekarchizadeh, M.; Mahoutian, M.; Soroushian, P. Mechanical properties of hybrid fiber reinforced lightweight aggregate concrete made with natural pumice. *Constr. Build. Mater.* **2011**, *25*, 2458–2464. [CrossRef]
25. Liu, X.; Wu, T.; Liu, Y. Stress-strain relationship for plain and fibre-reinforced lightweight aggregate concrete. *Constr. Build. Mater.* **2019**, *225*, 256–272. [CrossRef]
26. Afroughsabet, V.; Ozbakkaloglu, T. Mechanical and durability properties of high-strength concrete containing steel and polypropylene fibers. *Constr. Build. Mater.* **2015**, *94*, 73–82. [CrossRef]
27. Li, J.J.; Niu, J.G.; Wan, C.J.; Jin, B.; Yin, Y.L. Investigation on mechanical properties and microstructure of high performance polypropylene fiber reinforced lightweight aggregate concrete. *Constr. Build. Mater.* **2016**, *118*, 27–35. [CrossRef]
28. Liu, Q.; Xiao, J.; Pan, Z.; Li, L. Modeling Research on Waste Concrete Aggregate and Waste Brick Aggregate Recycled Concrete. *J. Build. Struct.* **2020**, *41*, 133–140.
29. Fu, F.; Parke, G.A.R. Assessment of the Progressive Collapse Resistance of Double-Layer Grid Space Structures Using Implicit and Explicit Methods. *Int. J. Steel Struct.* **2018**, *18*, 831–842. [CrossRef]
30. Fu, F.; Lam, D.; Ye, J. Moment resistance and rotation capacity of semi-rigid composite connections with precast hollowcore slabs. *J. Constr. Steel Res.* **2010**, *66*, 452–461. [CrossRef]
31. Wang, L.; Shen, N.; Zhang, M.; Fu, F.; Qian, K. Bond performance of Steel-CFRP bar reinforced coral concrete beams. *Constr. Build. Mater.* **2020**, *245*, 118456. [CrossRef]
32. Chen, X.; Wan, D.-W.; Jin, L.-Z.; Qian, K.; Fu, F. Experimental studies and microstructure analysis for ultra high-performance reactive powder concrete. *Constr. Build. Mater.* **2019**, *229*, 116924. [CrossRef]
33. Xu, M.; Gao, S.; Guo, L.; Fu, F.; Zhang, S. Study on collapse mechanism of steel frame with CFST-columns under column-removal scenario. *J. Constr. Steel Res.* **2018**, *141*, 275–286. [CrossRef]
34. Fu, F. *Advanced Modeling Techniques in Structural Design*; John Wiley & Sons: Hoboken, NJ, USA, 2015; ISBN 978-1-118-82543-3.
35. Fu, F. *Fire Safety Design for Tall Buildings*; Taylor Francis: Abingdon, UK, 2021; ISBN 978-0-367-44452-5.
36. Han, B.; Xiang, T.-Y. Axial compressive stress-strain relation and Poisson effect of structural lightweight aggregate concrete. *Constr. Build. Mater.* **2017**, *146*, 338–343. [CrossRef]

Article

Mechanical and Impact Properties of Engineered Cementitious Composites Reinforced with PP Fibers at Elevated Temperatures

Raad A. Al-Ameri [1], Sallal Rashid Abid [2],* and Mustafa Özakça [1]

[1] Department of Civil Engineering, Gaziantep University, Gaziantep 27310, Turkey; raada.alameri@gmail.com (R.A.A.-A.); ozakca@gantep.edu.tr (M.Ö.)
[2] Department of Civil Engineering, Wasit University, Kut 52003, Iraq
* Correspondence: sallal@uowasit.edu.iq

Abstract: The repeated impact performance of engineered cementitious composites (ECCs) is not well explored yet, especially after exposure to severe conditions, such as accidental fires. An experimental study was conducted to evaluate the degradation of strength and repeated impact capacity of ECCs reinforced with Polypropylene fibers after high temperature exposure. Compressive strength and flexural strength were tested using cube and beam specimens, while disk specimens were used to conduct repeated impact tests according to the ACI 544-2R procedure. Reference specimens were tested at room temperature, while three other groups were tested after heating to 200 °C, 400 °C and 600 °C and naturally cooled to room temperature. The test results indicated that the reference ECC specimens exhibited a much higher failure impact resistance compared to normal concrete specimens, which was associated with a ductile failure showing a central surface fracture zone and fine surface multi-cracking under repeated impacts. This behavior was also recorded for specimens subjected to 200 °C, while the exposure to 400 °C and 600 °C significantly deteriorated the impact resistance and ductility of ECCs. The recorded failure impact numbers decreased from 259 before heating to 257, 24 and 10 after exposure to 200 °C, 400 °C and 600 °C, respectively. However, after exposure to all temperature levels, the failure impact records of ECCs kept at least four times higher than their corresponding normal concrete ones.

Keywords: repeated impact; ACI 544-2R; high temperatures; fire; ECC; impact ductility

Citation: Al-Ameri, R.A.; Abid, S.R.; Özakça, M. Mechanical and Impact Properties of Engineered Cementitious Composites Reinforced with PP Fibers at Elevated Temperatures. *Fire* **2022**, *5*, 3. https://doi.org/10.3390/fire5010003

Academic Editor: Maged A. Youssef

Received: 28 November 2021
Accepted: 26 December 2021
Published: 30 December 2021

Publisher's Note: MDPI stays neutral with regard to jurisdictional claims in published maps and institutional affiliations.

1. Introduction

Regardless of the function and type of occupation of any structural facility, it is still probable to be subjected to unfavorable extreme or accidental loads. Most of the modern reinforced concrete structures are designed to withstand the usual design gravity loads in addition to lateral loads, such as wind and seismic loads. However, considering the accidental loading cases in design is not a typical procedure required by building design codes because this action would distend the construction cost. Among the most probable types of accidental loads are fires and impact loads. The rapid increase of temperature due to the combustion of furniture, nonstructural materials and electrical wiring can noticeably degrade the structural capacity of slabs, beams and columns. On the other hand, sudden impact loads can cause serious concentrated damage that may affect the integrity of the structure.

Although there are great advantages in fire resisting systems and materials in the construction industry, fires keep occurring every day. Large numbers of fire accidents are reported every year [1], where approximately half a million accidental fires were reported between 2013 and 2014 in the USA, while more than 150,000 fire accidents were reported in the UK during the same period. From these fires, 40% were recognized as structural fires [1,2]. Between 1993 and 2016, approximately 90 million fire accidents were recorded in 39 countries with more than a million death incidences [2]. The crucial question after each structural fire is whether the concrete structure can continue to be occupied as usual, should be rehabilitated before reoccupied or must be demolished [3]. Such a decision needs an accurate estimation of the residual properties of concrete, especially the mechanical strength to withstand the design loads. The physical and chemical actions that take place within the microstructure of concrete depend mainly on the temperature level reached and the fire exposure duration. Yet, the composition of the mixture, its porosity and the thermal properties of aggregate are also leading factors that determine the thermal resistance of concrete structures [3–6]. With the increase of temperature, several chemical and physical changes take place and affect the concrete strength owing to its heterogeneous state [7]. The first physical action of fire occurs at approximately 80 °C to 120 °C, where the contained free water in the concrete evaporates [7–9]. This action has a minor effect on concrete degradation, while the following action that usually occurs at temperatures higher than 300 °C and lower than 450 °C represents the starting point of the serious material degradation. This action is the dehydration of the C-S-H gel from the hydrated cement matrix [9–12]. The following actions depend not only on cement but also on the aggregate type [13–16]. The differences in thermal actions between the cement matrix and aggregate, due to the different thermal properties, result in breaking the bond at higher temperatures, which further weakens the concrete structure and deteriorates its residual strength [7,9]. Previous researchers showed that the tensile strength of concrete deteriorates at a faster rate compared to compressive strength [17,18]. Similarly, mechanical properties such as flexural strength, shear strength and modulus of elasticity showed significant deteriorations after exposure to 500 °C [19–23].

On the other hand, some parts of some structures are frequently subjected to the impact of falling objects or the lateral collision of moving vehicles, which are types of repeated accidental impact loads [24]. Other examples of repeated impacts are the offshore structures, where the waves of ocean water repeatedly subject these structures to hydraulic impacts. In hydraulic structures, such as stilling basins, the water acts as an impacting force on the downstream runway. Other examples of repeated impacts can be the forces subjected by airplane wheels on the airport runways [25–27]. The impact strength of concrete can be investigated using several techniques. However, the repeated impact test introduced by ACI 544-2R "Measurement of Properties of Fiber Reinforced Concrete" [28] is the simplest impact test and the only one that simulates the repeated impact case.

In recent years, several significant studies were conducted to evaluate the repeated impact strength of different concrete types using this testing technique. Mastali et al. [29] investigated the effect of the length and dosage of recycled carbon fiber reinforced polymer on the repeated impact performance of Self-Compacting Concrete (SCC). Ismail and Hassan [30] conducted experimental tests using the ACI 544-2R procedure to evaluate the impact resistance of SCC mixtures that include different contents of Steel Fibers (SF) and crumb rubber. The test results showed that the impact numbers increased by up to 30% and the impact ductility enhanced when crumb rubber was utilized, while the incorporation of 1% of SF significantly improved the retained impact numbers by more than 400%. The mono and dual effects of hooked-end and crimped SF on the ACI 544-2R impact resistance of SCC were investigated by Mahakavi and Chithra [31], where significant impact resistance improvement was reported when the two fiber types were hybridized. Jabir et al. [32] investigated the influence of single and hybrid micro SF and Polypropylene (PP) fibers on the impact resistance of ultra-high performance concrete. Abid et al. conducted ACI 544-2R [33] and flexural [34] repeated impact tests on SCC with micro SF contents of 0.5%, 0.75% and 1.0%. The results indicated that 1.0% of SF could increase the impact resistance by more than 800% compared to the reference plain specimens, while in another study [35], a percentage increase of approximately 1200% was recorded. Murali et al. [36–40] and others [41–43] conducted a series of experimental works that explored the repeated impact capabilities of multi-layered fibrous concrete. Double and triple layered concrete with preplaced aggregate and fibers with grouted cement paste were tested using the ACI 544-2R. Works on this material [36,37,41] showed that using intermediate fibrous meshes can improve the impact resistance at cracking and failure stages. However, the most influential contribution to impact strength development was attributed to the steel fibers.

Compared to conventional concrete that have similar strength and fiber content, Engineered Cementitious Composites (ECCs) are a type of high-performance SCC concrete that possess extraordinary ductility with strain hardening and multiple cracking under tensile and flexural stresses. ECCs were first introduced by Li in 1993 [44] and used in several applications [45]. Since that time, numerous studies have been conducted to introduce different ECC mixtures with different fiber types and contents. Plenty of research is available in literature on the different mechanical properties of ECCs. However, research on ECC repeated impact behavior is rare. The performance of ECCs under repeated impact was experimentally investigated by Ismail et al. [46] using the ACI 544-2R technique. Different ECC mixtures were introduced using fixed contents of binder, water, sand and fiber. The results indicated that using 15% to 20% metakaolin with fly ash significantly enhanced the impact performance. Similarly, some studies that evaluate the performance and residual mechanical properties of different ECC mixtures after fire exposure are available in literature [47–50].

It is obvious from the introduced literature that very rare experimental works are available in literature on the repeated impact strength of ECCs. Similarly, there is a serious gap of knowledge about the residual impact strength of fibrous concrete after fire temperatures. To the best of the authors' knowledge, no previous research was conducted to study the residual repeated impact strength of ECCs after high temperature exposure. To fill this gap of knowledge, an experimental program was directed in this research to investigate the cracking and failure repeated impact performances and impact ductility of PP-based ECCs after exposure to high temperatures reaching 600 °C. Such type of research is required because both accidental fire and impact loading are expected along the lifespan of structures. Hence, the research outputs can be utilized to evaluate the residual material and structural response of structural members made of ECCs under such accidental cases.

2. Materials and Methods

2.1. Mixtures and Materials

The aim of this study is to evaluate the residual repeated impact performance of ECCs after exposure to elevated temperatures, which can be considered as a type of new concrete that includes no aggregate particles and a high content of fine cementitious and filler materials. The M45 is a typical ECC mixture introduced by leading researchers, which was the base and most widely used mixture with proven characteristics [44,45]. This mixture was used in this study but with PP fiber instead of the typical and much more expensive polyvinyl alcohol fiber (PVA). On the other hand, a normal strength conventional concrete mixture (NC) with an approximately comparative compressive strength was used for comparison purposes. The mix design proportions of both mixtures are detailed in Table 1.

Table 1. Material contents in the ECC and NC mixtures (kg/m^3).

Mixture	Cement	Fly Ash	Sand	Silica Sand	Gravel	Water	SP	Fiber
NC	410	NA	787	NA	848	215	NA	NA
ECC	570	684	NA	455	NA	330	4.9	18.2 (2% PP)

A single type of Portland cement (Type 42.5) was used for both mixtures, while fly ash was used as a second cementitious material in the ECC mixture. The chemical composition and physical properties for both cement and fly ash are listed in Table 2. As preceded, the ECC mixture included no sand or gravel, where the filler of the mixture was composed of a single type of very fine silica sand with a grain size of 80 to 250 micrometer and a bulk density of 1500 kg/m^3. On the other hand, local sand and crushed gravel from the central region of Iraq were used as fine and coarse aggregates for the NC mixture. The grading of the sand and gravel are shown in Table 3, while the maximum size of the gravel particles was 10 mm. For the ECC mixture, a super plasticizer (SP) type ViscoCrete 5930-L from Sika® was used to assure the required workability due to the large amount of fine materials, while 2% by volume of PP fiber was used with the properties shown in Table 4.

Table 2. Properties of cement and fly ash.

Oxide (%)	Cement	Fly Ash
SiO_2	20.08	56.0
Fe_2O_3	3.6	24.81
Al_2O_3	4.62	5.3
CaO	61.61	4.8
MgO	2.12	1.48
SO_3	2.71	0.36
Loss on ignition (%)	1.38	5.78
Specific surface (m^2/kg)	368	-
Specific gravity	3.15	2.20
Fineness (% retain in 45 μm)	-	28.99

Table 3. Grading of sand and gravel used for NC specimens.

Sieve Size (mm)	Sand % Passing	Gravel % Passing
19	100	100
12.5	100	100
10	100	95.1
4.75	90.7	33.5
2.36	77.9	1.1
1.18	53.5	0
0.6	28.7	0
0.3	7.5	0
0.15	0	0

Table 4. Properties of polypropylene fiber.

Property	Density	Length	Diameter	Tensile Strength	Elastic Modulus
Value	910 kg/m^3	12 mm	0.032 mm	400 MPa	4000 MPa

2.2. Test Program and Heating Regime

From each mixture and for each temperature level, six 150 mm diameter and 64 mm thick disk specimens were used to evaluate the repeated impact test under the free drop-weight procedure of ACI 544-2R. On the other hand, six 100 mm cube specimens were used for a compressive strength test according to BS EN 12390-3: 2009 [51], while six beam specimens with 100 × 100 mm cross section and 400 mm length were tested under four-point bending test (300 mm span) for flexural strength according to BS EN 12390-5: 2009 [52]. All of the disk, cube and beam specimens were cured under the same standard conditions in temperature-controlled water tanks for 28 days.

After the curing period, the specimens were dried in the laboratory environment for 24 h. Previous researchers and trial tests conducted in this study showed that the heating of specimens without initial drying may lead to the explosive failure of some specimens at high temperatures. Therefore, all specimens were pre-dried using an electrical oven at a temperature of approximately 105 °C for 24 h. Afterwards, the specimens were heated using the electrical furnace shown in Figure 1a at a constant rate of approximately 4 °C/min to three levels of high temperatures of 200 °C, 400 °C and 600 °C. When the specified temperature level was reached, the temperature was kept constant for 60 min to assure the thermal saturation at this temperature. Finally, the furnace door was opened and the specimens were left to cool slowly at the laboratory temperature until testing time. The heating regime of the three temperature levels is described in Figure 1b. In addition to the three groups of heated specimens, a fourth group was tested at room temperature without heating as a reference group.

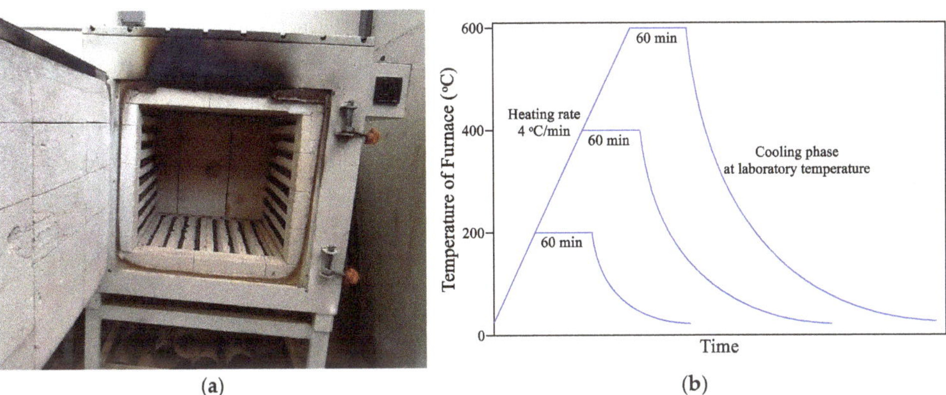

Figure 1. Heating of test specimens; (**a**) the electrical furnace; (**b**) the heating regime curve.

2.3. Repeated Drop-Weight Impact Test

The impact response of materials and structures can be experimentally evaluated using different types of tests, among which is the drop-weight one. ACI 544-2R [28] addressed two types of drop-weight tests. The first is the instrumented drop-weight test which is the most commonly used technique to evaluate the impact response of structural members. This test is mostly used for reinforced beam and slab elements and requires expensive sensor instrumentation and data acquisition equipment. On the other hand, the alternative drop-weight impact test is a very simple one that is conducted on small size specimens and requires no instrumentation or any sophisticated measurement systems. This test requires that a drop weight of 4.54 kg is dropped repeatedly on the test specimen from a height of 457 mm until a surface crack becomes visible, then the repeated impacts are resumed until the fracture failure of the specimen. The numbers of the impacts at which the first crack and failure occur are recorded as the cracking impact number and failure impact number. The test is generally considered as a qualitative evaluation technique, which compares the impact resistance of different concrete mixtures based on their ability to absorb higher or lower cracking and failure impact numbers.

The standard test specimen is a cylindrical (disk) with a diameter of approximately 150 mm and a thickness of approximately 64 mm. The standard test is operated manually by hand-lifting the drop weight to the specified drop height and releasing it to be freely dropped by gravity on a steel ball, which rests on the center of the specimen's top surface. The steel ball is used as a load distribution point and is held in place using a special framing system that also holds the concrete disk specimen, as illustrated in Figure 2a. However, it was found in previous works [27,32] that the manual operation requires significant effort and is time consuming, especially because at least 6 replication specimens are required to assess the test records due to the high dispersion of this test's results [27]. Therefore, an automatic repeated loading machine was manufactured to apply the standard dropping weight from the standard dropping height with a better accuracy and much less effort. The manufactured machine was provided with a high accuracy digital camera to observe the surface cracking and failure in addition to a special isolation cabin to reduce the test noise. The manufactured repeated drop weight impact testing machine is shown in Figure 2b.

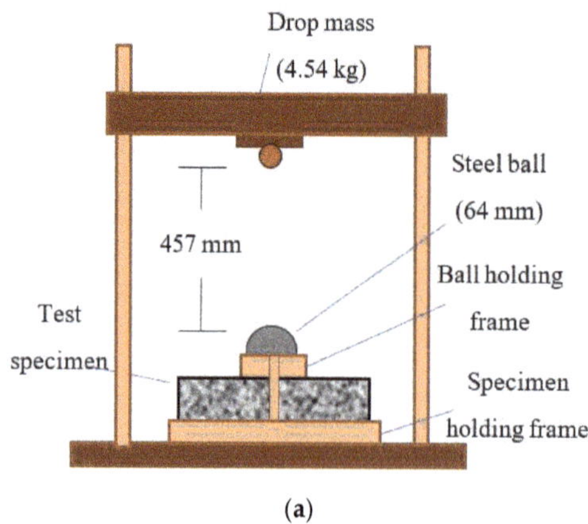

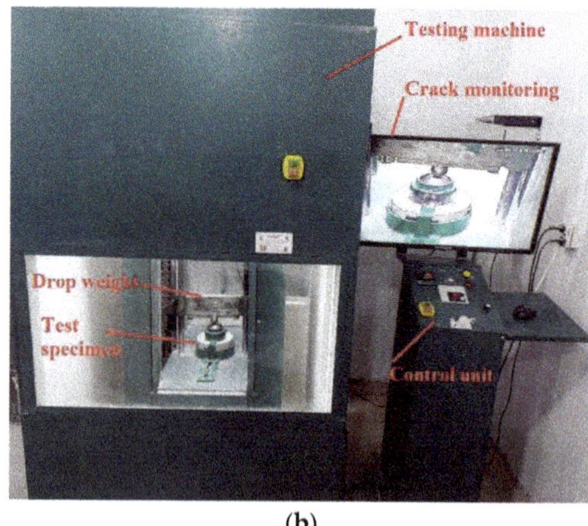

Figure 2. The drop-weight impact test; (**a**) schematic diagram of the test setup; (**b**) the automatic testing machine.

3. Results of Control Tests

3.1. Compressive Strength

The residual compressive strength–temperature relationship of the ECC tested cubes is shown in Figure 3, while Figure 4 shows that of the NC. It is clear in Figure 3 that the ECC strength reduced after exposure to 200 °C by approximately 22% compared to the reference unheated specimens, where the reference strength was 57.5 MPa, while it was 44.8 MPa after heating to 200 °C. A further decrease was recorded when the heating temperature was increased to 400 °C. However, this additional decrease was small compared to the initial one, where the residual strength percentages after exposure to 200 °C and 400 °C were approximately 78% and 70%, respectively. When the specimens were heated to 600 °C, a significant strength degradation was noticed with a residual compressive strength of 29.5 MPa, which means that the strength loss was approximately 49% compared to the strength of the unheated specimens. On the other hand, the percentage strength reduction of NC was less than that of the ECC after exposure to 200 °C and 400 °C. The residual compressive strength of the NC cubes after exposure to 200 °C and 400 °C was approximately 81% at both temperatures compared to the reference cubes as shown in Figure 4. However, the percentage residual compressive strength of the NC at 600 °C was approximately 50%, which was almost equal to that of the ECC (51.4%).

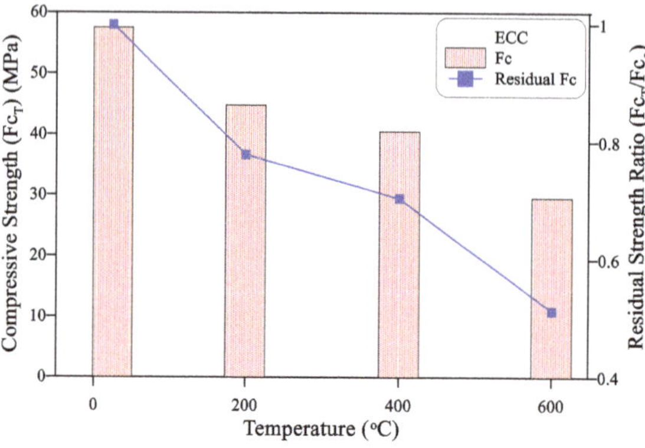

Figure 3. Residual compressive strength of ECC at different temperatures.

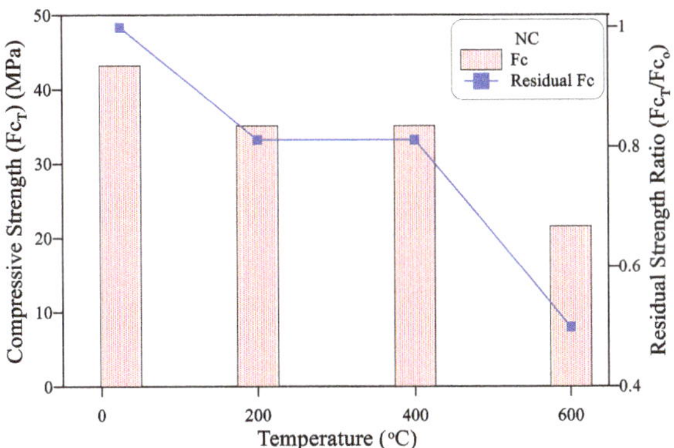

Figure 4. Residual compressive strength of NC at different temperatures.

The denser microstructure of ECCs compared to NC is considered as the main cause of the further strength reduction between 200 °C and 400 °C. ECCs comprise a much larger amount of very fine binder, fine silica sand, no coarse aggregate and lower water/cementitious material content, which in turn lowers the porosity of the ECC compared to the NC. The evaporation of the free pore water below 200 °C induces a pore pressure inside the microstructure. The dissipation of this pressure in the NC specimens due to the higher porosity relieves the internal thermal stresses, while these stresses are higher in the ECC due to the denser microstructure. As a result, the ECC suffered higher compressive strength losses at 200 °C and 400 °C. Previous researchers [53] reported that the total volume of the 0.1 micrometer and larger pores in the ECC reduced after exposure to 400 °C, which is attributed to the pozzolanic reaction of the unhydrated fly ash and other cementitious materials. Such a reaction would induce unfavorable volume changes due to the production of more C-S-H gel, which results in microstructural cracking leading to further strength degradation. The dehydration of hydrated products after exposure to temperatures higher than 400 °C is the main cause of the steep strength reduction at 600 °C, where this process leads to the degradation of the microstructure due to the increase of pore size and number and the further volume changes' micro-cracking. Sahmaran et al. [47] reported a significant increase in the volume and size of the pores of the ECC after exposure to 600 °C, where the porosity increased by 9% after exposure to 600 °C, which is large enough compared to 5% after exposure to 400 °C, while the pore size increased by at least 300% after 600 °C exposure.

3.2. Flexural Strength

As shown in Figure 5, the flexural strength of the ECC followed a continuous decrease behavior with temperature up to 600 °C. The reference flexural strength of the ECC at room temperature was 6.94 MPa, while it reduced to 5.75 MPa, 4.32 MPa and 2.31 MPa after exposure to 200 °C, 400 °C and 600 °C, respectively. This means that the strength respective reductions at these temperatures were approximately 17%, 38% and 67%. Similarly, the NC showed a continuous steep decrease in flexural strength with temperature increase as shown in Figure 6. The residual flexural strength records of the NC after heating to 200 °C, 400 °C and 600 °C were 2.87 MPa, 2.16 MPa and 0.32 MPa, while the reference unheated specimens recorded a flexural strength of 3.70 MPa. Hence, the percentage reductions were approximately 22%, 42% and 91% at 200 °C, 400 °C and 600 °C, respectively.

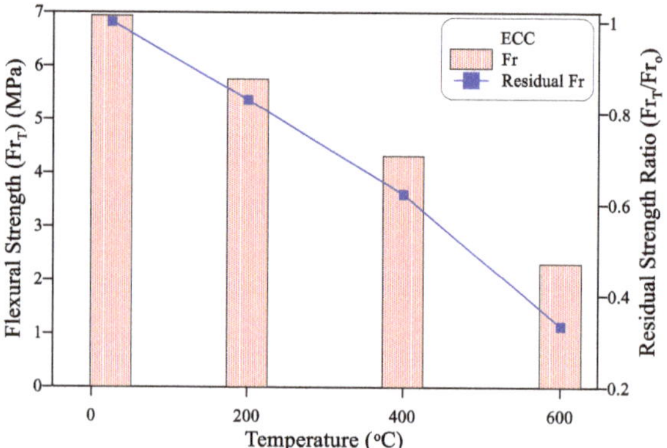

Figure 5. Residual flexural strength of ECC at different temperatures.

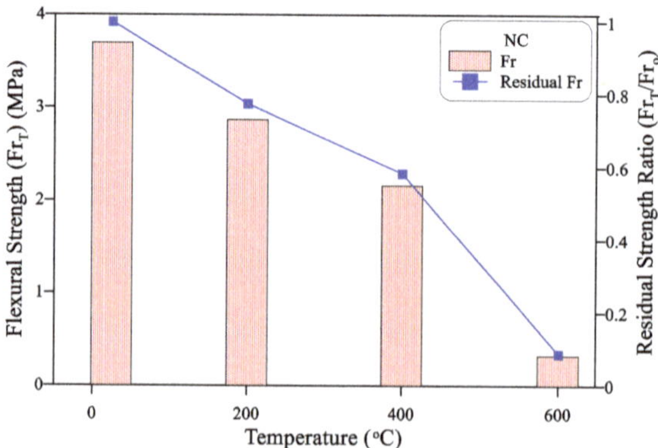

Figure 6. Residual flexural strength of NC at different temperatures.

The continuous decrease in the flexural strength after high temperature exposure is generally attributed to the volumetric changes in the cement matrix due to vapor movements beyond 100 °C and the bond loss between binder and filler after 400 °C due to their different thermal properties. In addition, most of the degradation at higher temperatures is attributed to the chemical reactions after 400 °C (dehydration of C-S-H) and the increased porosity as discussed in the previous section. As the flexural strength depends on the capability of concrete to withstand tensile stresses, the initial flexural strength was apparently higher for the ECC owing to the crack bridging activity of PP fibers, in addition to the higher content of cementitious materials. However, this bridging activity diminished after exposure to temperatures higher than 200 °C due to the melting of PP fibers. The better performance of the ECC at high temperatures compared to the NC might be attributed to the finer mixture constituents and the absence of coarse aggregate in the ECC, which minimized the effect of bond degradation. Wang et al. [54] showed that the residual flexural strength of PVA-based ECC after exposure to 400 °C was approximately 58% of the unheated strength, which is quite comparable to the obtained result in this study, while Yu et al. [55] reported that PVA-based ECC exhibited flexural strength reductions of more than 50% and more than 40% after exposure to temperatures of 400 °C and 600 °C, respectively.

4. Results of Repeated Impact Test

4.1. Description of Heated Specimens

Figure 7 shows the appearance of the external surfaces of a reference impact disk specimen and others heated to 200 °C, 400 °C and 600 °C before testing. No significant changes in the specimens' appearance were noticed after high temperature exposure. However, it was observed that the gray color became lighter after 200 °C and small yellow areas were noticed on the surface of specimens exposed to 600 °C. This slight color change might be due to the decomposition of C-S-H gel particles [56–58]. It should also be noticed that PP fibers cannot sustain high temperatures where its melting point is less than 200 °C. As shown in Figure 8a, the presence of PP fibers had a significant impact in bridging the crack's opposite sides, resulting in a more gradual and ductile failure of the reference unheated specimens. On the other hand, the complete melting of fibers after exposure to 400 °C and higher eliminated this effect and created a more porous media. The channels left after fiber melting would connect and produce continuous porous networks, which have a positive effect by relieving the internal stresses due to the vapor pressure dissipation. On the other hand, these channels may have a negative effect by making the media more porous and hence more brittle under loads. Figure 8b shows that after exposure to 600 °C, the vaporization of PP fibers changed the internal color of the specimen to a dark gray and left a very porous structure behind.

(**a**) R (**b**) 200 °C

(**c**) 400 °C (**d**) 600 °C

Figure 7. Impact test specimens subjected to different temperatures.

(**a**) R (**b**) 600 °C

Figure 8. Physical appearance of PP fibers in the impact specimens before and after heating.

4.2. Cracking and Failure Impact Numbers

The recorded cracking numbers (Ncr) of the ECC and NC are shown in Figure 9 at different levels of high temperatures, while the results of failure numbers (Nf) are shown in Figure 10. It is worthy to mention that the ACI 542-2R test is known for the high despersion of test results, where the Coefficient of Variation (COV) of the Ncr records of the ECC was in the range of 42% to 68.8%, while the COV of the recorded Nf results of the ECC specimens was in the range of 30.9% to 61.8%.

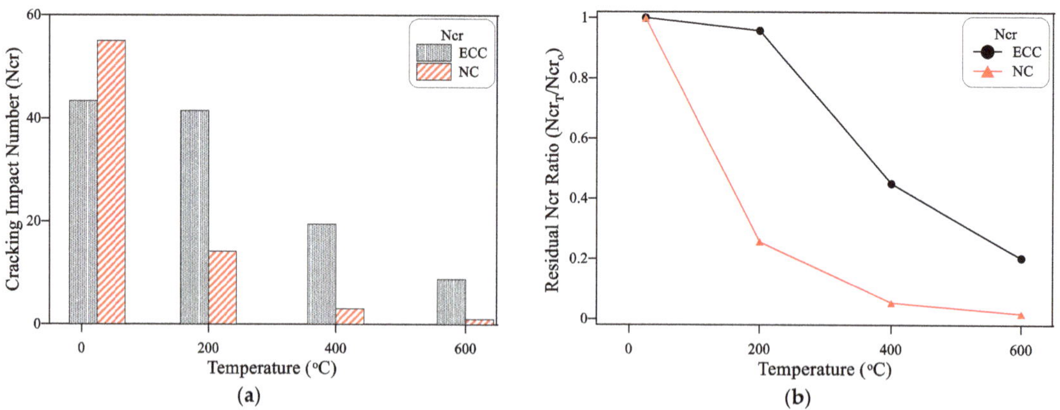

Figure 9. Residual cracking impact numbers of ECC and NC at different temperatures; (**a**) cracking number; (**b**) residual ratio of cracking number.

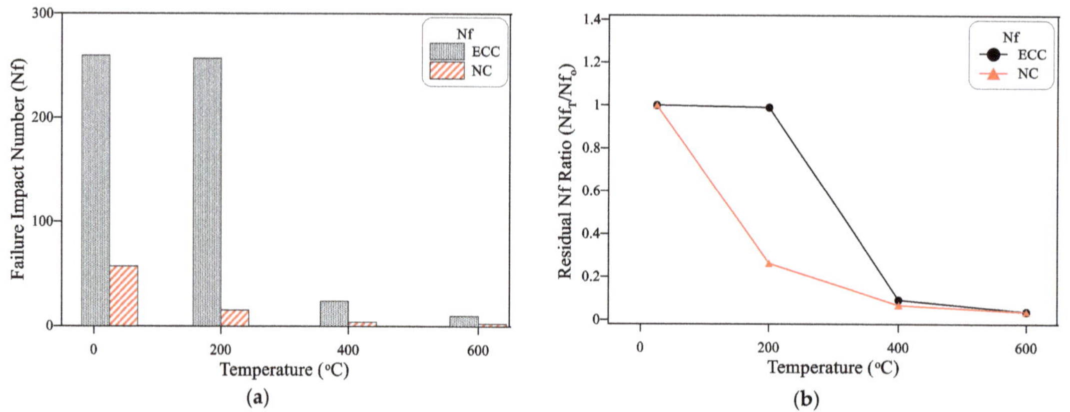

Figure 10. Residual failure impact numbers of ECC and NC at different temperatures; (**a**) failure number; (**b**) residual ratio of failure number.

Figure 9 shows that the reference unheated cracking number of the NC was higher than that of the ECC, which is attributed to the presence of gravel in the NC that enabled it to absorb a higher initial number of impacts before cracking. However, after high temperature exposure, the NC specimens showed much weaker response and deteriorated at much higher rate compared to the corresponding ECC specimens as shown in Figure 9a,b. The unheated Ncr of the ECC and NC were 43.3 and 55, respectively, noting that each impact number represents the average of six specimen records. On the other hand, the residual ECC cracking numbers were 41.5, 19.5 and 8.8 after exposure to 200 °C, 400 °C and 600 °C, respectively, while those of the NC specimens were 14.2, 3 and 1, respectively.

The results reveal a steep drop in the cracking impact numbers of the NC, where the percentage residual Ncr values were only 25.8%, 5.5% and 1.8%, respectively, compared to the reference unheated number as shown in Figure 9b. On the other hand, the ECC showed an insignificant decrease (less than 5%) after exposure to 200 °C, while the percentage residual Ncr values were 45% and 20.4% after exposure to 400 °C and 600 °C, respectively. The rapid decrease of the Ncr of the NC is attributed to the discussed physical and chemical changes that occur after exposure to high temperatures, especially the dehydration of C-S-H, which deteriorates the cement matrix, in addition to the different thermal movements of cement paste and aggregate. Consequently, the internal structure becomes more and more brittle as the temperature increases, which leads to the loss of impact energy absorption capacity and hence to rapid cracking. On the other hand, the higher cementitious materials content, the finer matrix and the absence of aggregate reduced these effects and enabled the ECC specimen to continue withstanding more impacts before cracking. It should be noticed that although the melting point of PP fibers is less than 200 °C, a significant amount of these fibers still existed in the specimens heated to 200 °C. These fibers helped maintain a significant impact number before cracking, which is approximately equal to that of the unheated specimens (95.8%). Aslani et al. [59] reported that PVA fibers did not melt completely after exposure to 300 °C, which is higher than the approximate melting point of PVA (200 °C to 230 °C).

ECCs are known for their high ability to withstand plastic deformation after cracking under tensile and flexural loads, which is attributed to their unique microstructure with high content of binder and fine filler in addition to the potential of fibers to withstand high tensile stresses across the cracks. These characteristics enabled the ECC specimens to absorb significantly higher energy compared to NC after cracking. The test results of this study showed that this potential is also valid under repeated impact loads. As shown in Figure 10, the failure impact number (Nf) of the unheated ECC specimens jumped to a very high limit compared to its corresponding Ncr, while that of NC was comparable to its cracking number, which duplicated the difference of Nf between the ECC and NC several times although the Ncr of the NC was higher than that of the ECC. The Nf of the unheated ECC was 259.3, while that of the NC was only 57.2. This means that the Nf of the NC was approximately equal to its Ncr with only 2.2 higher impacts, while the ECC sustained 216 more impacts after cracking.

After exposure to 200 °C, the NC specimens lost approximately 73% of their initial failure impact performance and retained only 15.2 impacts at failure. Oppositely, the ECC specimens kept approximately the same failure strength of the unheated specimens due to the same reasons discussed above. As shown in Figure 10b, the residual Nf of the ECC after exposure to 200 °C was 99% of the corresponding unheated Nf with 256.7 impacts. As discussed previously, the PP fibers did not melt completely at 200 °C, which means that the fiber bridging activity was still partially effective after cracking. The hydration of the unhydrated products at this temperature might be another reason that enabled the specimens to sustain high impact numbers before cracking. On the other hand, as temperature increased beyond 200 °C, the microstructure of the ECC deteriorated steeply after the complete melting and vaporization of the PP fibers (around 340 °C [60]) and the decomposition of C-S-H gel, which resulted in a weak microstructure. Therefore, the impact strength deteriorated sharply after exposure to 400 °C and 600 °C. As shown in Figure 10b, the percentage residuals of the Nf after exposure to these temperatures were only 9.2% and 3.8%, respectively.

4.3. Failure Patterns of Impact Specimens

The post-failure appearance of a reference ECC specimen and others heated to different high temperatures after repeated impact loading are shown in Figure 11. It is clear in Figure 11a that the central loading area of the top surface of the reference specimen was fractured due to the damage. This fracture zone occurred under the effect of the repeated concentrated compressive stresses from the steel ball, which reflects the ability of the material to absorb significant impact energy under the concentrated impact loading. After the fracture of the surface layer, the PP fibers kept bridging the internal micro-cracks where the compressive impacts try to split the cylinder and hence induce internal tensile stresses, see Figure 8a. However, the continuous impacting could finally break the fibers or their bond with the surrounding media resulting in a progressive crack widening and propagation. Hence, the surface cracks become visible. As shown in Figure 11a, the reference specimens exhibited a ductile failure behavior with central fracture zone and multi-surface cracking.

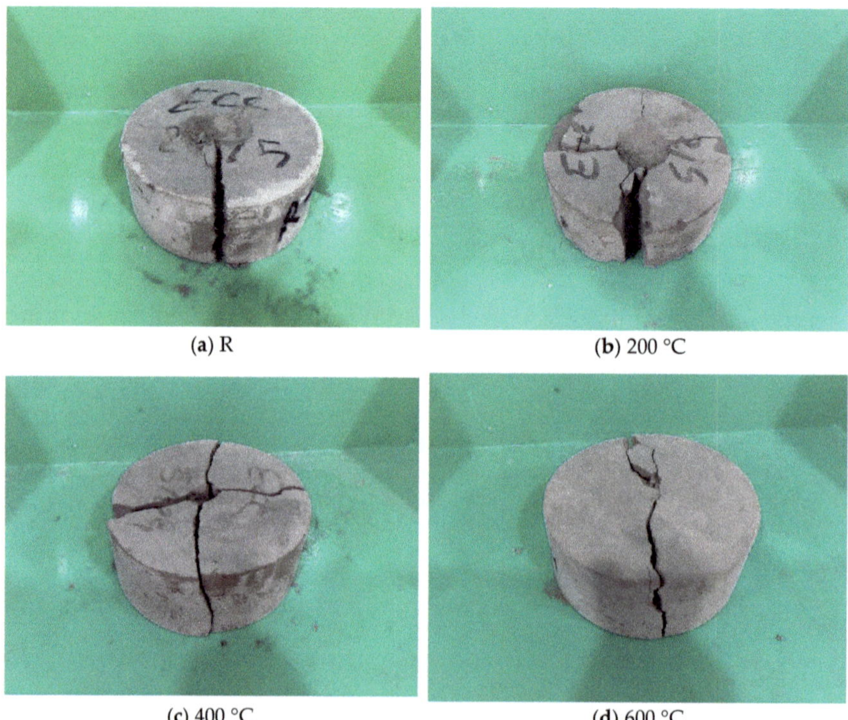

(a) R

(b) 200 °C

(c) 400 °C

(d) 600 °C

Figure 11. Failure patterns of tested impact specimens heated to different temperatures.

Referring to the impact response of the ECC specimens after exposure to 200 °C, the failure pattern at this temperature was similar to that of the reference unheated specimens, but with a lower number of standing fibers across the mouth of the main crack. It should also be noticed that the other minor cracks were wider at this temperature (Figure 11b) compared to those of the specimens, which discloses the lower ductility and higher brittleness of the heated specimens. As previously disclosed, the heating to 400 °C and 600 °C caused serious damage to the microstructure of the ECC and vaporized the reinforcing elements (PP fibers), which was approved by the brittle and sudden failure of the specimens to two, three or four pieces with wide cracks. This failure was not associated with central fracturing as in the case of the reference and 200 °C specimens, where the thermally weakened structure could not absorb significant concentrated impacts, as shown in Figure 11c,d.

5. Strength Correlation with Temperature

In some cases, it is required to evaluate the residual strength of a material after exposure to a specific temperature. If sufficient experimental data are not available, extrapolation from other existing data may be considered satisfactory for a quick primary evaluation.

Despite the limited number of points for each fit, simplified correlations were introduced, as shown in Figure 12, to describe the relation of the strength and impact numbers of the PP-based ECC after exposure to high temperatures. Figure 12a shows that the relations of both compressive strength and flexural strength with temperature can be represented using linear fits with good determination coefficients (R^2) of 0.96 and 0.99, respectively. Referring to Figure 12c, it can be said that a multilinear relation would better describe the reduction of compressive strength with temperature. However, a determination coefficient of 0.96 is good enough to accept the simpler linear correlation.

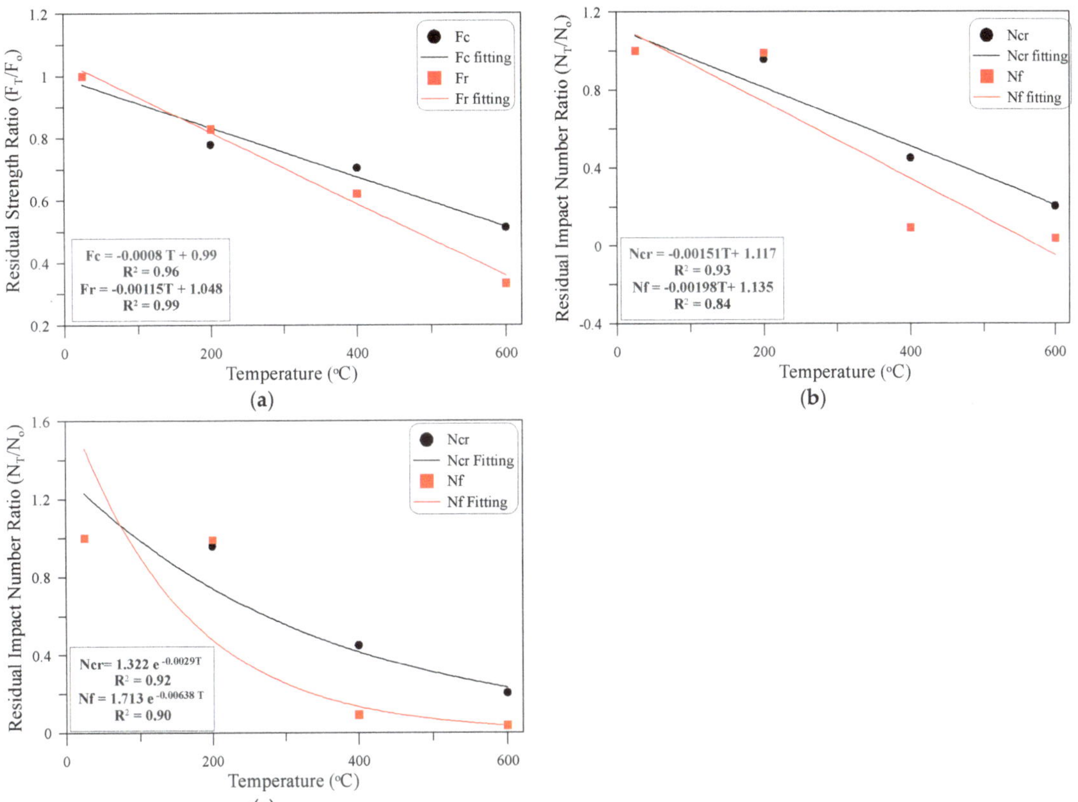

Figure 12. Correlation with temperature; (**a**) linear correlation of strength; (**b**) linear correlation of impact numbers; (**c**) exponential correlation of impact numbers.

The impact numbers showed a weaker linear correlation degree with temperature than those of compressive strength and flexural strength. As shown in Figure 12b, the linear relations of Ncr and Nf with temperature underestimate the retained impact numbers at 200 °C, while that of Nf overestimates the experimental failure impact number recorded at 400 °C. The deviations from the experimental records at these temperatures impacted the degree of the linear correlation, especially for Nf, where the R^2 of the linear correlation was 0.84, which is the lowest among the obtained ones. To avoid such a low degree of correlation, nonlinear correlations were tried and the exponential one was found to give a coefficient of determination of 0.9, which is quite acceptable as an indication of a good correlation. As shown in Figure 12c, the exponential correlations could acceptably estimate the degradation of Ncr and Nf after exposure to the highest temperatures (400 °C and

600 °C). However, these correlations significantly underestimated the residual impact numbers after exposure to 200 °C.

6. Conclusions

Compressive, flexural and repeated impact tests were conducted in this study to evaluate the residual strength of PP fiber-based ECCs after exposure to high temperatures up to 600 °C. Based on the results obtained from the experimental work of this study, the following are the most important conclusions:

1-The compressive strength of the ECC decreased with temperature increase. However, the residual strength at 400 °C was close to that at 200 °C, while exposure to 600 °C led to a significant strength reduction. The percentage residual compressive strengths of the tested ECC cubes after exposure to 200 °C, 400 °C and 600 °C were approximately 78%, 70% and 51%, respectively. The reason for the strength deterioration after 400 °C is attributed to the chemical and physical changes within the material microstructure due to the temperature exposure, which include the decomposition of C-S-H gel and the increase of porosity owing to the vaporization of PP fibers. The linear correlation could effectively describe the degradation of compressive strength after high temperature exposure with an R^2 of 0.96.

2-The flexural strength of the ECC showed a clear continuous reduction with temperature compared to that of compressive strength and higher percentage reductions at 400 °C and 600 °C. Therefore, the linear correlation with temperature was the most accurate one among the conducted tests with an R^2 of 0.99. The residual flexural strengths were reduced to approximately 62 and 33% after heating to 400 °C and 600 °C, respectively.

3-The ECC specimens exhibited minor reductions in the cracking number (Ncr) after exposure to 200 °C with a residual percentage of approximately 96%. The reduction in Ncr was much higher after exposure to the higher temperatures. However, the deterioration of normal concrete (NC) was much faster. ECCs retained percentage residual Ncr values of approximately 45% and 20% after exposure to 400 °C and 600 °C, respectively, while the corresponding percentages of NC were approximately 5% and 2%. The much higher binder content, finer matrix and the absence of aggregate enabled the heated ECC specimen to continue absorbing higher impacts till cracking compared to NC.

4-The failure impact number of the unheated ECC specimens jumped several times higher than the corresponding Ncr, which assured the ability of the dense and fine microstructure of the ECCs, with the help of the PP-fibers crack bridging elements, to amplify the capacity impact energy absorption at failure. The retained Nf was 259.3, which was approximately 4.5 times that of NC although of the higher Ncr of NC. After exposure to 200 °C, the ECC retained almost the same unheated Nf number (99%), while NC retained only 27% of its unheated failure number. Oppositely, both ECC and NC sharply lost their impact resistances after exposure to 400 °Cand 600 °C with percentage residual Nf values of less than 10%, and 4%, respectively.

5-The linear correlation was found suitable to describe the reduction of Ncr with temperature with a good R^2 of 0.93. However, such correlation noticeably underestimated the recorded Nf at 200 °C and overestimated that at 400 °C, which decreased its R^2 to 0.84. On the other hand, the exponential relation was found to better describe the deterioration of Nf after high temperature exposure, where R^2 was 0.9.

Author Contributions: Conceptualization, S.R.A.; methodology, R.A.A.-A. and S.R.A.; validation, S.R.A. and M.Ö.; formal analysis, S.R.A. and R.A.A.-A.; resources, R.A.A.-A.; data curation, S.R.A. and R.A.A.-A.; writing—original draft preparation, S.R.A. and R.A.A.-A.; writing—review and editing, S.R.A. and M.Ö.; visualization, S.R.A. and R.A.A.-A.; supervision, S.R.A. and M.Ö.; project administration, S.R.A. and M.Ö. All authors have read and agreed to the published version of the manuscript.

Funding: This research received no external funding.

Institutional Review Board Statement: Not Applicable.

Informed Consent Statement: Not Applicable.

Data Availability Statement: Data are available upon request from the corresponding author.

Acknowledgments: The authors acknowledge the support from Al-Sharq Lab., Kut, Wasit, Iraq and Ahmad A. Abbas.

Conflicts of Interest: The authors declare no conflict of interest.

References

1. Arna'ot, F.H.; Abid, S.R.; Özakça, M.; Tayşi, N. Review of concrete flat plate-column assemblies under fire conditions. *Fire Saf. J.* **2017**, *93*, 39–52. [CrossRef]
2. Brushlinsky, N.N.; Ahrens, M.; Sokolov, S.V.; Wagner, P. World fire statistics, Center of Fire Statistics of CTIF. *Int. Assoc. Fire Rescue Serv.* **2018**, *21*, 11.
3. Albrektsson, J.; Flansbjer, M.; Lindqvist, J.E.; Jansson, R. *Assessment of Concrete Structures after Fire*; SP Report 19; SP Technical Rsearch Institute of Sweden: Boras, Sweden, 2011; p. 93.
4. Guo, Y.; Zhang, J.; Chen, G.; Xie, Z. Compressive behaviour of concrete structures incorporating recycled concrete aggregates, rubber crumb and reinforced with steel fibre, subjected to elevated temperatures. *J. Clean. Prod.* **2014**, *72*, 193–203. [CrossRef]
5. Tufail, M.; Shahzada, K.; Gencturk, B.; Wei, J. Effect of elevated temperature on mechanical properties of limestone, quartzite and granite concrete. *Int. J. Concr. Struct. Mater.* **2017**, *11*, 17–28. [CrossRef]
6. Babalola, O.E.; Awoyera, P.O.; Le, D.-H.; Bendezú Romero, L.M. A review of residual strength properties of normal and high strength concrete exposed to elevated temperatures: Impact of materials modification on behaviour of concrete composite. *Constr. Build. Mater.* **2021**, *296*, 123448. [CrossRef]
7. Roufael, G.; Beaucour, A.-L.; Eslami, J.; Hoxha, D.; Noumowe, A. Influence of lightweight aggregates on the physical and mechanical residual properties of concrete subjected to high temperatures. *Constr. Build. Mater.* **2021**, *268*, 121221. [CrossRef]
8. Drzymala, T.; Jackiewicz-Rek, W.; Tomaszewski, M.; Kus, A.; Galaj, J.; Sukys, R. Effects of high temperatures on the properties of high performance concrete (HPC). *Procedia Eng.* **2017**, *172*, 256–263. [CrossRef]
9. Abrams, M.S. *Compressive Strength of Concrete at Temperatures to 1600 °F. ACI SP 25, Temperature and Concre*; American Concrete Institute ACI: Detroit, MI, USA, 1971.
10. DüGenci, O.; Haktanir, T. Experimental research for the effect of high temperature on the mechanical properties of steel fiber-reinforced concrete. *Constr. Build. Mater.* **2015**, *75*, 82–88. [CrossRef]
11. Arna'ot, F.H.; Abbass, A.A.; Abualtemen, A.A.; Abid, S.R.; Özakça, M. Residual strength of high strength concentric column-SFRC flat plate exposed to high temperatures. *Constr. Build. Mater.* **2017**, *154*, 204–218. [CrossRef]
12. Chu, H.-Y.; Jiang, J.-Y.; Sun, W.; Zhang, M. Mechanical and physicochemical properties of ferro-siliceous concrete subjected to elevated temperatures. *Constr. Build. Mater.* **2016**, *122*, 743–752. [CrossRef]
13. Phan, L.T.; Carino, N.J. Code provisions for high strength concrete strength-temperature relationship at elevated temperatures. *Mater. Struct.* **2003**, *36*, 91–98. [CrossRef]
14. Netinger, I.; Kesegic, I.; Guljas, I. The effect of high temperatures on the mechanical properties of concrete made with different types of aggregates. *Fire Saf. J.* **2011**, *46*, 425–430. [CrossRef]
15. Deng, Z.H.; Huang, H.Q.; Ye, B.; Wang, H.; Xiang, P. Investigation on recycled aggregate concretes exposed to high temperature by biaxial compressive tests. *Constr. Build. Mater.* **2020**, *244*, 118048. [CrossRef]
16. Phan, L.T.; Carino, N.J. Review of mechanical properties of HSC at elevated temperatures. *J. Mater. Civ. Eng.* **1998**, *10*, 58–64. [CrossRef]
17. Al-Owaisy, S.R. Effect of high temperatures on shear transfer strength of concrete. *J. Eng. Sustain. Dev.* **2007**, *11*, 92–103.
18. Sultan, H.K.; Alyaseri, I. Effects of elevated temperatures on mechanical properties of reactive powder concrete elements. *Constr. Build. Mater.* **2020**, *261*, 120555. [CrossRef]
19. Cheng, F.P.; Kodur, V.K.R.; Wang, T.C. Stress-strain curves for high strength concrete at elevated temperatures. *J. Mater. Civ. Eng.* **2004**, *16*, 84–94. [CrossRef]
20. Husem, M. The effects of high temperature on compressive and flexural strengths of ordinary and high-performance concrete. *Fire Saf. J.* **2006**, *41*, 155–163. [CrossRef]
21. Al-Owaisy, S.R. Strength and elasticity of steel fiber reinforced concrete at high temperatures. *J. Eng. Sustain. Dev.* **2007**, *11*, 125–133.
22. Toric, N.; Boko, I.; Peroš, B. Reduction of postfire properties of high-strength concrete. *Adv. Civ. Eng.* **2013**, *2013*, 712953.
23. Alimrani, N.; Balazs, G.L. Investigations of direct shear of one-year old SFRC after exposed to elevated temperatures. *Constr. Build. Mater.* **2020**, *254*, 119308. [CrossRef]
24. Nili, M.; Afroughsabet, V. Combined effect of silica fume and steel fibers on the impact resistance and mechanical properties of concrete. *Int. J. Impact Eng.* **2010**, *37*, 879–886. [CrossRef]
25. Salaimanimagudam, M.P.; Suribabu, C.R.; Murali, G.; Abid, S.R. Impact response of hammerhead pier fibrous concrete beams designed with topology optimization. *Period. Polytech. Civ. Eng.* **2020**, *64*, 1244–1258. [CrossRef]
26. Wang, W.; Chouw, N. The behavior of coconut fibre reinforced concrete (CFRC) under impact loading. *Constr. Build. Mater.* **2017**, *134*, 452–461. [CrossRef]

27. Abid, S.R.; Abdul-Hussein, M.L.; Ali, S.H.; Kazem, A.F. Suggested modified testing techniques to the ACI 544-R repeated drop-weight impact test. *Constr. Build. Mater.* **2020**, *244*, 118321. [CrossRef]
28. ACI 544-2R. *Measurement of Properties of Fiber Reinforced Concrete*; American Concrete Institute: Detroit, MI, USA, 1999.
29. Mastali, M.; Dalvand, A. The impact resistance and mechanical properties of self-compacting concrete reinforced with recycled CFRP pieces. *Compos. B Eng.* **2016**, *92*, 360–376. [CrossRef]
30. Ismail, M.K.; Hassan, A.A. Impact resistance and mechanical properties of self-consolidating rubberized concrete reinforced with steel fibers. *ASCE J. Mater. Civ. Eng.* **2017**, *29*, 04016193. [CrossRef]
31. Mahakavi, P.; Chithra, R. Impact resistance, microstructures and digital image processing on self-compacting concrete with hooked end and crimped steel fiber. *Constr. Build. Mater.* **2019**, *220*, 651–666. [CrossRef]
32. Jabir, H.A.; Abid, S.R.; Murali, G.; Ali, S.H.; Klyuev, S.; Fediuk, R.; Vatin, N.; Promakhov, V.; Vasilev, Y. Experimental Tests and Reliability Analysis of the Cracking Impact Resistance of UHPFRC. *Fibers* **2020**, *8*, 74. [CrossRef]
33. Abid, S.R.; Abdul-Hussein, M.L.; Ayoob, N.S.; Ali, S.H.; Kadhum, A.L. Repeated drop-weight impact tests on self-compacting concrete reinforced with micro-steel fiber. *Heliyon* **2020**, *6*, e03198. [CrossRef]
34. Abid, S.R.; Murali, G.; Ali, S.H.; Kadhum, A.L.; Al-Gasham, T.S.; Fediuk, R.; Vatin, N.; Karelina, M. Impact performance of steel fiber-reinforced self-compacting concrete against repeated drop weight impact. *Crystals* **2021**, *11*, 91. [CrossRef]
35. Abid, S.R.; Ali, S.H.; Goaiz, H.A.; Al-Gasham, T.S.; Kadhum, A.L. Impact resistance of steel fiber-reinforced self-compacting concrete. *Mag. Civ. Eng.* **2021**, *105*, 10504.
36. Murali, G.; Abid, S.R.; Mugahed Amran, Y.H.; Abdelgader, H.S.; Fediuk, R.; Susrutha, A.; Poonguzhali, K. Impact performance of novel multi-layered prepacked aggregate fibrous composites under compression and bending. *Structures* **2020**, *28*, 1502–1515. [CrossRef]
37. Murali, G.; Abid, S.R.; Karthikeyan, K.; Haridharan, M.K.; Amran, M.; Siva, A. Low-velocity impact response of novel prepacked expanded clay aggregate fibrous concrete produced with carbon nano tube, glass fiber mesh and steel fiber. *Constr. Build. Mater.* **2021**, *284*, 122749. [CrossRef]
38. Murali, G.; Abid, S.R.; Abdelgader, H.S.; Amran, M.Y.H.; Shekarchi, M.; Wilde, K. Repeated projectile impact tests on multi-layered fibrous cementitious composites. *Int. J. Civ. Eng.* **2021**, *19*, 635–651. [CrossRef]
39. Murali, G.; Abid, S.R.; Amran, M.; Fediuk, R.; Vatin, N.; Karelina, M. Combined effect of multi-walled carbon nanotubes, steel fibre and glass fibre mesh on novel two-stage expanded clay aggregate concrete against impact loading. *Crystals* **2021**, *11*, 720. [CrossRef]
40. Murali, G.; Asrani, N.P.; Ramkumar, V.R.; Siva, A.; Haridharan, M.K. Impact Resistance and Strength Reliability of Novel Two-Stage Fibre-Reinforced Concrete. *Arab. J. Sci. Eng.* **2019**, *44*, 4477–4490. [CrossRef]
41. Ramkumar, V.R.; Murali, G.; Asrani, N.P.; Karthikeyan, K. Development of a novel low carbon cementitious two stage layered fibrous concrete with superior impact strength. *J. Build. Eng.* **2019**, *25*, 100841. [CrossRef]
42. Prasad, N.; Murali, G. Exploring the impact performance of functionally-graded preplaced aggregate concrete incorporating steel and polypropylene fibres. *J. Build. Eng.* **2021**, *35*, 102077. [CrossRef]
43. Ramakrishnan, K.; Depak, S.; Hariharan, K.; Abid, S.R.; Murali, G.; Cecchin, D.; Fediuk, R.; Amran, Y.M.; Abdelgader, H.S.; Khatib, J.M. Standard and modified falling mass impact tests on preplaced aggregate fibrous concrete and slurry infiltrated fibrous concrete. Constr. Build. Mater. **2021**, *298*, 123857. [CrossRef]
44. Li, V.C. From micromechanics to structural engineering: The design of cementitious composites for civil engineering applications. *J. Struct. Mech. Earthq. Eng.* **1993**, *10*, 37–48. [CrossRef]
45. Li, V.C. *Engineering Cementitious Composites (ECC)-Materials, Structural, and Durability Performance*; University of Michigan: Ann Arbor, MI, USA, 2007.
46. Ismail, M.K.; Hassan, A.A.A.; Lachemi, M. Performance of self-consolidating engineered cementious composite under drop-weight impact loading. *ASCE J. Mater. Civ. Eng.* **2019**, *31*, 04018400. [CrossRef]
47. Sahmaran, M.; Lachemi, M.; Li, V. Assessing mechanical properties and microstructure of fire-damaged engineered cementitious composites. *ACI Mater. J.* **2010**, *107*, 297–304.
48. Çavdar, A. A study on the effects of high temperature on mechanical properties of fiber reinforced cementitious composites. *Compos. B* **2012**, *43*, 2452–2463. [CrossRef]
49. Shang, X.; Lu, Z. Impact of high temperature on the compressive strength of ECC. *Adv. Mater. Sci. Eng.* **2014**, *2014*, 919078. [CrossRef]
50. Rafiei, P.; Shokravi, H.; Mohammadyan-Yasouj, S.E.; Koloor, S.S.R.; Petru, M. Temperature impact on engineered cementitious composite containing basalt fibers. *Appl. Sci.* **2021**, *11*, 6848. [CrossRef]
51. *BS EN 12390-3: 2009*; Testing Hardened Concrete-Part 3: Compressive Strength of Test Specimens. British Standards: London, UK, 2009.
52. *BS EN 12390-5: 2009*; Testing Hardened Concrete-Part 5: Flexural Strength of Test Specimens. British Standards: London, UK, 2009.
53. Şahmaran, M.; Özbay, E.; Yücel, H.E.; Lachemi, M.; Li, V.C. Effect of fly ash and PVA fiber on microstructural damage and residual properties of engineered cementitious composites exposed to high temperatures. *J. Mater. Civ. Eng.* **2011**, *23*, 1735–1745. [CrossRef]

54. Wang, Z.-B.; Han, S.; Sun, P.; Liu, W.-K.; Wang, Q. Mechanical properties of polyvinyl alcohol-basalt hybrid fiber engineered cementitious composites with impact of elevated temperatures. *J. Cent. South Univ.* **2021**, *28*, 1459–1475. [CrossRef]
55. Yu, Z.; Yuan, Z.; Xia, C.; Zhang, C. High temperature flexural deformation properties of engineered cementitious composites (ECC) with hybrid fiber reinforcement. *Res. Appl. Mater. Sci.* **2020**, *2*, 17–26. [CrossRef]
56. Li, Q.-H.; Sun, C.-J.; Xu, S.-L. Thermal and mechanical properties of ultrahigh toughness cementitious composite with hybrid PVA and steel fibers at elevated temperatures. *Compos. B* **2019**, *176*, 107201. [CrossRef]
57. Li, Q.; Gao, X.; Xu, S.; Peng, Y.; Fu, Y. Microstructure and mechanical properties of high-toughness fiber-reinforced cementitious composites after exposure to elevated temperatures. *J. Mater. Civ. Eng.* **2016**, *28*, 04016132. [CrossRef]
58. Liu, J.-C.; Tan, K.H.; Fan, S.-X. Residual mechanical properties and spalling resistance of strain-hardening cementitious composite with Class C fly ash. *Constr. Build. Mater.* **2017**, *181*, 253–265. [CrossRef]
59. Aslani, F.; Wang, L. Fabrication and characterization of an engineered cementitious composite with enhanced fire resistance performance. *J. Clean. Prod.* **2019**, *221*, 202–214. [CrossRef]
60. Poon, C.S.; Shui, Z.H.; Lam, L. Compressive behavior of fiber reinforced high-performance concrete subjected to elevated temperatures. *Cem. Concr. Res.* **2004**, *34*, 2215–2222. [CrossRef]

Article

Residual Axial Behavior of Restrained Reinforced Concrete Columns Damaged by a Standard Fire

Mohamed Monir A. Alhadid and Maged A. Youssef *

Department of Civil and Environmental Engineering, Western University, London, ON N6A 5B9, Canada;
majjanal@uwo.ca
* Correspondence: youssef@uwo.ca

Abstract: A simplified procedure to predict the residual axial capacity and stiffness of both rectangular and circular reinforced concrete (RC) columns after exposure to a standard fire provides the means to replace the current descriptive methods. The availability of such a procedure during the design phase provides engineers with the flexibility to come up with better designs that ensure safety. In this paper, finite difference heat transfer and sectional analysis models are combined to determine the axial behavior of RC columns with various end-restraint conditions at different standard fire durations. The influence of cooling phase on temperature distribution and residual mechanical properties is considered in the analysis. The ability of the model to predict the axial behavior of the damaged columns is validated in view of related experimental studies and shown to be in very good agreement. A parametric study is then conducted to assess the axial performance of fire-damaged RC columns. A procedure is proposed to determine the residual strength and stiffness of fire-damaged RC columns in typical frame structures.

Keywords: reinforced concrete; columns; standard fire; cooling phase; axial capacity; temperature-stress history

Citation: Alhadid, M.M.A.; Youssef, M.A. Residual Axial Behavior of Restrained Reinforced Concrete Columns Damaged by a Standard Fire. *Fire* **2022**, *5*, 42. https://doi.org/10.3390/fire5020042

Academic Editor: Wojciech Węgrzyński

Received: 24 February 2022
Accepted: 17 March 2022
Published: 23 March 2022

Publisher's Note: MDPI stays neutral with regard to jurisdictional claims in published maps and institutional affiliations.

1. Introduction

Reinforced concrete (RC) structures are widely used in construction due to their outstanding structural performance and design flexibility [1]. The behavior of RC members at ambient conditions is addressed by various building codes and standards [2–4]. However, when exposed to elevated temperatures, the capacity and deformation of such members change due to material degradation, residual strains, and stress redistribution [1,5,6]. In addition, the temperature–load history and the interaction between mechanical and thermal stresses significantly affect the residual properties of the members [1,7,8].

The structural integrity and mechanical properties of most fire-exposed concrete members are either fully or partially restored after the fire incident. Many design codes and standards [9–12] adopt a prescriptive approach through providing data related to the anticipated fire resistance of various RC members based on their geometrical properties and fire exposure conditions. This approach is easy to implement but usually results in bigger sections than what is required to support the loads. The prescriptive approach also overlooks the influence of temperature–load history despite its important role in determining the residual performance of the members. In practice, a preliminary assessment of the damaged members is performed immediately after the structure is exposed to elevated temperatures [13]. This includes visual inspection, hammer tapping, determination of fire propagation route and residual strength of concrete, cracking and spalling schemes, color changes, and smoke deposits to identify the fire duration and maximum temperatures reached [14]. After that, the structure is evaluated according to the relevant design code based on the extent of damage and the affordability of the required work. Load-bearing members, such as columns, should maintain their structural integrity to sustain the applied load without failure or excessive deflections.

This study is an attempt to propose an analytical procedure to supplement the in situ preliminary assessment after a fire incident in RC structures considering standard fire. A model utilizing both heat transfer analysis and sectional analysis is developed to evaluate the residual axial behavior of rectangular and circular RC columns. Temperature–load history is explicitly considered in the analysis. The various strain components developed during and after fire are calculated and their influence on changing the residual performance of the damaged members under various restraining conditions is evaluated. The validity of the proposed model is assessed in view of relevant experimental results obtained from the literature. The validated model is then utilized to perform a parametric study aiming at investigating the influence of mechanical properties, cross-sectional dimensions, fire exposure and support conditions on the residual performance of RC columns. A simplified procedure is then proposed to predict the residual axial capacity and stiffness of RC columns in typical frame structures. The outcomes of the current study provide a solid basis for a more comprehensive work that accounts for other fire types and exposure conditions.

The determination of the residual axial capacity of RC columns subjected to elevated temperatures is not practical in design offices due to the complexity associated with performing comprehensive thermal and structural analyses. The proposed simplified method allows engineers to utilize the commercially available structural analysis software to predict the residual axial capacity and deformation behavior of fire-exposed structures. This can be performed by considering the residual axial stiffness as an input in the definition of the mechanical properties of the affected members to account for the deterioration they exhibited due to elevated temperatures. The internal forces in all the members can then be obtained due to the load redistribution triggered by the fire scenario. The capability of the structural members to resist the applied loads is determined in view of their residual capacity. The residual axial properties used in performing the structural analysis study are obtained from the proposed models in the current article.

2. Proposed Analytical Approach

Assessment of the post-fire behavior of RC columns in typical frame structures requires the consideration of not only the residual mechanical properties of the composing materials but also the temperature–load interaction before and during fire. Figure 1 illustrates the influence of heating and loading history on the total strains (ε_t) induced in concrete. Path 1 shows the case where the column supports a load that causes a mechanical strain (ε_m)$_1$ before heat exposure. By heating the column, a combination of thermal and transient strains (ε_{th})$_1$ is induced. On the other hand, path 2 shows the development of total strains under a successive application of temperature and load. In this case, the column experiences thermal strains (ε_{th})$_2$ followed by mechanical strains (ε_m)$_2$ due to the loads applied on the fire-damaged member. Transient strains are not considered as the column is unloaded during heating. Although the column is supporting the same load and is exposed to the same maximum temperature in both cases, the total strain differs significantly. In real structures, the total strain can be somewhere in between the two previously mentioned extreme cases. Since the free thermal strain is partially irrecoverable and the transient strain is irreversible [7,15], detailed examination of the actual load–temperature path must be considered in the analysis. Guo and Shi [1] experimentally demonstrated the variation in the deformation behaviors of RC columns when subjected to different heating–loading paths.

The analytical approach performed in this study encompasses three main stages that describe the structural variations in the exposed member throughout the heating–cooling cycle. Firstly, the structural performance of the intact member is determined in terms of its capacity and stiffness considering the relevant material models at ambient conditions. The obtained structural characteristics act as a basis to calculate the initial axial load level (λ) and to determine the extent of deterioration in the member after exposure to fire. The second stage involves thermal and structural analyses of the exposed member during the heating and cooling cycles. Heat transfer analysis is carried out using the finite difference method to determine the maximum temperature distribution within the member based on concrete thermal and physical properties. In Figure 2, the residual properties of the member at the final stage (point 2) are highly dependent on the temperature–load path followed. Therefore, at each time increment, the change in the applied load level ($\Delta\sigma$) associated with the restraint conditions is considered. Both thermal and transient strains are calculated at each time increment, as represented by the step function shown in Figure 2. The residual capacity of the member during fire is calculated based on the relevant material models to check if failure occurs during fire. The third analysis stage commences after the member is completely cooled down to room temperature. In this stage, sectional analysis is carried out to determine the residual capacity and stiffness of the fire-damaged member in view of the maximum temperature reached and residual strain distribution. The analysis is performed by applying uniform strain increments until failure occurs considering the post-fire mechanical properties and material models.

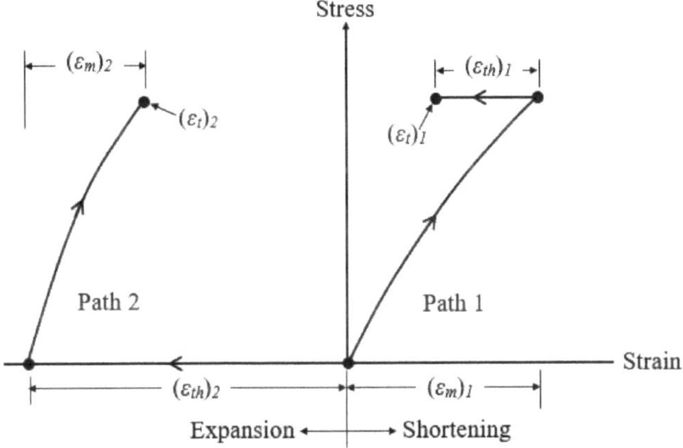

Figure 1. Influence of temperature–stress interaction on the concrete strains.

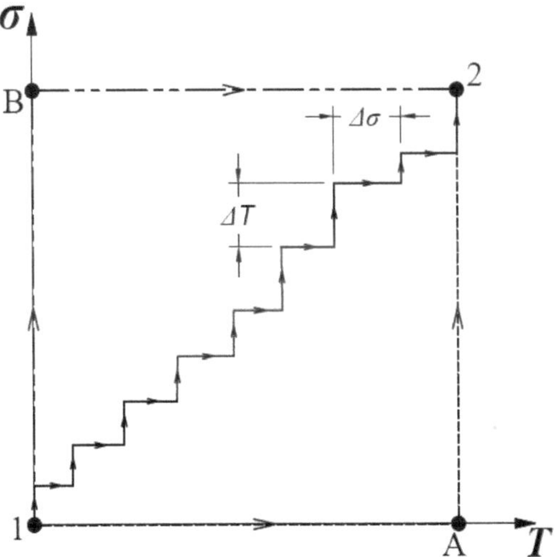

Figure 2. Potential temperature–stress paths between initial and final conditions of the fire-exposed members.

The current study focuses on the axial behavior of rectangular and circular RC members exposed to fire from all sides. The restraint condition is determined by performing structural analysis of the entire frame, Figure 3a, with the aid of a suitable commercially available software. The first iteration is performed considering the mechanical properties of the section at ambient conditions. At any specific time during fire, the columns' axial capacity and stiffness are reduced as a function of the temperature distribution within the section. The fire-exposed column can be isolated as shown in Figure 3b. A pin support is assigned to one end of the column, while the other end is attached to a roller support and a spring with an axial stiffness (k_δ) that represents the axial constraints provided by the adjacent frame members. The value of k_δ can be obtained based on structural analysis, as will be discussed later in Section 11 of this paper. Springs act in resisting the columns' expansion but not contraction. When the column expands, the magnitude of the axial load acting on the column encompasses both the initial applied load (P_i) and the restraining force caused by thermal expansion. The axial stiffness (EA) of the columns varies at each time step during fire and is considered in the calculation of the restraining force. The mutual dependency is considered in the proposed model, as discussed in the subsequent sections.

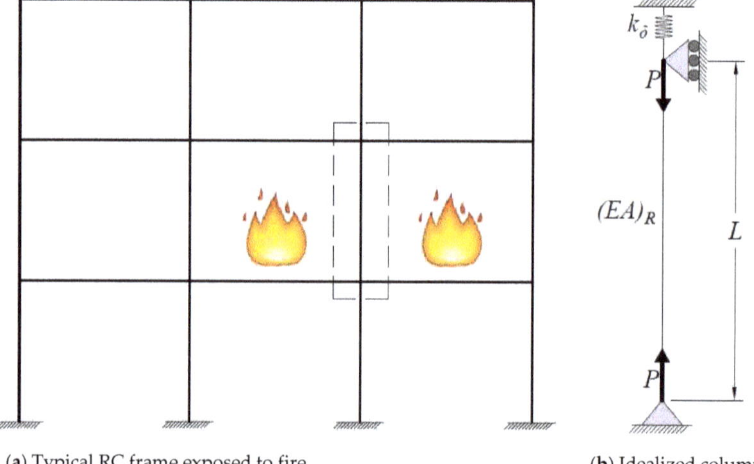

(a) Typical RC frame exposed to fire   (b) Idealized column

Figure 3. Isolation of a fire-exposed column in a typical RC frame.

The proposed analysis of the fire-damaged RC members is carried out based on the following assumptions:

1. Cross-sections remain plane before and after fire. The validity of this assumption was validated for temperatures up to 1200 °C [6].
2. Perfect bond exists between the steel reinforcement and the surrounding concrete.
3. Spalling of concrete is not considered as the analysis is limited to normal-weight concrete.
4. Two-dimensional heat transfer analysis is considered. Thus, heat flow is uniform along the member length.
5. Geometrical nonlinearity is not considered in the analysis.
6. Failure of the compression members is not governed by buckling.

3. Definition of Cross-Sections

The residual axial capacity and stiffness of fire-exposed RC rectangular and circular columns subjected to standard fire from all sides are considered in the analysis. The geometrical properties and reinforcement distribution of a typical cross-section are defined in Figures 4a and 5a for rectangular and circular sections, respectively. Rectangular sections are defined in terms of section width (b), section height (h), steel reinforcement ratio (ρ), top steel reinforcement (A_{st}) and bottom steel reinforcement (A_{sb}), whereas circular columns are defined in terms of cross-sectional diameter (D), steel reinforcement ratio (ρ) where steel reinforcement (A_s) is assumed to be uniformly distributed along the circumference. Table 1 details the mechanical and geometrical properties of the selected rectangular and circular sections discussed in this paper.

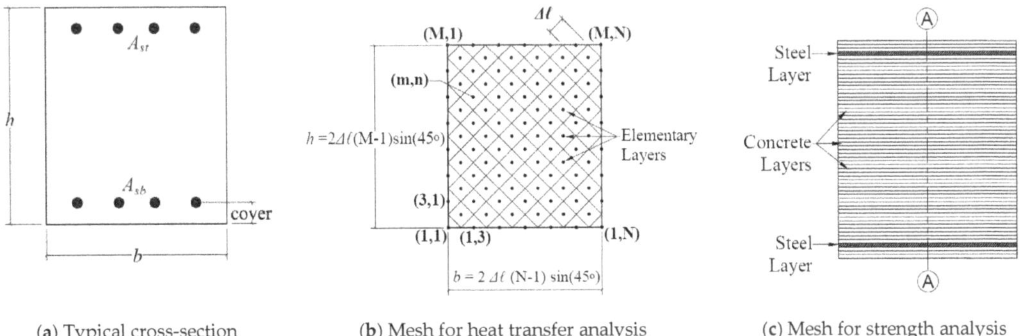

(**a**) Typical cross-section   (**b**) Mesh for heat transfer analysis   (**c**) Mesh for strength analysis

Figure 4. Geometry and meshing of rectangular sections.

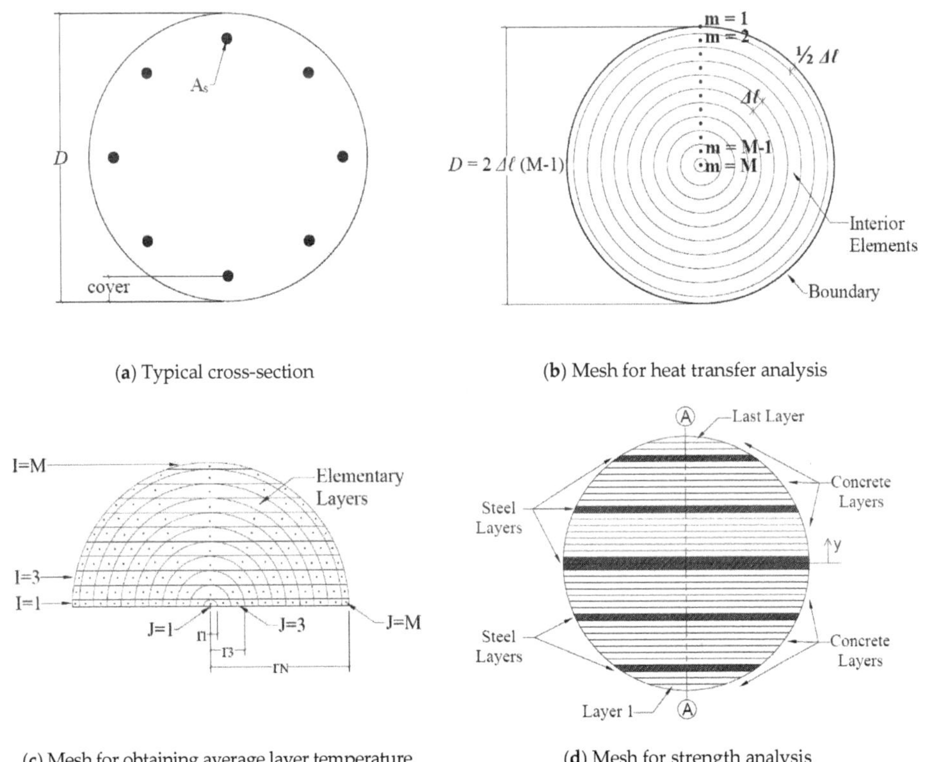

(**a**) Typical cross-section   (**b**) Mesh for heat transfer analysis

(**c**) Mesh for obtaining average layer temperature   (**d**) Mesh for strength analysis

Figure 5. Geometry and meshing of circular sections.

Table 1. Properties of the considered rectangular and circular column sections.

Rectangular Sections							Circular Sections						
Case	t (hr)	f_c' (MPa)	f_y (Mpa)	b (mm)	h (mm)	ρ	R_D	Case	t (hr)	f_c' (MPa)	f_y (Mpa)	D (mm)	ρ	R_D
R1	1.5	35	400	400	500	0.04	0.0	C1	1.5	35	400	500	0.04	0.0
R2	0.5	35	400	400	500	0.04	0.0	C2	0.5	35	400	500	0.04	0.0
R3	2.5	35	400	400	500	0.04	0.0	C3	2.5	35	400	500	0.04	0.0
R4	1.5	25	400	400	500	0.04	0.0	C4	1.5	25	400	500	0.04	0.0
R5	1.5	35	300	400	500	0.04	0.0	C5	1.5	35	300	500	0.04	0.0
R6	1.5	35	400	250	500	0.04	0.0	C6	1.5	35	400	310	0.04	0.0
R7	1.5	35	400	600	500	0.04	0.0	C7	1.5	35	400	400	0.04	0.0
R8	1.5	35	400	400	300	0.04	0.0	C8	1.5	35	400	780	0.04	0.0
R9	1.5	35	400	400	800	0.04	0.0	C9	1.5	35	400	500	0.02	0.0
R10	1.5	35	400	400	500	0.02	0.0	C10	1.5	35	400	500	0.04	0.5
R11	1.5	35	400	400	500	0.04	0.5	C11	1.5	35	400	500	0.04	1.0
R12	1.5	35	400	400	500	0.04	1.0							

4. Thermal Analysis

Temperature distribution at any section along the member is determined based on the finite difference method described by Lie [16]. The physical and thermal properties of both concrete and steel are provided by Lie [16]. For each time increment, the temperature distribution within the section is obtained by solving the heat balance equations [16]. In the current study, the columns are exposed to an ASTM E119 [17] standard fire along their perimeter during the heating phase, as approximated by Equation (1).

$$T_f - T_o = 750 \left[1 - e^{(-3.79553\,\sqrt{t}\,)}\right] + 170.41\sqrt{t} \tag{1}$$

where T_f is the fire temperature (°C), T_o is the room temperature (°C) and t is the time after the start of the fire (hr). During the cooling phase, the rate of decrease in temperature per minute is calculated according to the ISO 834 [18] specifications as provided in Equation (2) in terms of fire duration at the end of the heating phase (t_{hot}).

$$\Delta T = \begin{cases} -10.417 & , & t < 30 \text{ min} \\ -4.167(3 - \frac{t_{hot}}{60}) & , & 30 \text{ min} \le t < 120 \text{ min} \\ -4.167 & , & t \ge 120 \text{ min} \end{cases} \tag{2}$$

Concrete thermal properties are assumed to be irreversible and maintain a constant value corresponding to the maximum temperature reached [1,15]. A distinction in the meshing procedure between rectangular and circular column sections is illustrated in Figures 4b and 5b, respectively.

4.1. Rectangular Sections

The analysis procedure begins by dividing the cross section into M × N 45° inclined square elements, as shown in Figure 4b. The point at the center of each internal element or on the hypotenuse of each boundary element represents the temperature of the entire element. Steel bars are considered as perfect conductors due to their high thermal conductivity, and their temperature is assumed to be identical to the adjacent concrete elements. Heat energy is transferred from the outer elements toward the concrete core, causing a subsequent increase in temperature depending on concrete thermal conductivity and moisture content. The influence of moisture is considered by assuming that when an element reaches a temperature of 100 °C, all the transferred heat causes the evaporation of water particles instead of raising the element's temperature. Heat transfer equations between the elements throughout the cross-section are given by Lie [16].

Having determined the temperature distribution within the cross-section, the section is divided into multiple horizontal layers, each having a thickness of $\Delta \ell \sin(45°)$, as shown in Figure 4c. Average temperature is then calculated in each layer considering two methods that result in different temperature distribution along the cross-section. In the first one, the temperature of each horizontal layer is calculated as the algebraic average temperature of the square elements composing it. The other calculation procedure is performed by first calculating the residual compressive strength of each square element, and then evaluating the temperature, which would result in the same average compressive strength in that layer. The first temperature distribution is utilized to calculate thermal and transient strains, whereas the second one is used in calculating the residual strength of each layer. The temperature of the steel layer is assumed to be similar to the temperature of the square mesh elements within which they are located. A similar procedure was performed and validated by El-Fitiany and Youssef [6].

4.2. Circular Sections

To determine the temperature within the circular cross-section along the RC columns, the area is first divided into M concentric layers as shown in Figure 5b. The change in temperature (T) in each circular layer is derived by solving the heat balance equations at each time increment, assuming that the column is exposed to heat along its circumference, as described by Lie [16]. The influence of steel bars and moisture contents is considered in the analysis in a similar manner to the rectangular sections.

In this study, a method is proposed and validated to transform the circular layers into equivalent horizontal layers that can be utilized in the sectional analysis procedure. The procedure commences by dividing the semi-circular section into M horizontal layers (I), each corresponding to a unique circular layer (J), as indicated in Figure 5c. The upper and lower boundaries of any horizontal layer (I) are taken as the tangents to the two circular layers denoted by ($J = I$) and ($J = I - 1$), respectively. The intersection between the horizontal and circular layers produces elementary layers whose temperatures represent the temperature of the circular element they are located in. The area (A) of each elementary layer is derived in terms of the distance (r) from the center of the circular cross-section to each layer, as given in Equation (3).

$$A_{I,J} = \begin{cases} \dfrac{\pi\, r_1^2}{2} & , I = J = 1 \\[2em] r_I^2 \times \dfrac{2\cos^{-1}\left(\frac{r_{I-1}}{r_I}\right) - \sin\left[2\cos^{-1}\left(\frac{r_{I-1}}{r_I}\right)\right]}{2} & , I = J \\[2em] \dfrac{\pi\, r_J^2}{2} - r_J^2 \times \dfrac{2\cos^{-1}\left(\frac{r_I}{r_J}\right) - \sin\left[2\cos^{-1}\left(\frac{r_I}{r_J}\right)\right]}{2} - \sum_{1,1}^{I,J} A_{i,j} & , I \neq J \end{cases} \qquad (3)$$

The temperature in each layer is calculated twice, similar to the procedure performed in rectangular sections. However, in the first case, the weighted average is calculated for each layer instead of calculating the normal average. This requires the determination of the area and temperature of each small element composing the horizontal layer. In the second case, the average temperature that would result in the same weighted average of residual compressive strength is determined. The temperature of each steel layer is taken as the maximum temperature reached at a distance equal to the provided concrete cover since all bars are uniformly distributed parallel to the circumference.

For both rectangular and circular columns, the temperature distribution within the section varies with the thermal properties of concrete and the cross-sectional dimensions. Figure 6 illustrates the change in temperature at different points along the mid-width of sections R3 and C3, whose characteristics are detailed in Table 1. The location of each point is defined as the distance from the face of the column in terms of section height (h) for rectangular sections and radius (r) for circular sections. Two main observations can be drawn from these figures. Firstly, curves representing the points further away from the surface show a continuous increase in temperature after the end of heating. This causes the maximum temperature in the interior elements to be reached during the cooling phase, indicating that heat flow propagates not only to the atmosphere, but also to the inner, colder portions of the member. The second observation shows that cooling continues for a considerable amount of time before heat flow starts to take one direction only toward the atmosphere. A distinction between the rectangular and circular sections is detected in terms of response to temperature variation. In the aforementioned two sections, the concrete in column C3 located at a distance of up to (0.5 *r*) responds faster to the increase in temperature than that in rectangular sections located at the same distance. However, at a greater depth within the section, temperature variation becomes less pronounced in the circular section compared to its rectangular counterpart. This change in behavior is attributed to the more concrete area acting as a protecting cover for points closer to the core in section C3 compared to section R3.

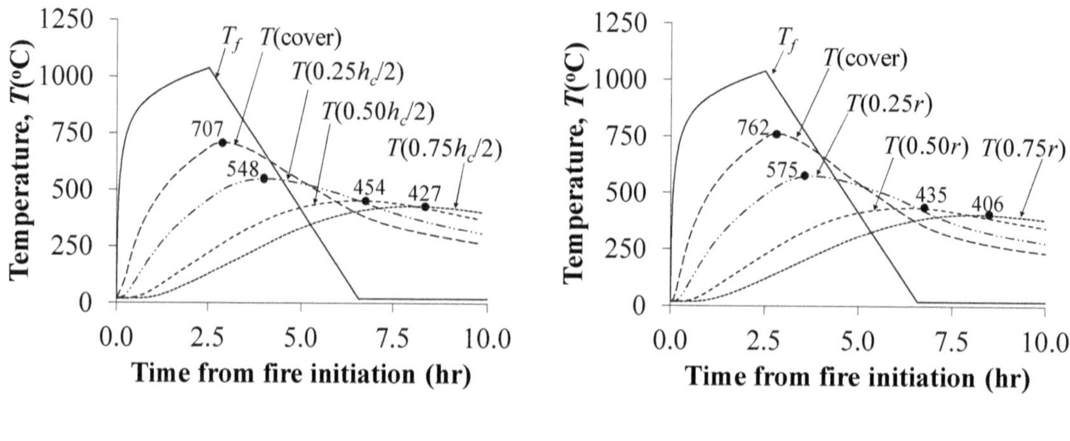

(**a**) Rectangular section (R3)

(**b**) Circular section (C3)

Figure 6. Temperature variation with time at different points along the cross-section.

Temperature distributions within sections R3 and C3 corresponding to maximum temperature reached as well as the end of both the heating and cooling phases are shown in Figures 7 and 8, respectively. As indicated in Figures 7a and 8a, heat flow is initiated from the section perimeter towards the inner core, resulting in the highest temperature rise near the exposed surfaces and the lowest values at the center point of section. During the gradual cooling phase, heat transfer takes place from the hot outer regions towards both the colder concrete zones and the surrounding air. This causes the temperature to keep increasing in the interior concrete elements for a certain period, beyond which heat transfer towards the atmosphere becomes predominant, as shown in Figures 7b and 8b, for the rectangular and circular sections, respectively. The maximum temperature distribution attained at each point within the section throughout the heating–cooling cycle is illustrated in Figures 7c and 8c for the same two sections, respectively. Maximum temperature distribution results in higher temperature values than those at the end of the heating phase. Hence, the residual mechanical properties and constitutive relationships of both concrete and steel are determined in the following sections based on the maximum temperature reached.

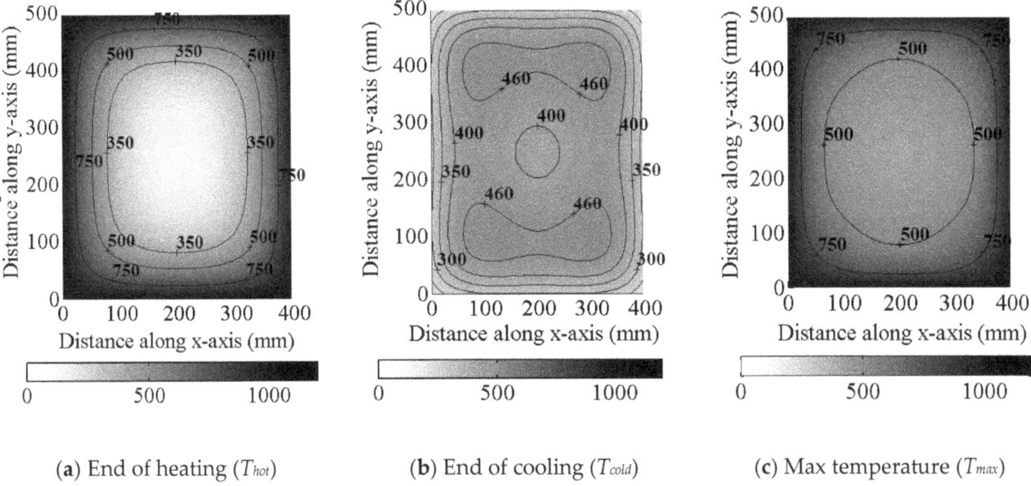

(**a**) End of heating (T_{hot})   (**b**) End of cooling (T_{cold})   (**c**) Max temperature (T_{max})

Figure 7. Temperature distribution within the rectangular cross-section of column (R3) at different time increments.

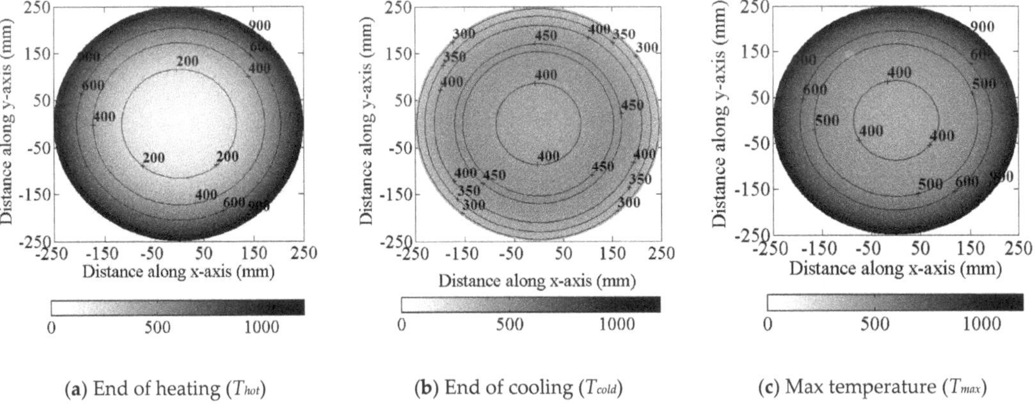

(**a**) End of heating (T_{hot})   (**b**) End of cooling (T_{cold})   (**c**) Max temperature (T_{max})

Figure 8. Temperature distribution within the circular cross-section of column (C3) at different time increments.

5. Material Models and Strain Components

The general form of the Tsai [19] model is adopted in this study to represent the compressive stress–strain relationship of concrete at all stages. During fire, the reduced compressive strength due to fire (f'_{cT}) proposed by Hertz [15] is used, whereas concrete strain at peak stress at elevated temperatures (ε_{0T}) is determined by the Terro [20] formula. The post-fire mechanical properties are calculated based on the expressions by Chang et al. [21].

Regarding steel, the constitutive model used by Karthik and Mander [22] is adopted for both ambient and post-fire conditions as it conveniently combines the initial elastic response, yield plateau and strain hardening stages. At elevated temperatures, the Lie [23] model is used as it implicitly includes the reduction in yield strength due to fire.

Total strain in concrete (ε_t) is calculated as the summation of stress-related strain (ε_σ), free thermal strain (ε_{th}), creep strain (ε_{cr}), and transient strain (ε_{tr}). The tendency of the structural members to deform due to external applied loads is described in terms of the stress-related strain component. Free thermal strain of both concrete and steel bars is determined from Eurocode [4]-proposed expressions. The residual free thermal strain (ε_{thR}) represents the irreversible part of the free expansion that occurred during fire. After a complete heating–cooling cycle, thermal strain is restored with a rate of $8 \times 10^{-6}/°C$ from the maximum temperature reached [1], while ε_{thR} for steel is set to zero. If the member is initially loaded or restrained, then transient strain is generated in concrete and maintains its maximum values after cooling [1]. The empirical model proposed by Terro [20] is adopted to calculate the transient creep strain as referred to by load-induced thermal strain (ε_{LITS}). Regarding steel bars, the residual thermal strain is brought back to zero at the end of the cooling phase. Both transient and creep strain are not applicable for steel during and after fire. Detailed descriptions of the material models and strain components during fire exposure are provided by Youssef and Moftah [5].

6. Strength Analysis

An iterative sectional analysis procedure is carried out to determine the residual P-ε behavior of the fire-damaged RC columns. The residual properties are determined in view of the temperature distribution obtained from thermal analysis. At every loading step, the axial strain is increased incrementally until reaching the total applied axial load. The kinematic and compatibility conditions are considered in view of the corresponding residual mechanical properties and stress–strain relationships of both concrete and steel. The strength analysis is performed by dividing the cross-section into multiple horizontal layers, as shown in Figures 4c and 5d, for the rectangular and circular cross-sections, respectively. To maintain the high accuracy while reducing the computation time, a sensitivity analysis was performed, and the maximum layer height was chosen as not to exceed 3 mm. The centroid of each concrete and steel layer is determined considering the appropriate geometrical expressions for both circular and rectangular sections. For concrete, temperature is obtained from the average distribution that would result in average compressive strength in each layer, whereas the maximum temperature reached is used directly for steel layers corresponding to the exact location of steel bars. The failure criterion of the RC element is defined by the crushing of concrete once the strain in any of the sectional layers reaches the residual ultimate strain (ε_{cuR}) proposed and validated by Alhadid and Youssef [24]. The restraining effect due to elevated temperature is considered in the analysis through calculating the axial restraint at each time increment depending on the assumed supporting condition. The axial force generated due to restraint is added to the initial applied load to determine the total axial load during fire exposure.

7. Equivalent Residual Strain

Residual stresses are induced in fire-damaged members for two main reasons:

(1) Thermal strain in concrete is partially reversible, while transient strain is completely irreversible [1]. At equilibrium, unloaded fire-damaged concrete tends to remain either expanded or contracted depending on the temperature–load history. On the other hand, thermal strain in steel is fully reversible and the embedded steel bars tend to restore their initial length after fire, provided that they did not reach the yield point at the elevated temperature. The variation in behavior between concrete and the embedded steel bars generates internal stresses.

(2) Both thermal and transient strain distributions along section height are nonlinear as they follow the nonlinear temperature profile. Therefore, internal stresses are developed in order to maintain the plane section assumption.

Figure 9 illustrates the development of the strain components along section (A-A) of Figure 4c for rectangular sections. The same analysis procedure is considered for circular sections while accounting for the modified location of the steel layers. The difference between the residual thermal strain (ε_{thR}) and the residual transient strain (ε_{trR}) is the total residual strain (ε_R), which can be either positive or negative depending on the temperature–load history and the magnitude of the developed transient strain. Due to the plane section assumption, the deformed section is represented by a uniform equivalent strain (ε_{eq}) along the cross-section. Residual stress-induced strain ($\varepsilon_{\sigma i}$) distribution is determined as the difference between an equivalent strain (ε_{eq}) and the total residual strain (ε_R). An iteration process is performed to evaluate the uniformly distributed equivalent strain (ε_{eq}) that satisfies the equilibrium condition of $\varepsilon_{\sigma i}$ distribution. The value of ε_{eq} is determined such that the total axial force in concrete and steel resulting from $\varepsilon_{\sigma i}$ distribution is equal to zero.

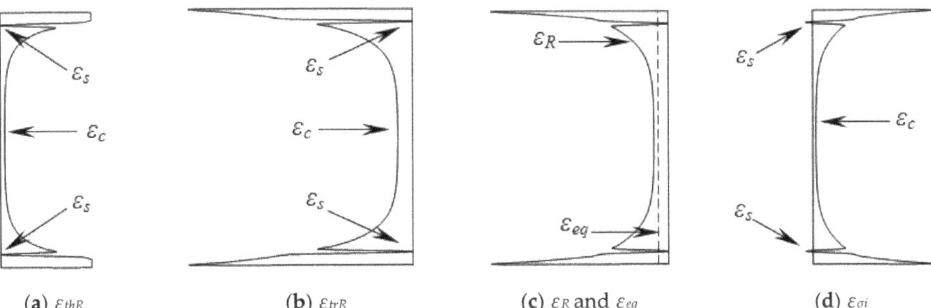

(**a**) ε_{thR} (**b**) ε_{trR} (**c**) ε_R and ε_{eq} (**d**) $\varepsilon_{\sigma i}$

Figure 9. Development of various strain components along the discretized cross-section.

Once equilibrium is achieved, $\varepsilon_{\sigma i}$ is applied as initial strains in the concrete and steel layers, whereas ε_{eq} results in shifting the *P-ε* curve, as illustrated in Figure 10 for both rectangular and circular sections. The residual and equivalent strain distribution along column R3 and C3 cross-sections are shown in Figure 11. If the column is not initially loaded during fire exposure ($\lambda = 0$), then the residual equivalent strain (ε_{eq1}) is always negative, causing the *P-ε* curve to shift to the expansion side. However, by imposing an initial load to the column during the heating phase, a transient strain component develops and counteracts the influence of the thermal strain. If the applied load is large enough, the column experiences residual contraction instead of expansion after the cooling, as indicated by the positive equivalent strain (ε_{eq2}). The change in stiffness is attributed to the elimination of the residual stress-induced strains. Restraining the column affects the magnitude of the generated transient strain, especially if the column is not subjected to initial load. When the column is restrained, part of the equivalent strain (ε_{eq}) induces stresses within the section depending on the considered degree of restraint while maintaining the equilibrium condition. By restraining the column, additional compressive forces are developed in the column as a result of preventing the column's tendency to expand.

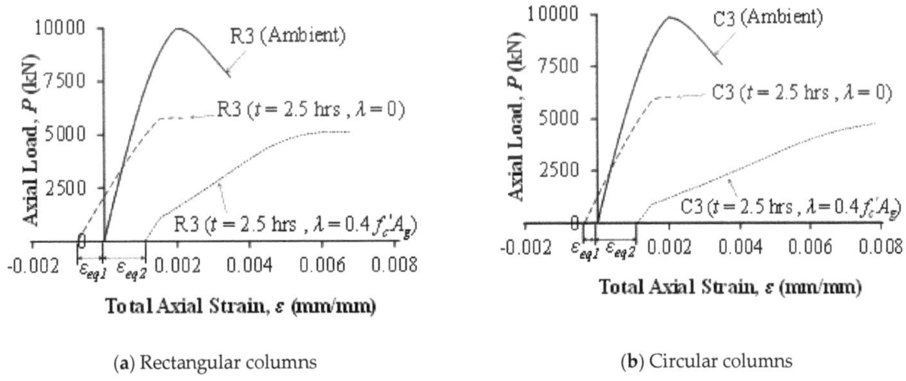

(**a**) Rectangular columns   (**b**) Circular columns

Figure 10. Influence of initial load level on the residual (*P-ε*) relationship.

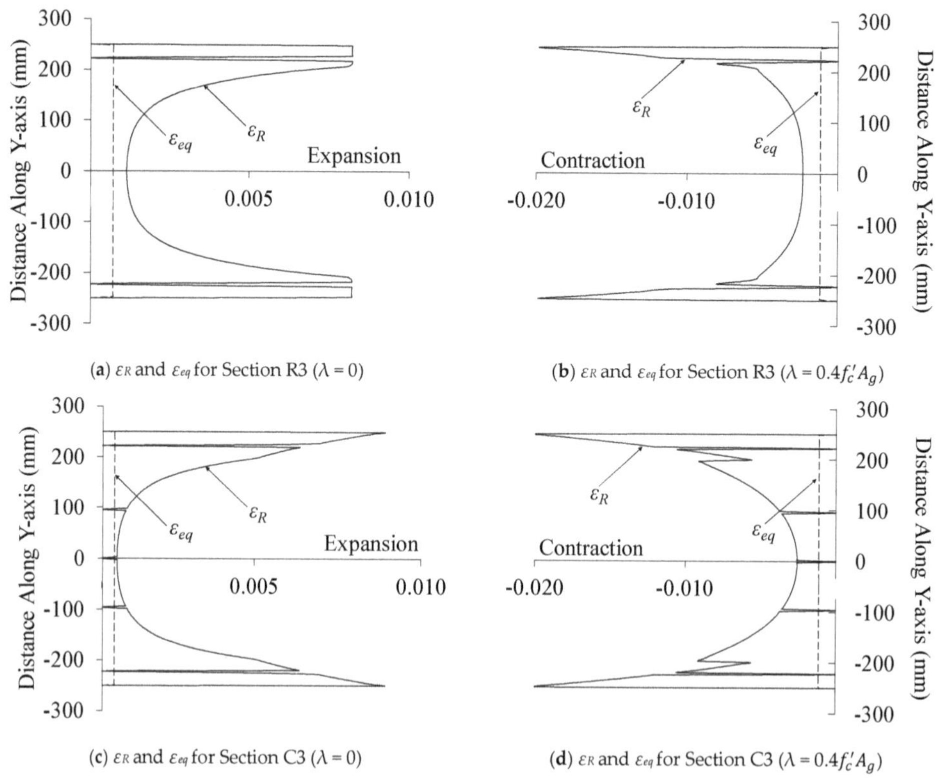

(**a**) ε_R and ε_{eq} for Section R3 ($\lambda = 0$)

(**b**) ε_R and ε_{eq} for Section R3 ($\lambda = 0.4f_c'A_g$)

(**c**) ε_R and ε_{eq} for Section C3 ($\lambda = 0$)

(**d**) ε_R and ε_{eq} for Section C3 ($\lambda = 0.4f_c'A_g$)

Figure 11. Residual and equivalent strain distribution along column R3 and C3 cross-sections.

8. Validation of the Proposed Analytical Model

The capability of the present model to predict the post-fire structural performance of axially loaded RC members is validated in view of the experimental results by Chen et al. [25], Jau and Huang [26], Yaqub and Bailey [27] and Elsanadedy et al. [28]. The validation is limited to structural members made of normal-strength concrete where spalling does not occur.

Chen et al. [25] carried out a full-scale experiment to investigate the performance of RC columns after exposure to different fire conditions. The results obtained from the proposed analytical model are compared with the measured data of columns FC06 and FC05. These columns are exposed to an ISO 834 (2014) standard fire curve from four sides for 2 hrs and 4 hrs, respectively. The tested columns have cross-sectional dimensions of 300 mm × 450 mm, concrete cover of 40 mm and an overall length of 3.0 m. The concrete compressive strength at ambient conditions is 29.5 MPa. The longitudinal reinforcement consists of 4 Φ 19 mm and 4 Φ 16 mm steel bars with yield strengths of 476 MPa and 479 MPa, respectively. Both columns were subjected to an initial axial load of 797 kN prior to heat exposure. The specimens were axially loaded during the whole heating and cooling cycle. After 30 days from the fire test, the columns were subjected to a constant initial concentric load of 797 kN and an additional eccentric load offset a distance of 650 mm from the cross-sectional centre along the y-axis (weak axis) producing the bending moment about the x-axis, while another eccentric load is applied at 600 mm from the centroidal axis. Figure 12a shows the analytical and experimental load–deflection curves at the column mid-span due to the eccentric load about the y-axis. A very good agreement between both curves can be shown with a percent difference of 3.8% and 4.6% in the ultimate capacity of columns FC06 and FC05, respectively, and a percent difference of 6.3% and 5.4% in the 40%

secant stiffness for the same two columns, respectively. This variation can be attributed to the sensitivity of the adopted thermal expansion model to the experimental conditions and concrete mix.

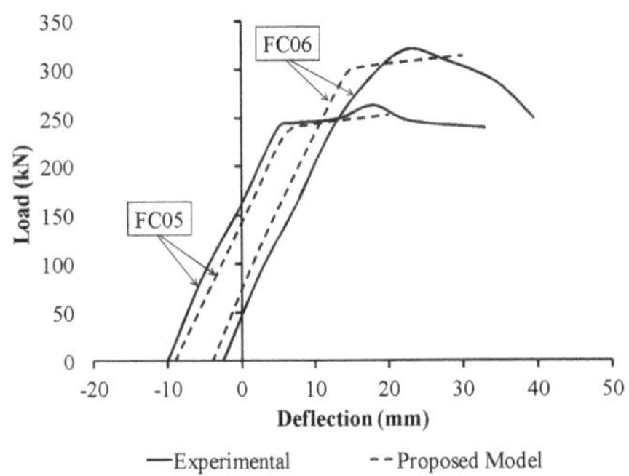

(**a**) Using experiment by Chen et al. (2009)

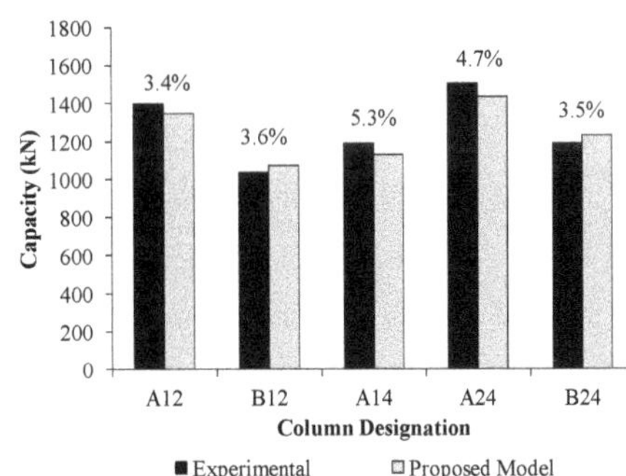

(**b**) Using experiment by Jau and Huang (2008)

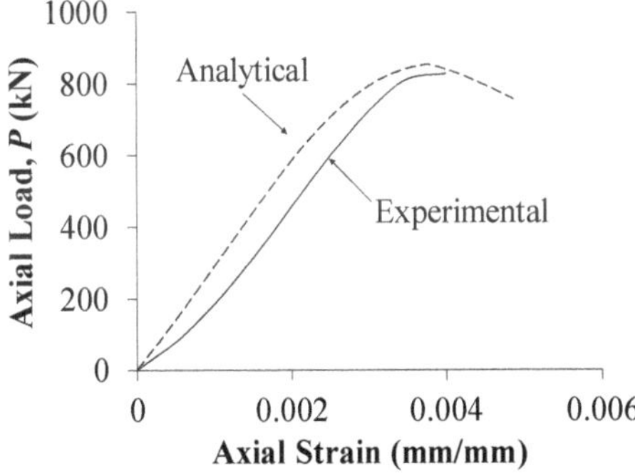

(**c**) Using experiment by Yaqub and Bailey (2011)

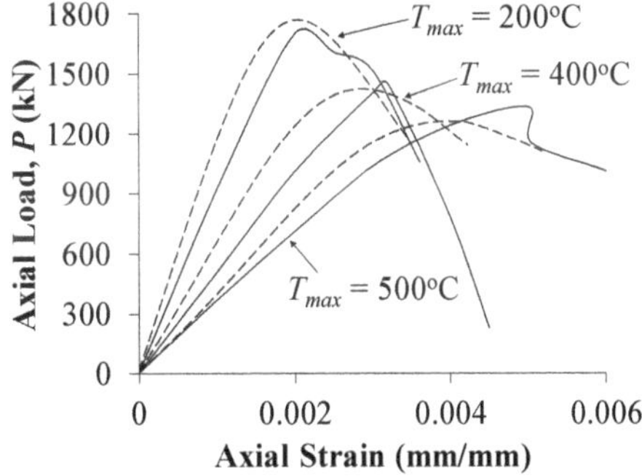

(**d**) Using experiment by Elsanadedy et al., (2016)

Figure 12. Validation of the proposed analytical model.

In another experimental study, Jau and Huang [26] investigated the residual behavior of initially loaded restrained RC columns subjected to heat from two adjacent sides. The cross-sectional dimensions of all columns are 300 mm × 450 mm, with an overall length of 2.7 m. The concrete cover varies between 50 mm or 70 mm, whereas the steel reinforcement ratio varies between 1.8% and 3.0%. Normal-strength concrete with a compressive strength of 33.7 MPa and steel bars with a yield strength of 475.8 MPa are used. The test setup allows the heat to flow through two adjacent surfaces only while the other two surfaces are insulated and not subjected to fire. The restrained columns are subjected to a 10% axial preloading of their ambient compressive strength during the 2 or 4 hr fire tests. After the columns naturally cooled down, the load is applied until failure occurs. Figure 12b shows both the experimental and predicted residual capacity of columns A12, B12, A14, A24 and B24 whose detailed geometrical and mechanical properties are provided by Jau and Huang [26]. The proposed model is found to predict the capacity of the tested columns with high accuracy, as indicated by the maximum percent error of 5.3% depicted for column A14 shown in Figure 12b. Overall, the agreement between the experimental and analytical results is very good. It is worth mentioning that exposing the specimens to fire from two adjacent sides only resulted in a lateral displacement in the range of 7 to 25 mm, depending on fire duration. Due to the influence of load–temperature interaction, the residual strength and stiffness varies within the cross-section. The lateral displacement and the associated curvature should be considered when analyzing the fire-exposed columns, especially those with high slenderness ratios.

The influence of elevated temperature on the residual axial capacity, axial stiffness and stress–strain behavior of circular columns strengthened with fiber-reinforced polymers (FRP) was experimentally investigated by Yaqub and Bailey [27]. The unwrapped control specimen (i.e., Specimen No. 2) is considered for comparison. The examined column has a diameter of 200 mm and an overall length of 1000 mm. The concrete cover to the centroid of the steel bars was taken as 30 mm. The reinforcement consisted of 6 Φ 10 mm steel bars, resulting in a reinforcement ratio of 1.5%. Normal-weight concrete with a compressive strength of 42.4 MPa and steel bars with a yield strength of 570 MPa were used. The column was exposed to a predefined heating–cooling cycle 9 months after casting until the entire cross-section reached a uniform temperature of 500 °C. After that, the column was subjected to a displacement-controlled uniaxial compression load until failure. The member temperature at the time of the compression test was 22 °C. Figure 12c presents both the experimental and analytical axial load–deformation curves for the specimen, which was exposed to a uniform temperature of 500 °C before cooling. The proposed model is found to provide very good prediction of the experimental results, as indicated by the 4.2% percent error. The strength of the heat-exposed columns was reduced by 41.8% after the heating–cooling cycle, as implied by the strength reduction from an average of 1418 kN for the intact columns to 826 kN of the heated column. Based on Figure 4.4.2.2.1c(b) of ACI 216-14 [9], the estimated residual strength of the specimen at 500 °C is determined as about 51.0% of the initial compressive strength. This represents an error of about 12.4% with respect to the actual residual strength of 58.2% determined in the lab. The incremental stiffness at service load is almost identical between the two curves. Additionally, the load–deformation behavior obtained from the proposed model is shown to be consistent with that obtained experimentally in terms of stiffness, peak strain and failure strain.

Elsanadedy et al. [28] examined the effect of high temperature on the residual capacity and deformation behavior of R.C. columns strengthened with FRP wraps. The control specimens, which were unwrapped, were tested at a room temperature of 26 °C and are considered for comparison in this paper. All the examined columns have a diameter of 242 mm and an overall length of 900 mm. The concrete cover to the centroid of the steel bars was taken as 41 mm. The reinforcement consisted of 4 Φ 10 steel bars, resulting in a reinforcement ratio of 0.68%. Normal-weight concrete with a compressive strength of 42 MPa and steel bars with a yield strength of 593 MPa were used. The columns were exposed to elevated temperature along their circumference under unstressed conditions with a heating rate ranging between 5 °C and 15 °C per minute. Specimens C-200, C-400 and C-500 are considered for comparison, where the letter "C" indicates un-strengthened specimens, and the number indicates the maximum temperature reached in the oven. The specimens were subjected to the specified maximum temperature for 3 h before shutting down the oven. The columns were then naturally cooled inside the oven to room temperature. After that, the columns were taken out of the oven and subjected to a displacement-controlled uniaxial compression load until failure. Figure 12d presents both the experimental and analytical axial load–deformation curves for the examined specimens. The residual strength of specimens C-200, C-400 and C-500 were found to be 1745 kN, 1490 kN and 1350 kN, respectively. These values represent a residual strength of 90.1%, 77.1% and 69.9% of the initial strength of the intact specimen, respectively. The capability of the proposed model to capture the residual capacity of the heat-exposed columns is very good, as indicated by the 4.7%, 3.7% and 6.5% percent errors for specimens C-200, C-400 and C-500, respectively. Considering Figure 4.4.2.2.1c(b) of ACI 216-14 [9], the residual strength of specimens C-200, C-300 and C-400 are 84.0%, 75.0% and 63.0% of the initial compressive strength, respectively. This represents a percent error of 7.0%, 2.8% and 9.8% relative to the tested specimens, respectively. Additionally, the load–deformation behavior obtained from the proposed model is shown to be consistent with that obtained experimentally in terms of stiffness, peak strain and failure strain. The error between the model and experimental results can be attributed to the variation in heating rate and the presence of residual surface cracks and initial misalignment, which are not accounted for in the model.

9. Parametric Study

The main parameters include the concrete compressive strength, f_c' (25 MPa and 35 MPa); steel yield strength, f_y (300 MPa and 400 MPa); fire duration, t (0.5 hr, 1.5 hrs and 2.5 hrs); initial load level, λ (0.0, 0.2 f_c', 0.4 f_c'); axial restraint stiffness ratio, R_D (0.0, 0.5 and 1.0); and steel reinforcement ratio, ρ (0.02 and 0.04). The cross-sectional dimensions of the rectangular sections are defined in terms of member height, h (400 mm and 800 mm) and width, b (300 mm and 600 mm), whereas for circular sections, the geometrical properties are determined in terms of their diameter, D (350 mm and 650 mm). A 30 mm clear concrete cover was considered for all specimens. The members are exposed to fire along their perimeters according to ASTM E119 [17] standard fire curve, followed by a cooling phase according to ISO 834 [18] recommendations. The influence of the considered factors on the post-fire behavior of both rectangular and circular RC axially loaded members is investigated in view of a parametric study. Based on these parameters, the analytical investigation consists of a total of 1728 different cases.

The effect of the aforementioned parameters on both the residual axial capacity and the residual 40% secant axial stiffness is illustrated in view of the members presented in Table 1. The variation in the residual capacity and stiffness in terms of the different parameters at different initial load levels is presented Figures 13 and 14 for both rectangular and circular sections, respectively.

9.1. Effect of Fire Duration

Fire duration has been found to have the most significant influence on reducing the post-fire capacity and stiffness of both rectangular and circular RC columns. The influence of increasing the fire duration on the residual flexural behavior is examined in view of the rectangular sections (R1, R2 and R3) and the circular sections (C1, C2 and C3), as shown in Figures 13a and 14a, respectively. Prolonged exposure to fire results in material strength degradation and softening, which adversely affect the stiffness and capacity of the fire-damaged section. The permanent strength and stiffness reductions in the circular columns are found to be slightly higher than those with rectangular sections. This can be attributed to the higher maximum temperature reached within the circular sections subjected to fire for the same fire duration, as was previously described in Figure 6. The additional deterioration in both concrete and steel residual mechanical properties caused by the longer duration of the heating–cooling cycle provides more time for heat to transfer to the inner elementary layers raising their temperatures.

9.2. Effect of Section Size

Increasing the cross-sectional dimensions of both rectangular and circular columns results in higher residual flexural strength and stiffness after fire, as indicated in Figures 13b and 14b. This larger residual capacity is caused by the lower temperature increase within the larger member as it requires more heat energy to increase its temperature. This is attributed to the additional concrete cover provided by the larger sections causing the hindrance of heat transfer from the column perimeter towards its core. Hence, internal concrete fibers experience lower temperatures and consequently higher residual compressive strength and stiffness than the inner elements of columns with smaller dimensions. For the same fire duration, concrete within the inner parts of the wider member experience a lower increase in temperature and consequently more recovery after a fire. The influence of strength recovery in steel bars is neglected since concrete cover is the same in all specimens causing the maximum temperature reached in all steel bars to be the same.

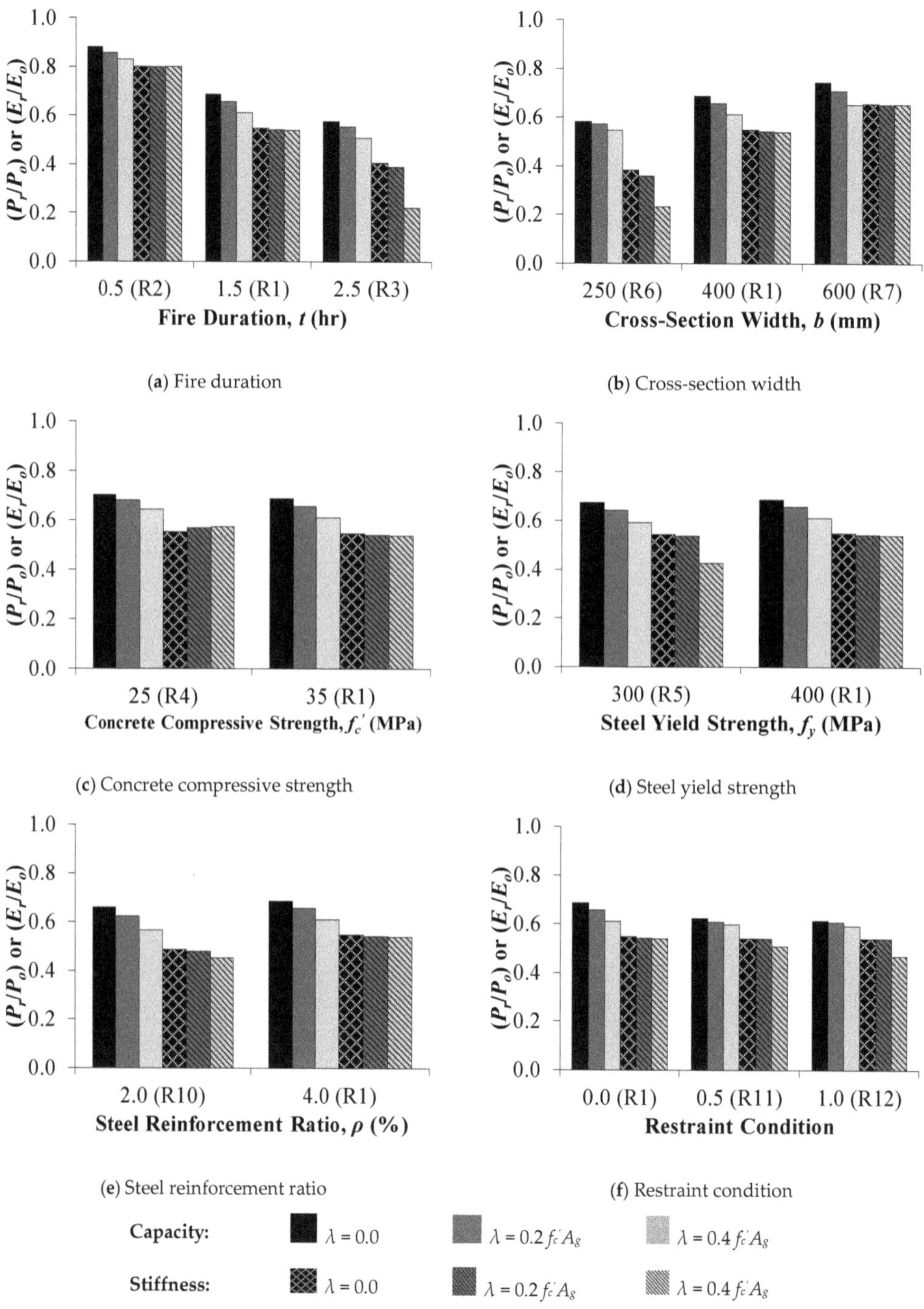

Figure 13. Influence of varying the examined parameters on the axial capacity and stiffness of rectangular columns.

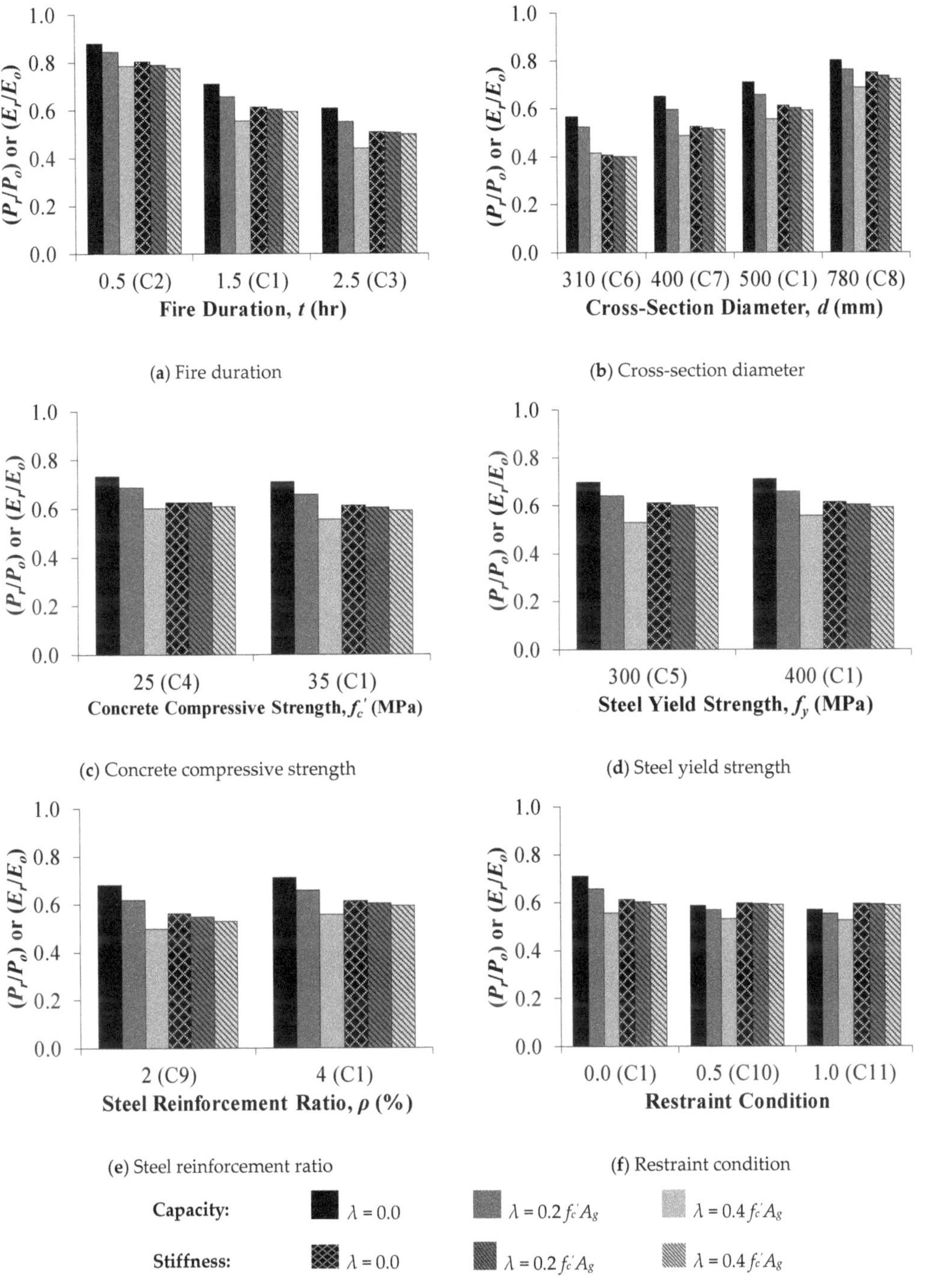

Figure 14. Influence of varying the examined parameters on the axial capacity and stiffness of circular columns.

9.3. Effect of Mechanical Properties

Increasing the concrete compressive strength is found to have an insignificant inverse relationship on the reduction ratio of both capacity and stiffness for all load levels in the examined range, as shown in Figures 13c and 14c, for rectangular and circular columns, respectively. The decreasing rate can be justified by the greater reduction in compressive strength of the stronger concrete after fire. Hence, the reduction in concrete contribution within the compression zone becomes more pronounced and results in the observed larger decrease relative to the original capacity. The use of normal-strength concrete infers that no spalling is encountered, which could otherwise significantly affect the residual capacity. The same observation can be drawn by varying the grade of the embedded steel bars from 300 MPa to 400 MPa, as shown Figures 13d and 14d, for rectangular and circular columns, respectively. This is attributed to the fact the steel bars restore a significant portion of their capacity and stiffness after fire as discussed previously.

9.4. Effect of Steel Reinforcement Ratio

Steel bars are located near the exposed surfaces of the columns and are subjected to relatively high temperatures. However, this has negligible impact on the overall axial capacity and stiffness reduction due to the significant recovery of mild steel bars after fire exposure [29–31]. Figures 13e and 14e show that increasing the reinforcement ratio results in an insignificant increase in both residual capacity and stiffness in the rectangular and circular columns, respectively. This is attributed to the higher impact of the larger steel area in replacing the fire-damaged concrete since the recovery of steel bars is very significant as opposed to concrete.

9.5. Effect of Restraint Conditions

The influence of restraining the member against thermal expansion during heating has been found to slightly decrease its post-fire stiffness and capacity, as shown in Figures 13f and 14f, for both rectangular and circular columns, respectively. The reduction in residual properties is more pronounced when comparing the fully unrestrained sections with the restrained ones. However, the reduction seems to be almost identical for columns that are fully restrained or 50% restrained. This is explained by the impact of transient strain in changing the deformation behavior of axially loaded members during fire exposure through alleviating the thermal expansion. As the stiffness of the supports provided by the adjacent frame members increases, more restraining forces are generated to counteract the tendency of the column to expand. This additional force results in transient creep strain, which reduces the thermal strain and consequently decreases the amount of restraining force required to overcome the expansion. These two processes occur simultaneously and have a negative influence on each other, causing them to reduce the impact of restrains.

During fire exposure, the column's tendency to undergo thermal expansion increases with time, causing the support to counteract this potential movement, depending on the column's stiffness. Initially, the member's stiffness remains close to that at ambient conditions as the temperature increase within the member is relatively low. Thus, an increase in restraining force results in significant hindrance of the column's deformation as the thermal strain component increases. However, after a certain period of time, the temperature within the member becomes relatively high, causing the stiffness degradation to become more pronounced. Thus, the forces required to resist the larger thermal expansion of the member drops. The axial force required to restrain the member keeps decreasing as a result of the continuous reduction in stiffness caused by elevated temperatures. Therefore, the change in the restraining load is characterized by a mild increase followed by a gradual decrease with time.

10. Proposed Simplified Expressions to Obtain Residual Axial Capacity and Stiffness

Prolonged exposure of RC columns to elevated temperatures according to a standard fire has a substantial influence on their axial capacity and deformation behavior.

The residual structural performance of such columns relies on the geometrical characteristics, mechanical properties, initial load, restraint conditions and fire duration that should be appropriately accounted for in the analysis. Accurate determination of temperature distribution and residual strain components developed within RC columns is tedious and requires detailed thermal and structural analyses that may not be convenient for design engineers. The proposed analytical model comprehensively addresses the influence of the aforementioned factors on determining the post-fire response of both rectangular and circular RC columns. Hence, based on the extensive parametric study conducted on the 1728 different cases, regression analysis is carried out to develop expressions for obtaining both the residual axial capacity and secant axial stiffness of fire-damaged rectangular and circular RC columns. These proposed expressions take into consideration the loading history, restraint conditions, fire duration, material strength and cross-sectional dimensions of the exposed members. The validity and accuracy of the proposed equations depend on the range of parameters considered in the parametric study. The proposed expressions provide a suitable approach for predicting the behavior of RC columns after exposure to an extreme standard fire scenario. This would be a valuable tool for both researchers and engineers to predict the post-fire performance of RC columns during the design phase.

10.1. Rectangular Sections

Linear multiple regression analysis is performed to propose an expression for both the residual capacity and axial stiffness ratios (ω), as given in Equation (4).

$$\omega = A_1 + A_2\lambda + A_3 f_c' + A_4 f_y + A_5\rho + A_6\frac{\rho f_y}{f_c'} + A_7 b + A_8 h \tag{4}$$

where λ is the initial load level relative to ambient capacity, f_c' is the concrete compressive strength (MPa), f_y is the steel yield strength (MPa), ρ is the steel reinforcement ratio, b is section width (m), and h is section height (m). The coefficients ($A_{i\,=\,1,2,3,4,5,6,7,8}$) are given in Table 2 in terms of the axial restraint ratio (R_D) and fire duration at the end of the heating phase (t) in hours. For values other than the listed t and R_D, linear interpolation of the upper and lower calculated ω should be performed. In Table 2, P_o and P_r are the axial capacities at ambient and post-fire conditions, respectively; EA_i and $(EA_i)_r$ are the initial axial stiffness at ambient and post-fire conditions, respectively; $EA_{0.4}$ and $(EA_{0.4})_r$ are the 40% axial stiffness at ambient and post-fire conditions, respectively; and $EA_{0.8}$ and $(EA_{0.8})_r$ are the 80% axial stiffness at ambient and post-fire conditions, respectively.

It is worth mentioning that although the rectangular column is exposed to fire from all sides, the coefficients of the section height (h) and section width (b) are different in Equation (4). This variation is attributed to the assumed reinforcement configuration where the steel bars lie in two opposite layers that are parallel to the section width as indicated in Figure 4a.

The applicability of the proposed expressions is assessed by comparing the values obtained using the proposed equations and the results obtained from the analytical analysis. A comparison between the values predicted from Equation (4) and the results determined through performing detailed analytical analysis for all examined cases revealed a very good agreement, as shown in Figures 15a and 16a, for both residual capacity and axial stiffness, respectively. The equality line denotes the location on the graph where the predictions from the proposed equations match those obtained from the proposed analytical model. As shown in the figure, the data points are uniformly distributed in the vicinity of the equality line.

The residual compressive strength of column R6 is determined from ACI 216.1-14 and compared to that obtained from Equation (4) considering an exposure of 1 h to ASTM E119 [17] standard fire. Figure 4.4.2.3a and 4.4.2.2.1c(b) of ACI 216.1-14 provide the temperature distribution and the residual compressive strength of concrete at various heating and loading conditions, respectively. The unstressed residual compressive strength obtained from the ACI procedure for column R6 is found to be just under 71.0% of its initial strength.

Considering the same parameters, the percentage of the residual strength obtained from the proposed Equation (4) is found to be 77.7%. In general, there is a good agreement between the two values considering the complex behavior of fire-exposed RC members. The variation between the two values is attributed mainly to the difference between the exposure and boundary conditions assumed in both methods. The residual strength is calculated from Equation (4) considering the reinforcement ratio, steel yield strength and section height, which is not specifically provided in the ACI procedure.

Table 2. Coefficient of Equation (4) for rectangular sections.

ω	A_i	$R_D = 0$ (Unrestrained)			$R_D = 0.5$ (Partially Restrained)			$R_D = 1$ (Fully Restrained)		
		$t = 0.5$ hr	$t = 1.5$ hrs	$t = 2.5$ hrs	$t = 0.5$ hr	$t = 1.5$ hrs	$t = 2.5$ hrs	$t = 0.5$ hr	$t = 1.5$ hrs	$t = 2.5$ hrs
$\dfrac{P_r}{P_o}$	A_1	6.90×10^{-1}	3.60×10^{-1}	2.06×10^{-1}	6.16×10^{-1}	3.28×10^{-1}	2.17×10^{-1}	5.96×10^{-1}	3.14×10^{-1}	2.18×10^{-1}
	A_2	-1.22×10^{-1}	-1.70×10^{-1}	-1.48×10^{-1}	-7.29×10^{-2}	-7.23×10^{-2}	-8.24×10^{-2}	-6.74×10^{-2}	-6.43×10^{-2}	-7.03×10^{-2}
	A_3	-7.09×10^{-4}	-1.28×10^{-3}	-1.58×10^{-3}	-1.08×10^{-3}	-1.97×10^{-3}	-2.16×10^{-3}	-1.16×10^{-3}	-1.94×10^{-3}	-2.26×10^{-3}
	A_4	5.01×10^{-5}	6.72×10^{-5}	7.83×10^{-5}	6.94×10^{-5}	8.57×10^{-5}	1.02×10^{-4}	7.58×10^{-5}	9.78×10^{-5}	1.04×10^{-4}
	A_5	1.31	1.90	2.34	1.82	2.46	3.05	2.09	2.65	3.17
	A_6	9.03×10^{-2}	1.16×10^{-1}	1.43×10^{-1}	1.14×10^{-1}	1.43×10^{-1}	1.41×10^{-1}	1.14×10^{-1}	1.42×10^{-1}	1.47×10^{-1}
	A_7	1.66×10^{-1}	3.45×10^{-1}	4.04×10^{-1}	1.54×10^{-1}	2.51×10^{-1}	3.00×10^{-1}	1.49×10^{-1}	2.37×10^{-1}	2.78×10^{-1}
	A_8	1.30×10^{-1}	2.23×10^{-1}	2.42×10^{-1}	1.78×10^{-1}	2.56×10^{-1}	2.43×10^{-1}	1.92×10^{-1}	2.61×10^{-1}	2.43×10^{-1}
$\dfrac{(EA_i)_r}{EA_i}$	A_1	5.36×10^{-1}	-1.38×10^{-1}	-3.18×10^{-1}	5.34×10^{-1}	-1.64×10^{-1}	-2.57×10^{-1}	5.32×10^{-1}	-2.41×10^{-1}	-2.75×10^{-1}
	A_2	-7.45×10^{-3}	-5.66×10^{-2}	-7.28×10^{-2}	-3.75×10^{-3}	-6.59×10^{-2}	-1.26×10^{-1}	-3.55×10^{-3}	-8.61×10^{-2}	-1.35×10^{-1}
	A_3	-3.90×10^{-4}	-2.01×10^{-3}	-3.58×10^{-3}	-4.00×10^{-4}	-3.97×10^{-3}	-2.77×10^{-3}	-3.56×10^{-4}	-2.69×10^{-3}	-4.56×10^{-3}
	A_4	-2.41×10^{-5}	1.79×10^{-4}	3.70×10^{-4}	-2.83×10^{-5}	3.14×10^{-4}	2.02×10^{-4}	-2.98×10^{-5}	2.99×10^{-4}	3.83×10^{-4}
	A_5	2.52	9.00	$1.49 \times 10^{+1}$	2.55	$1.16 \times 10^{+1}$	$1.12 \times 10^{+1}$	2.56	9.90	$1.38 \times 10^{+1}$
	A_6	4.68×10^{-2}	-1.37×10^{-1}	-3.52×10^{-1}	5.55×10^{-2}	-2.70×10^{-1}	-6.62×10^{-2}	5.88×10^{-2}	-3.94×10^{-2}	-2.54×10^{-1}
	A_7	2.11×10^{-1}	8.16×10^{-1}	7.84×10^{-1}	2.04×10^{-1}	8.70×10^{-1}	7.46×10^{-1}	2.01×10^{-1}	9.32×10^{-1}	7.42×10^{-1}
	A_8	1.64×10^{-1}	2.63×10^{-1}	2.56×10^{-1}	1.68×10^{-1}	2.49×10^{-1}	2.56×10^{-1}	1.69×10^{-1}	2.35×10^{-1}	2.59×10^{-1}
$\dfrac{(EA_{0.4})_r}{EA_{0.4}}$	A_1	5.74×10^{-1}	-1.53×10^{-1}	-2.88×10^{-1}	5.82×10^{-1}	-3.85×10^{-1}	-3.70×10^{-1}	5.81×10^{-1}	-4.48×10^{-1}	-4.43×10^{-1}
	A_2	1.37×10^{-3}	-1.58×10^{-1}	-3.34×10^{-1}	-5.76×10^{-4}	-1.49×10^{-1}	-3.58×10^{-1}	-1.15×10^{-3}	-1.38×10^{-1}	-3.08×10^{-1}
	A_3	-5.86×10^{-4}	8.44×10^{-4}	6.42×10^{-4}	-6.26×10^{-4}	1.16×10^{-3}	5.47×10^{-4}	-5.61×10^{-4}	1.94×10^{-3}	2.56×10^{-3}
	A_4	-1.88×10^{-5}	-4.47×10^{-5}	5.78×10^{-5}	-4.12×10^{-5}	1.41×10^{-4}	1.81×10^{-4}	-4.48×10^{-5}	1.61×10^{-4}	1.50×10^{-4}
	A_5	2.19	1.29	2.18	2.08	-1.56	6.68×10^{-1}	2.07	-3.09	-3.46
	A_6	7.94×10^{-2}	4.52×10^{-1}	4.49×10^{-1}	1.25×10^{-1}	7.39×10^{-1}	6.88×10^{-1}	1.33×10^{-1}	8.25×10^{-1}	1.06
	A_7	2.10×10^{-1}	9.25×10^{-1}	8.43×10^{-1}	1.89×10^{-1}	1.15	8.45×10^{-1}	1.83×10^{-1}	1.19	8.45×10^{-1}
	A_8	1.69×10^{-1}	2.78×10^{-1}	2.88×10^{-1}	1.77×10^{-1}	2.86×10^{-1}	2.83×10^{-1}	1.79×10^{-1}	2.97×10^{-1}	2.65×10^{-1}
$\dfrac{(EA_{0.8})_r}{EA_{0.8}}$	A_1	5.96×10^{-1}	-2.33×10^{-1}	-3.10×10^{-1}	5.11×10^{-1}	-4.88×10^{-1}	-4.27×10^{-1}	4.57×10^{-1}	-4.96×10^{-1}	-4.32×10^{-1}
	A_2	1.79×10^{-2}	-3.16×10^{-1}	-4.64×10^{-1}	-3.54×10^{-2}	-1.35×10^{-1}	-2.16×10^{-1}	-3.89×10^{-2}	-1.15×10^{-1}	-1.58×10^{-1}
	A_3	-8.59×10^{-4}	1.51×10^{-3}	1.54×10^{-3}	-1.37×10^{-3}	3.87×10^{-3}	3.30×10^{-3}	-1.11×10^{-3}	4.18×10^{-3}	3.29×10^{-3}
	A_4	2.76×10^{-5}	1.55×10^{-4}	1.06×10^{-4}	2.29×10^{-4}	2.43×10^{-4}	9.84×10^{-5}	3.04×10^{-4}	2.38×10^{-4}	8.56×10^{-5}
	A_5	1.88	-1.67	-9.57×10^{-1}	1.60	-8.52	-6.22	1.03	-8.62	-6.28
	A_6	1.44×10^{-1}	5.98×10^{-1}	5.20×10^{-1}	2.43×10^{-1}	9.97×10^{-1}	7.77×10^{-1}	3.07×10^{-1}	9.58×10^{-1}	7.35×10^{-1}
	A_7	2.03×10^{-1}	9.81×10^{-1}	8.88×10^{-1}	2.25×10^{-1}	1.09	8.52×10^{-1}	2.46×10^{-1}	1.08	8.43×10^{-1}
	A_8	1.72×10^{-1}	3.18×10^{-1}	3.08×10^{-1}	1.95×10^{-1}	3.54×10^{-1}	3.11×10^{-1}	2.03×10^{-1}	3.52×10^{-1}	3.04×10^{-1}

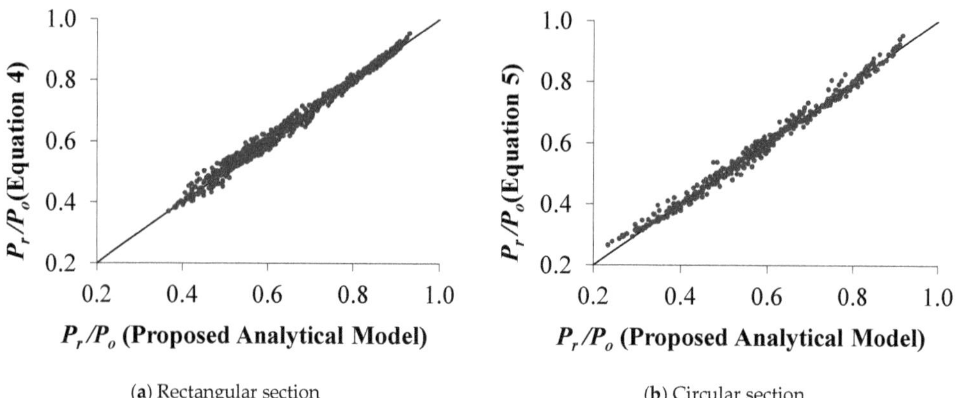

(**a**) Rectangular section

(**b**) Circular section

Figure 15. Validation of the proposed Equations (4) and (5) for residual capacity.

10.2. Circular Sections

Multiple linear regression analysis is also performed to propose a similar expression for the residual capacity and stiffness of the axially loaded circular RC columns, as shown in Equation (5).

$$\omega = B_1 + B_2\lambda + B_3 f'_c + B_4 f_y + B_5\rho + B_6\frac{\rho f_y}{f'_c} + B_7 D \tag{5}$$

where D is the diameter of the cross-section (m). The coefficients ($B_i = {}_{1,2,3,4,5,6,7}$) are given in Table 3 in a similar manner to the coefficients of the rectangular section. The line of equality plot reveals that the proposed expressions provide an excellent prediction of the capacity and stiffness compared to the results obtained from the analytical model, as illustrated in Figures 15b and 16b, respectively. The presence of outliers is almost negligible, which enhances the confidence of using the proposed expressions. The simplicity and robustness of the proposed expressions is an advantage for increasing their applicability during the design phase.

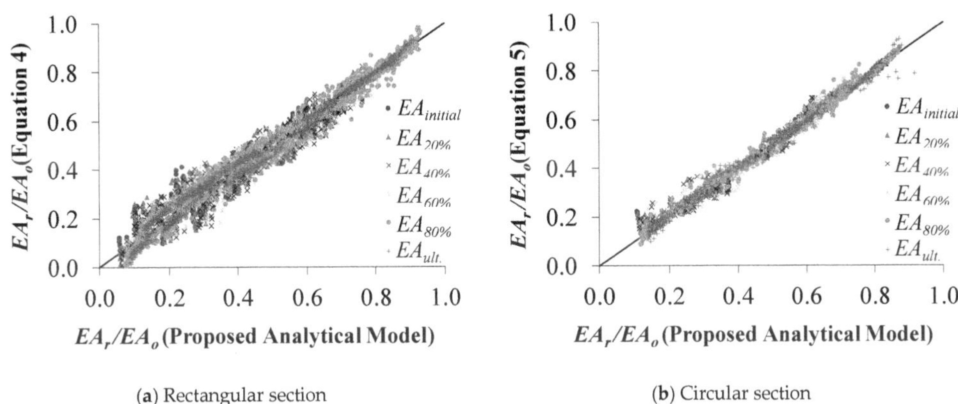

(a) Rectangular section

(b) Circular section

Figure 16. Validation of the proposed Equations (4) and (5) for residual axial stiffness.

Table 3. Coefficient of Equation (5) for circular sections.

ω	B_i	$R_D = 0$ (Unrestrained)			$R_D = 0.5$ (Partially Restrained)			$R_D = 1$ (Fully Restrained)		
		$t = 0.5$ hr	$t = 1.5$ hrs	$t = 2.5$ hrs	$t = 0.5$ hr	$t = 1.5$ hrs	$t = 2.5$ hrs	$t = 0.5$ hr	$t = 1.5$ hrs	$t = 2.5$ hrs
$\frac{P_r}{P_o}$	B_1	6.70×10^{-1}	3.41×10^{-1}	1.96×10^{-1}	4.87×10^{-1}	1.74×10^{-1}	1.04×10^{-1}	4.41×10^{-1}	1.71×10^{-1}	4.63×10^{-2}
	B_2	-2.52×10^{-1}	-3.97×10^{-1}	-3.96×10^{-1}	-1.44×10^{-1}	-1.44×10^{-1}	-1.43×10^{-1}	-1.17×10^{-1}	-1.24×10^{-1}	-8.38×10^{-2}
	B_3	-1.76×10^{-3}	-1.91×10^{-3}	-1.86×10^{-3}	-1.77×10^{-3}	-2.32×10^{-3}	-2.40×10^{-3}	-1.92×10^{-3}	-2.62×10^{-3}	-7.54×10^{-4}
	B_4	5.55×10^{-6}	9.80×10^{-5}	1.05×10^{-4}	1.14×10^{-4}	1.28×10^{-4}	1.49×10^{-4}	1.25×10^{-4}	7.86×10^{-5}	1.02×10^{-4}
	B_5	5.75×10^{-1}	1.30	1.43	1.32	1.60	1.81	1.54	1.66	8.87×10^{-1}
	B_6	7.74×10^{-2}	8.00×10^{-2}	8.96×10^{-2}	7.71×10^{-2}	1.22×10^{-1}	1.06×10^{-1}	7.86×10^{-2}	1.37×10^{-1}	1.62×10^{-1}
	B_7	3.88×10^{-1}	6.00×10^{-1}	6.53×10^{-1}	4.62×10^{-1}	6.24×10^{-1}	5.74×10^{-1}	4.95×10^{-1}	6.32×10^{-1}	6.12×10^{-1}
$\frac{(EA_i)_r}{EA_i}$	B_1	4.51×10^{-1}	7.79×10^{-2}	-1.64×10^{-1}	4.14×10^{-1}	7.25×10^{-2}	-1.59×10^{-1}	4.00×10^{-1}	7.57×10^{-2}	-1.59×10^{-1}
	B_2	-5.91×10^{-2}	-4.15×10^{-2}	-6.81×10^{-2}	-3.41×10^{-2}	-1.79×10^{-2}	-4.96×10^{-2}	-2.83×10^{-2}	-1.56×10^{-2}	-5.38×10^{-2}
	B_3	1.38×10^{-4}	-2.76×10^{-4}	-1.75×10^{-3}	6.76×10^{-4}	-2.02×10^{-5}	-1.79×10^{-3}	9.26×10^{-4}	-7.20×10^{-6}	-3.19×10^{-3}
	B_4	-3.46×10^{-5}	-5.30×10^{-5}	1.60×10^{-4}	-4.79×10^{-5}	-7.07×10^{-5}	1.50×10^{-4}	-5.48×10^{-5}	-8.46×10^{-5}	2.40×10^{-4}
	B_5	1.33	2.41	4.78	1.40	2.41	4.78	1.41	2.42	5.64
	B_6	3.36×10^{-2}	5.16×10^{-2}	-5.01×10^{-2}	5.10×10^{-2}	7.11×10^{-2}	-4.03×10^{-2}	5.80×10^{-2}	7.49×10^{-2}	-9.85×10^{-2}
	B_7	4.98×10^{-1}	7.76×10^{-1}	9.21×10^{-1}	4.92×10^{-1}	7.42×10^{-1}	8.97×10^{-1}	4.94×10^{-1}	7.37×10^{-1}	9.04×10^{-1}
$\frac{(EA_{0.4})_r}{EA_{0.4}}$	B_1	4.95×10^{-1}	8.07×10^{-2}	-1.97×10^{-1}	4.42×10^{-1}	3.46×10^{-2}	-3.02×10^{-1}	4.23×10^{-1}	1.80×10^{-3}	-3.47×10^{-1}
	B_2	-7.98×10^{-2}	-7.01×10^{-2}	-1.66×10^{-1}	-3.87×10^{-2}	-5.52×10^{-2}	-1.28×10^{-1}	-3.26×10^{-2}	-5.93×10^{-2}	-1.04×10^{-1}
	B_3	4.74×10^{-5}	-1.17×10^{-3}	1.20×10^{-3}	8.94×10^{-4}	-1.66×10^{-3}	-3.48×10^{-4}	1.19×10^{-3}	-2.38×10^{-3}	2.62×10^{-4}
	B_4	-3.24×10^{-5}	3.93×10^{-5}	-6.58×10^{-5}	-7.04×10^{-5}	1.35×10^{-4}	1.55×10^{-4}	-7.71×10^{-5}	2.33×10^{-4}	1.32×10^{-4}
	B_5	1.31	3.24	1.31	1.33	4.32	3.06	1.37	5.19	2.26
	B_6	3.86×10^{-2}	1.76×10^{-2}	2.43×10^{-1}	6.98×10^{-2}	-1.77×10^{-2}	1.77×10^{-1}	7.61×10^{-2}	-6.54×10^{-2}	2.60×10^{-1}
	B_7	4.69×10^{-1}	7.98×10^{-1}	1.03	4.74×10^{-1}	7.77×10^{-1}	1.07	4.78×10^{-1}	7.91×10^{-1}	1.11
$\frac{(EA_{0.8})_r}{EA_{0.8}}$	B_1	5.44×10^{-1}	-1.01×10^{-2}	-2.63×10^{-1}	4.82×10^{-1}	-2.81×10^{-1}	-5.45×10^{-1}	4.53×10^{-1}	-3.12×10^{-1}	-5.66×10^{-1}
	B_2	-8.70×10^{-2}	-2.37×10^{-1}	-3.61×10^{-1}	-5.11×10^{-2}	-1.22×10^{-1}	-1.19×10^{-1}	-4.65×10^{-2}	-9.72×10^{-2}	-8.29×10^{-2}
	B_3	3.60×10^{-5}	9.20×10^{-5}	1.51×10^{-3}	3.15×10^{-4}	2.29×10^{-3}	3.44×10^{-3}	3.48×10^{-4}	2.34×10^{-3}	3.04×10^{-3}
	B_4	-4.74×10^{-5}	1.39×10^{-4}	8.88×10^{-5}	-7.26×10^{-5}	2.60×10^{-4}	2.96×10^{-4}	-5.30×10^{-5}	2.62×10^{-4}	3.22×10^{-4}
	B_5	1.22	1.16	-3.62×10^{-1}	1.55	-4.88×10^{-1}	-2.10	1.84	-5.61×10^{-1}	-1.37
	B_6	6.96×10^{-2}	2.09×10^{-1}	3.45×10^{-1}	1.00×10^{-1}	3.91×10^{-1}	4.48×10^{-1}	9.48×10^{-2}	3.92×10^{-1}	3.57×10^{-1}
	B_7	4.22×10^{-1}	9.14×10^{-1}	1.14	4.47×10^{-1}	1.06	1.25	4.60×10^{-1}	1.09	1.31

11. Application of the Proposed Procedure

The proposed method is suitable to be implemented by engineers during the preliminary design phase for estimating the residual performance of RC frames exposed to extreme standard fire conditions. The current study represents a step toward developing an integrated approach for considering all the components of the RC frames subjected to different loading conditions and exposed to various fire curves. This research assumes that the global behavior of the frame system is merely affected by the deterioration taking place in columns subjected to pure axial loads. This implies that beams and eccentrically loaded columns are either perfectly insulated against fire or are not exposed to critical temperatures capable of affecting their residual performance. The proposed procedure considers the interaction between the entire frame system and the fire-damaged columns in terms of connections' stiffness and load path. The fire-exposed columns are considered in the analysis as isolated members using an equivalent spring model whose stiffness is determined from the stiffness of the entire frame.

The steps required to adopt the proposed procedure are discussed in view of the 20-storey frame structure shown in Figure 17. The frame is composed of 8 m-long 300×450 mm RC beams made of normal-weight concrete with f_c' of 35 MPa and reinforced with grade 400 MPa steel bars. The 300×400 mm columns are 3.6 m long with the reinforcement ratio of 0.04 and are constructed of the same materials as the beams. The moment of inertia of both member types is determined assuming cracked cross-sections (i.e., $I_{beam} = 0.35I_g$ and $I_{column} = 0.7I_g$), where I_g is the gross moment of inertia of the considered member. The frame is loaded by subjecting the beams to a uniformly distributed load of 33 kN/m along the entire span. ASTM E119 standard fire is assumed to spread in the first floor of the building for 1.5 h, followed by a gradual cooling phase, according to ISO 834 specifications. Beams and corner columns are assumed to not be significantly influenced by fire, while the interior columns (i.e., columns IC1 and IC2) are exposed to fire from all sides. To determine the residual performance of the frame, the proposed procedure is discussed with reference to column IC1 in Figure 17. The structural analysis is performed using the commercially available ETABS [32] finite element software.

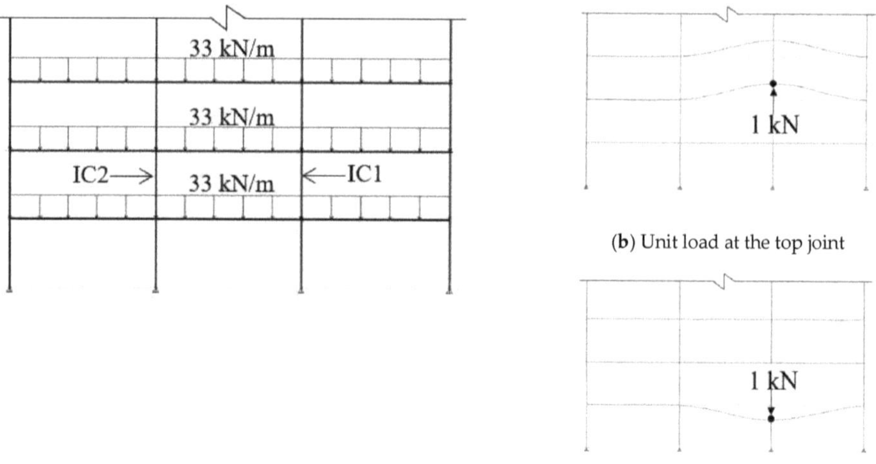

(a) Part of the considered loaded frame   (c) Unit load at bottom joint

Figure 17. Description of the proposed analysis procedure.

(1) Determine the equivalent axial stiffness (k_δ) of the spring shown in Figure 3b that represents the vertical stiffness of the structural system at that point. This is performed by replacing the examined column with a unit load acting at each joint individually, as shown in Figure 17b,c. The structural analysis is then performed on the frame to find the corresponding displacement of the considered joint. k_δ for each joint is calculated as the ratio between the unit load to the induced displacement. The total equivalent

axial stiffness (k_δ) is then determined by considering the two joints as springs in series according to Equation (6).

$$k_\delta = \frac{(k_\delta)_1 (k_\delta)_2}{(k_\delta)_1 + (k_\delta)_2} \tag{6}$$

In this example, $(k_\delta)_1$ is determined as 10,000 kN/m, while $(k_\delta)_2$ is found to be 829,187 kN/m. Thus, k_δ for the isolated column model is 9881 kN/m.

(2)   Calculate the axial restraint ration (R_D) as the ratio between k_δ calculated in step 1 and the axial stiffness of column per unit length (EA/L). In this example, R_D is found to be 0.012.

(3)   Determine the axial force acting on the considered column by performing structural analysis on the entire frame while the actual loads are added. Column IC1 in this example is subjected to an axial load of 2383 kN.

(4)   Calculate the applied load level (λ) as the ratio between the applied load and the column axial capacity. In this example, λ is determined as 0.4.

(5)   Determine the residual axial capacity (P_r) and axial stiffness (EA)$_r$ of the considered column in view of the proposed expressions provided in Equation (4) along with Table 2 for rectangular sections. In this case, ω corresponding to the capacity and axial stiffness is 0.531 and 0.311, respectively. For columns IC1, this would be translated into a residual capacity and an axial stiffness of 3161 kN and 995,923,429 kN, respectively.

(6)   Repeat the same procedure for all other axially loaded columns. In this example, the only other affected column is IC2.

(7)   Adjust the axial capacity and stiffness of the considered columns in the structural program and repeat the analysis. Repeat steps 1 through 6 until the obtained variation in both capacity and stiffness for each column is within an acceptable tolerance.

(8)   Once the residual behavior of all fire-damaged columns is adjusted in the program, the engineer can check the stresses, straining actions and deformation behavior of the frame in both the local and global levels.

12. Conclusions

In this paper, both thermal and sectional analyses are performed to determine the residual capacity and stiffness of fire-damaged rectangular and circular columns in typical RC frames. The temperature–load history experienced by the exposed members is considered in detail in the analytical study. The model is validated against relevant experimental studies found in the literature. A parametric study is carried out to determine the influence of various loading conditions and fire scenarios on the residual properties of the members. An objective-based method is then proposed to assist the engineers in evaluating the residual behavior of axially loaded RC columns considering an extreme standard fire scenario. The applicability of the proposed procedure is limited to RC columns made of normal-weight concrete and siliceous aggregate with fire durations up to 2.5 h, initial load level up to $0.4\,f_c'$, section height between 400 mm and 800 mm, section width between 300 mm and 600 mm, section diameter between 350 mm and 650 mm, and steel reinforcement ratio between 2% and 4%. The main findings are as follows:

- The deterioration in both concrete and steel residual mechanical properties continues during the cooling phase as heat transfers not only to the atmosphere, but also to the colder inner elementary layers, raising their temperatures.
- Fire duration and cross-sectional dimensions are found to be the main parameters affecting the residual stiffness and capacity of the fire-exposed members.
- The influence of the initial load level on the residual stiffness and deformation behavior is noticeable as opposed to that on the residual capacity of the fire-exposed members.
- The permanent strength and stiffness reductions in the circular columns are found to be slightly higher than those with rectangular sections due to the higher maximum temperature reached within the circular sections.

Author Contributions: Conceptualization, M.A.Y.; data curation, M.M.A.A.; investigation, M.M.A.A.; writing—original draft preparation, M.M.A.A.; writing—review and editing, M.A.Y.; acquisition, M.A.Y. All authors have read and agreed to the published version of the manuscript.

Funding: This research was funded by the Natural Sciences and Engineering Research Council of Canada (NSERC), and Western University.

Conflicts of Interest: The authors declare that they have no known competing financial interest or personal relationships that could have appeared to influence the work reported in this paper.

Nomenclature

A	area of the cross-section (mm^2)
A_{sb}	area of bottom steel reinforcement (mm^2)
A_{st}	area of top steel reinforcement (mm^2)
b	width of rectangular cross-section (mm)
D	diameter of circular cross-section (mm)
E	modulus of elasticity of concrete (MPa)
h	height of rectangular cross-section (mm)
k_δ	axial stiffness of the equivalent spring (N/mm)
P_i	initial applied load (N)
r	distance from the center of the circular cross-section to each layer (mm)
R_D	axial restraint ratio
t	time after the start of fire (hr)
T_f	fire temperature (°C)
T_o	room temperature (°C)
$\Delta\sigma$	change in the applied load level
ε_c	strain in concrete (mm/mm)
ε_{cr}	creep strain (mm/mm)
ε_{cuR}	residual ultimate strain of concrete (mm/mm)
ε_{eq}	equivalent strain (mm/mm)
ε_{LITS}	load-induced thermal strain (mm/mm)
ε_m	mechanical strain (mm/mm)
ε_R	total residual strain (mm/mm)
ε_s	strain in steel bars (mm/mm)
ε_t	total strain (mm/mm)
ε_{th}	free thermal strain (mm/mm)
ε_{thR}	residual free thermal strain (mm/mm)
ε_{tr}	transient strain (mm/mm)
ε_{trR}	residual transient strain (mm/mm)
ε_σ	stress-related strain (mm/mm)
$\varepsilon_{\sigma i}$	residual stress-induced strain (mm/mm)
λ	axial load level
ρ	steel reinforcement ratio
ω	residual capacity and axial stiffness ratio

References

1. Guo, Z.; Shi, X. *Experimental and Calculation of Reinforced Concrete at Elevated Temperatures*; Butterworth-Heinemann: Oxford, UK, 2011. [CrossRef]
2. *CAN/CSA A23.3-14*; Design of Concrete Structures. Cement Association of Canada: Ottawa, ON, Canada, 2014.
3. *ACI Committee 318*; Building Code Requirements for Structural Concrete and Commentary (ACI 318-14). American Concrete Institute: Farmington Hills, MI, USA, 2014.
4. *Eurocode2*; Design of Concrete Structures-Part 1.1: General Rules and Rules for Buildings. European Committee for Standardization (CEN): Brussels, Belgium, 2004.
5. Youssef, M.; Moftah, M. General Stress-Strain Relationship for Concrete at Elevated Temperatures. *Eng. Struct.* **2007**, *29*, 2618–2634. [CrossRef]
6. El-Fitiany, S.; Youssef, M. Assessing the Flexural and Axial Behaviour of Reinforced Concrete Members at Elevated Temperatures Using Sectional Analysis. *Fire Saf. J.* **2009**, *45*, 691–703. [CrossRef]
7. Schneider, U. *Properties of Materials at High Temperatures: Concrete*; RILEM: Kassel, Germany, 1985.

8. Anderberg, Y. *Stress and Deformation Characteristics of Concrete at High Temperatures*; Lund Institute of Technology: Stockholm, Sweden, 1976.
9. *ACI Committee 216*; Code Requirements for Determining Fire Resistance of Concrete and Masonry Construction Assemblies (ACI 216.1-14). American Concrete Institute: Farmington Hills, MI, USA, 2007.
10. National Research Council Canada (NRCC). *National Building Code of Canada*; Associate Committee on the National Building Code: Ottawa, ON, Canada, 2015.
11. AS-3600; Concrete Structures. Committee BD-002. 2001. Available online: https://www.saiglobal.com/PDFTemp/Previews/OSH/as/as3000/3600/3600-2001(+A2).pdf (accessed on 12 January 2022).
12. *Eurocode 2*; Eurocode 2: Design of Concrete Structures-Part 1-2: General Rules-Structural Fire Design. European Committee for Standardization (CEN): Brussels, Belgium, 2004.
13. Concrete Society. *Assessment, Design and Repair of Fire-Damaged Concrete Structures*; The Concrete Society: Camberley, UK, 2008.
14. *ACI Committee 201*; Guide for Conducting a Visual Inspection of Concrete in Service. American Concrete Institute: Farmington Hills, MI, USA, 2008.
15. Hertz, K.D. Concrete Strength for Fire Safety Design. *Mag. Concr. Res.* **2005**, *57*, 445–453. [CrossRef]
16. Lie, T.T. *Structural Fire Protection, ASCE Manuals and Reports on Engineering Practice*; No. 78; American Society of Civil Engineers: New York, NY, USA, 1992.
17. ASTM. *Standard Methods of Fire Test of Building Construction and Materials, Test Method E119-01*; American Society for Testing and Materials: West Conshohocken, PA, USA, 2001.
18. *ISO 834*; Fire Resistance Tests, Elements of Building Construction. International Organization for Standardization: London, UK, 2014.
19. Tsai, W.T. Uniaxial Compressional Stress-Strain Relation of Concrete. *J. Struct. Eng.* **1988**, *114*, 2133–2136. [CrossRef]
20. Terro, M.J. Numerical Modeling of the Behavior of Concrete Structures in Fire. *ACI Struct. J.* **1998**, *52*, 183–193.
21. Chang, Y.F.; Chen, Y.H.; Sheu, M.S.; Yao, G.C. Residual Stress-Strain Relationship for Concrete after Exposure to High Temperatures. *Cem. Concr. Res.* **2006**, *36*, 1999–2005. [CrossRef]
22. Karthik, M.; Mander, J. Stress-Block Parameters for Unconfined and Confined Concrete based on a Unified Stress-Strain Model. *J. Struct. Eng.* **2011**, *137*, 270–273. [CrossRef]
23. Lie, T.T.; Stanzak, W.W. Structural Steel and Fire: More Realistic Analysis. *Eng. J.* **1976**, *13*, 35–42. Available online: https://www.aisc.org/Structural-Steel-and-Fire-More-Realistic-Analysis (accessed on 12 January 2022).
24. Alhadid, M.; Youssef, M. Analysis of Reinforced Concrete Beams Strengthened Using Concrete Jackets. *Eng. Struct.* **2017**, *132*, 172–187. [CrossRef]
25. Chen, Y.H.; Chang, Y.F.; Yao, G.C.; Sheu, M.S. Experimental Research on Post-Fire Behaviour of Reinforced Concrete Column. *Fire Saf. J.* **2009**, *44*, 741–748. [CrossRef]
26. Jau, W.C.; Huang, K.L. A Study of Reinforced Concrete Corner Columns after Fire. *Cement Concr. Comp.* **2008**, *30*, 622–638. [CrossRef]
27. Yaqub, M.; Bailey, C.G. Repair of Fire Damaged Circular Reinforced Concrete Columns with FRP Composites. *Constr. Build. Mater.* **2011**, *25*, 359–370. [CrossRef]
28. Elsanadedy, H.; Almusallam, T.; Al-Salloum, Y.; Iqbal, R. Effect of High Temperature on Structural Response of Reinforced Concrete Circular Columns Strengthened with Fiber Reinforced Polymer Composites. *J. Compos. Mater.* **2016**, *51*, 333–355. [CrossRef]
29. Neves, I.C.; Rodrigues, J.P.C.; Loureiro, A.D.P. Mechanical Properties of Reinforcing and Prestressing Steels after Heating. *J. Mater. Civil. Eng.* **1996**, *8*, 189–194. [CrossRef]
30. Edwards, W.T.; Gamble, W.L. Strength of Grade 60 Reinforcing Bars After Exposure to Fire Temperatures. *Concr. Int.* **1986**, *8*, 17–19.
31. Qiang, X.; Bijlaard, F.S.K.; Kolstein, H. Post-Fire Mechanical Properties of High Strength Structural Steels S460 and S690. *Eng. Struct.* **2012**, *35*, 1–10. [CrossRef]
32. ETABS, 15.0.0. Computers and Structures Inc.: Walnut Creek, CA, USA, 2015. Available online: https://www.advanceduninstaller.com/ETABS-2015-64-bit-cc2613d7b22a7d65c7b8ba662ef17719-application.htm. (accessed on 12 January 2022).

Article

Nonlinear Analysis of a Steel Frame Structure Exposed to Post-Earthquake Fire

Alaa T. Alisawi *, **Philip E. F. Collins** and **Katherine A. Cashell**

Department of Civil and Environmental Engineering, Brunel University London, London UB8 3PN, UK;
Philip.Collins@brunel.ac.uk (P.E.F.C.); Katherine.Cashell@brunel.ac.uk (K.A.C.)
* Correspondence: Alaa.Al-isawi@brunel.ac.uk

Abstract: The probability of extreme events such as an earthquake, fire or blast occurring during the lifetime of a structure is relatively low but these events can cause serious damage to the structure as well as to human life. Due to the significant consequences for occupant and structural safety, an accurate analysis of the response of structures exposed to these events is required for their design. Some extreme events may occur as a consequence of another hazard, for example, a fire may occur due to the failure of the electrical system of a structure following an earthquake. In such circumstances, the structure is subjected to a multi-hazard loading scenario. A post-earthquake fire (PEF) is one of the major multi-hazard events that is reasonably likely to occur but has been the subject of relatively little research in the available literature. In most international design codes, structures exposed to multi-hazards scenarios such as earthquakes, which are then followed by fires are only analysed and designed for as separate events, even though structures subjected to an earthquake may experience partial damage resulting in a more severe response to a subsequent fire. Most available analysis procedures and design codes do not address the association of the two hazards. Thus, the design of structures based on existing standards may contribute to a significant risk of structural failure. Indeed, a suitable method of analysis is required to investigate the behaviour of structures when exposed to sequential hazards. In this paper, a multi-hazard analysis approach is developed, which considers the damage caused to structures during and after an earthquake through a subsequent thermal analysis. A methodology is developed and employed to study the nonlinear behaviour of a steel framed structure under post-earthquake fire conditions. A three-dimensional nonlinear finite element model of an unprotected steel frame is developed and outlined.

Keywords: fire; earthquake; finite element analysis; Abaqus; multi hazard analysis

Citation: Alisawi, A.T.; Collins, P.E.F.; Cashell, K.A. Nonlinear Analysis of a Steel Frame Structure Exposed to Post-Earthquake Fire. *Fire* **2021**, *4*, 73. https://doi.org/10.3390/fire4040073

Academic Editor: Maged A. Youssef

Received: 2 September 2021
Accepted: 6 October 2021
Published: 15 October 2021

Publisher's Note: MDPI stays neutral with regard to jurisdictional claims in published maps and institutional affiliations.

1. Introduction

Extreme events such as earthquakes, fires or blasts have a low likelihood of incidence during a structure's lifecycle but they can have tremendous after-effects with regard to the safety of any inhabitants and the integrity of the structure. In addition, there may be a higher risk of a second extreme event occurring, owing to any damage that occurs during the initial event, for example, a fire after an earthquake [1,2]. In such a case, the structure is exposed to multiple hazards. The current paper is concerned with the response of steel framed structures when subjected to an earthquake that is followed by a fire. This particular multi-hazard event is known as a post-earthquake fire (PEF). Most structures are required to satisfy 'life safety' design criteria as specified in design standards. These codes guarantee that structures remain stable and continue to carry gravity loads, dead loads and a percentage of live loads during extreme events, thus allowing the building's occupants to evacuate the buildings safely [3,4]. Based on the function of the structure and its importance, the allowable rate and type of damage that is tolerable during an extreme loading is typically specified during its design. The design codes ensure building safety under a variety of load combinations that represent different extreme loading scenarios.

However, the load combination of an earthquake followed by a fire has yet to be included in international design standards although the forces and moments applied to a structure during a PEF are likely to be much greater than for individual extreme events [4,5].

Mitigating the effects of PEF on buildings during the design process in order to ensure the safety of occupants and emergency service personnel is a crucial aspect to consider for any PEF safety strategy. The effects of a PEF can be diminished by controlling and determining the status of structural stresses after the first event (the earthquake) and also designing and/or strengthening the building to withstand and survive the fire loading. Eurocode 8 Part 1 [6] provides a design load combination for a set of different actions (Equation (1)). These actions must be combined with those from other loads, such as permanent loads (G), pre-stressing loads (P), seismic actions ($A_{E,d}$) and a proportion of the variable (live) loads (Q). A specific reduction factor ($\Psi_{2,i}$) is provided in Eurocode 8 and the recommended values of factors for buildings are specified in Eurocode 3 Part 1–2 [7].

$$\sum_{f \geq 1} G_{k,j} + P + A_{E,d} + \sum_{i \geq 1} \Psi_{2,i}\ Q_{k,i} \tag{1}$$

There are two important concepts that should be considered when designing a structure that can resist different magnitudes of earthquakes, which are frequent earthquakes and design earthquakes. The return period of a frequent earthquake is lower than that of a design or 'maximum considered' earthquake. A design earthquake is characterized by a return period R of 475 years, which corresponds to a 10% probability of exceedance in 50 years. As shown in Figure 1, a usual building must be operational for a frequent return period, and safe in the zone of a design earthquake. For very important structures, the critical components must remain operational for a 'maximum considered' earthquake [8,9].

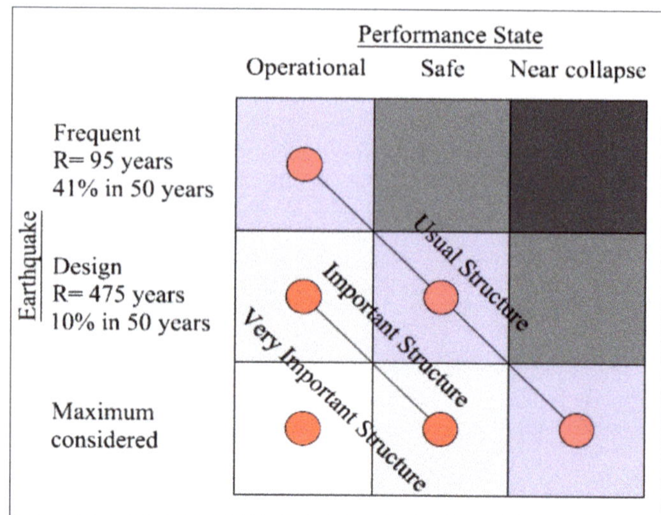

Figure 1. Requirements for structural performance during different types of earthquake in accordance with EN 1998-1 [6].

In this context, it is clear that in order to develop appropriate design methods for a PEF event, it is critical to first develop a good understanding of the complex structural behaviour that occurs in this scenario. The structural behaviour and material properties of the remaining parts of the structure after the first hazard are classified as the input properties of the structure during the fire, and it is important that these are accurately represented. For this reason, in the current paper, a multi-hazard analysis approach is presented for steel-framed buildings. The paper proceeds with an overview of the state-of-the-art, which is followed by a description of the three-dimensional (3D) finite element (FE) model that was developed using the Abaqus software [10]. The damage caused to a structure during and after an earthquake is included in the sequential thermal analysis. This methodology is developed and employed to study the nonlinear behaviour of a steel-framed structure when subjected to the PEF loading condition.

2. State of the Art

There has been limited research into multi-hazard extreme events and their effects on building structures compared with single extreme events such as a fire or an earthquake. Nevertheless, as more has become known and understood about single hazard events, researchers have begun to study the more complex case of PEF [1–3,11]. Della Corte et al. [12] investigated the fire resistance rating for unprotected steel frames for the PEF condition, assuming elastic-perfectly plastic steel behaviour. This study considered second-order effects, whereby the lateral displacements caused by the stresses and strains resulting from the earthquake, reduce the structural stability under gravity loads. However, this study did not include stiffness degradation in the analysis.

Ali et al. [13] conducted a comprehensive study that considered the effects of geometry and stiffness degradation in the PEF condition, in which they also developed of a 3D numerical model. The behaviour of an unprotected, single-storey, multi-bay steel frame was analysed after its exposure to a seismic load followed by a sizeable uncontrolled fire. It was shown that the PEF resistance is significantly dependent on both the particular fire scenario as well as the gravity loads that are applied to the structure. Mousavi et al. [14] presented a review on the key issues and hazards related to PEF for a building and found that the principle influential factors are the intensity and duration of the earthquake and fire, the level of protection included in the original design and the structural materials used. Zaharia and Pintea [15] examined two types of steel frames which were designed for different return periods of ground motion (2475 and 475 years, respectively). The seismic response of the system was evaluated by conducting a nonlinear static analysis, i.e., a pushover analysis. The structure that was designed for a return period of 475 years suffered from a more significant inter-storey drift in the plastic range after the earthquake event, whilst the frame designed for the longer return period continued to respond in the elastic range. A fire analysis was then performed for both frames and the results showed that the fire resistance of the frame with a shorter return period, which had experienced greater deformations during the earthquake, was less than for the other frame, which did not have a history of plastic deformations before the fire.

Ghoreishi et al. [16] presented a review of the existing experimental and numerical studies on structural systems when subjected to fire, which included a multi hazard analysis of PEF. This study revealed that traditional design methods based on the concept of fire resistance ratings do not consider many of the significant typical structural conditions such as size, control conditions and loading. Moreover, the fire resistance of a singular structural element is different to that of the overall structure, due to the influences of continuity, interaction between elements and load and stress redistribution. Memari et al. [17] presented their insights into the consequences of PEF on low-, medium- and high-rise steel moment-resisting frames, using FE and nonlinear time-history analysis. An uncoupled thermal-mechanical analysis was conducted and a fire was applied at the reduced beam section connections (RBS). The material properties were assumed to be elastic-perfectly plastic in this analysis, but it is noteworthy that one-dimensional beam elements were employed to represent the structure's components that were incapable of depicting local buckling failure in the members.

Chicchi and Varma [1] published a state-of-the-art review for the analysis and the design of moment-resisting framed structures subjected to PEF, which was largely focused on events in the USA. This review included an assessment of the consequences of non-structural damage produced through earthquakes on the subsequent structural fire resistance. A methodology was proposed for analysing and designing these types of structures, so that they may resist a PEF event using incremental dynamic and fire analyses. Zhou et al. [18] proposed an integrated multi-hazard analysis framework using FEA and the OpenSEES software. This framework provides a practical solution for measuring the residual fire resistance of a system with cementitious passive fire protection (PFP) subjected to fire following a moderate earthquake. However, it is noteworthy that this study analysed individual structural members rather than the overall structure.

The research that has been conducted to date generally illustrates that the behaviour of a building subjected to a PEF is not significantly affected by the nonlinear geometric effects caused by an earthquake if the initial design of the structure complies with the serviceability limit state requirements. However, there are shortcomings in some of the assumptions that have been made in the available research, including simplifications of the element types, methods of analysis and the applied input motions. The nonlinear geometric effects are generally assumed without considering the influence of structural resonance and the frequency effect. Moreover, if an inaccurate design spectrum is determined, in accordance with Eurocode 8, the acceleration time history applied during the seismic stage of the multi-hazard event could lead to an underestimation of the stresses and strains experienced in the structure. Such is the basis for this work, which provides a novel approach to quantifying the effect of a PEF event on structural behaviour, using a coupled nonlinear sequential analysis. The study highlights the unique relationship between the geotechnical and geological properties of the applied motion during the earthquake stage and the system behaviour during a multi-hazard event. The coupled nonlinear time-history analysis is used to identify the residual material properties of the subsequent fire analysis.

3. Basis of the Analysis

It is clear that an accurate evaluation of a structure's response following an earthquake, which serves as the input data in the fire analysis for a PEF event, is critically important. Its response is influenced by many factors including the level of certainty of the material properties and the mechanical behaviour of the structural components as well as the intensity of the seismic action (e.g., [19]). These difficulties and uncertainties have led researchers to adopt simplified approaches for assessing the seismic structural behaviour and damage in PEF analyses [20,21]. However, simplified methods may not present an accurate depiction of the actual structural behaviour following an earthquake, particularly for the stress redistributions that occur and are likely to be quite influential in its fire performance (e.g., [22,23]). The key problem lies in the appraisal of the physical condition of the structure following the earthquake, or the 'initial condition' for the subsequent fire action.

During most major earthquakes, structures are required to withstand significant levels of plastic deformation. The availability of reliable analytical methods, including sophisticated numerical models, may facilitate a more realistic reflection of the performance and damage of a structure when subjected to an earthquake. The structural damage experienced can be classified as either geometric, whereby the initial geometry is altered due to plastic deformations that occur during the earthquake, or mechanical, i.e., the degradation of the mechanical properties of the structural components that are in the plastic range of deformation during the earthquake.

3.1. Seismic Analysis for PEF

Traditionally, the effects of an earthquake on a structure are studied using either approximate methods, such as a pushover analysis, or a time-history analysis. A pushover analysis is a nonlinear static analysis procedure used to estimate the strength of a structure beyond its elastic limit but does not induce actual plastic damage in the structure and does not require a ground motion time history. On the other hand, a time-history analysis is a nonlinear dynamic response analysis performed using an actual or artificial earthquake to evaluate the response of the system. A time-history analysis usually takes significantly longer to complete compared to a pushover analysis and is also more computationally demanding. However, it provides a more accurate depiction of the structural response to a seismic event, which is imperative in a PEF assessment. When the damage from an earthquake is underestimated, a structure can be highly vulnerable to failure even if it has been rigorously designed for an isolated fire condition. It is in this context, that this study applies a time-history analysis to assess the structural response to the seismic excitation.

3.2. Input Data

The earthquake input data is generated in accordance with the structure frequency modes, geotechnical and geological site properties, and the design response spectrum characteristics. In a performance-based design, a structure subjected to a design earthquake should maintain the required design-level performance [24]. Eurocode 8 specifies two types of earthquakes, namely Type 1 and Type 2 spectra and also four different importance classifications for buildings, depending on their function. In the current work, it is assumed that the structure being analysed has an importance classification of III (i.e., buildings with a seismic resistance that is of importance due to the consequences associated with a collapse, e.g., schools, assembly halls, cultural institutions, etc.) and is therefore subjected to a Type 2 earthquake. The ground conditions are Type E as defined in Eurocode 8, described by various stratigraphic profiles and parameters and with viscous damping set at 5%. For these conditions, the peak ground acceleration (PGA) that occurs during the earthquake is 0.35 g.

The design response spectrum is also developed in accordance with Eurocode 8 for selected targeted time histories. The user-selected time histories are subjected to a scaling and matching procedure to derive earthquake input data within the spectrum periods of interest. The spectral scaling method used in the current study employs a computer algorithm—using SeismoSignal and SeismoMatch software [25]—to modify the real and artificial time histories in order to closely match the target design response spectrum. Using these procedures, data from a real earthquake are modified to a PGA of 0.35 g and a frequency content according to the design conditions.

To examine the seismic structural response, two predominant periods are selected for the modified real earthquake, namely 0.24 sec and 0.36 sec, in addition to one predominant period of 0.16 sec for the artificial motion. For the latter, a MATLAB algorithm has been developed to create the white noise artificial earthquake to satisfy the Eurocode 8 value of the structural natural period; there are more details on this later. The SeismoSignal and SeismoMatch software are combined with data from the U.S. Geological Survey (USGS) peer database [26] to meet the spectral design requirements. Figure 2a illustrates the Eurocode 8 design response spectrum with the modified real earthquake spectra with predominant periods of 0.24 sec and 0.36 sec, respectively, and the corresponding acceleration time histories are shown in Figure 2b. Figure 3 represents the corresponding data for a spectrum with a predominant period of 0.16 sec, for the artificial motion.

3.3. Thermal Stress Analysis in PEF Analysis

In the post-earthquake fire analysis, the deformed or damaged structural configuration that occurs following the earthquake event is employed as the input for the application of the thermal loads [27,28]. For the fire load, a uniform standard ISO-834 fire exposure [29] is applied to all the components of the frame, as shown in Figure 4. The frame is made from mild steel with a yield and ultimate strength, at the ambient temperature, of 385 N/mm^2 and 450 N/mm^2, respectively. The steel has a density of 7850 kg/m^3 and a coefficient of thermal expansion (α_s) of 1.4×10^{-5}. The changes in material properties resulting from increasing levels of elevated temperature are obtained from the reduction factors provided in Eurocode 3 Part 1–2 [7].

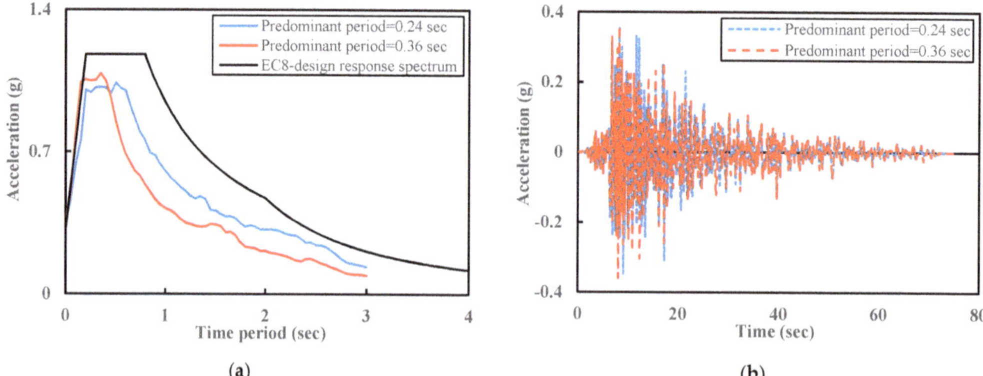

(a)   (b)

Figure 2. Comparison of the real and design earthquake input data including (**a**) the design response spectrum and (**b**) the acceleration time histories.

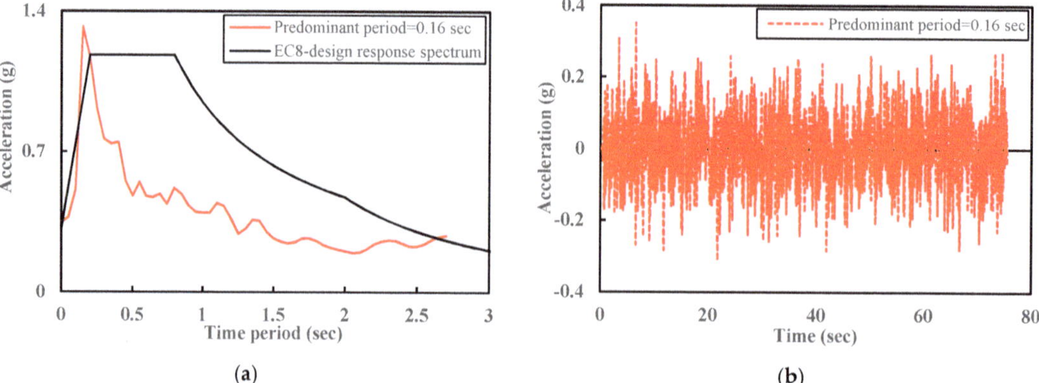

(a)   (b)

Figure 3. Comparison of the artificial and design earthquake input data including (**a**) the design response spectrum and (**b**) the acceleration time histories.

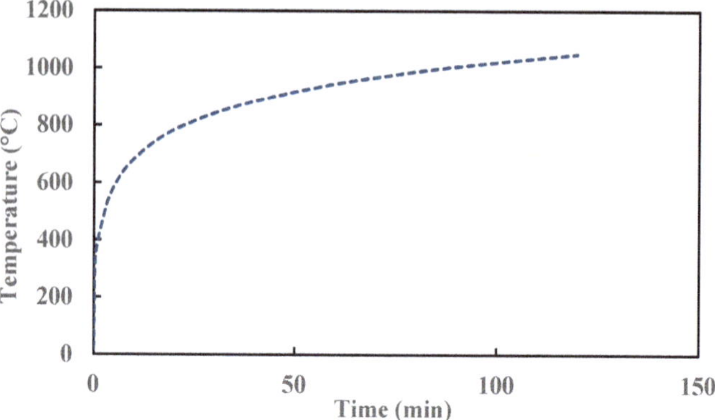

Figure 4. Standard fire curve [7].

4. Development of the Numerical Model

4.1. General

A geometrically and materially nonlinear three-dimensional model of an unprotected single-storey steel frame has been developed using the Abaqus software, in order to analyse the behaviour of the given structure during a post-earthquake fire (PEF). The frame is fabricated from beams and columns of the same I-shaped cross-section, which are connected with rigid joints. The frame is 5720 mm in length, 5370 mm in width and has a height of 4050 mm. The cross-section has a depth (D) of 350 mm, flange width (B) of 170 mm, identical web (t) and flange (T) thicknesses of 10 mm each, root radius (r) of 12 mm and a depth between the flange fillets (d) of 306 mm. The frame is designed to withstand gravity and seismic loads in accordance with Eurocode 8 Part 1 [6]. In accordance with the basis for design information provided in EN 1990 [30] and the guidance on actions in EN 1991 [31], the frame has been designed for a load combination comprising of 100% of the permanent actions and 60% of the variable actions during the PEF event, as discussed later.

4.2. Elements, Meshing and Boundary Conditions

The steel sections are modelled through the finite element model using general purpose linear brick elements with reduced integration, referred to as C3D8R in the Abaqus library [32]. A mesh sensitivity study was conducted to achieve the optimal combination of accuracy and computational efficiency, which resulted in element sizes ranging between 10×20 mm and 20×20 mm at the beam-column connections and 10×100 mm and 20×100 mm for the rest of the beam/column steel sections. The steel is represented using a nonlinear elastoplastic material model which has a yield and an ultimate strength of 385 N/mm^2 and 450 N/mm^2, respectively. These properties degrade with an elevated temperature in accordance with the reduction factors provided in Eurocode 3 Part 1–2 [7]. The beam-column connection is achieved using the tie condition. The base of the columns are assumed to rest on a rigid foundation system, so the earthquake boundary condition is applied at the base of all the columns. A roller support is used to constrain the displacement, placed vertically at the bottom of the model. The horizontal boundary conditions permit 'free' horizontal shaking in the direction/directions of the applied seismic load.

4.3. Loading and Solution Procedure

The analysis is performed sequentially, comprising of static, dynamic and thermal analysis steps, as illustrated in Figure 5. The analysis is carried out in three main multi-hazard analysis steps, as well as an initial sub-step. Firstly, a linear perturbation–frequency step is conducted to identify the structural modal analysis (as discussed in more detail later) and frequency content window of the dynamic system. Then, in the first analysis stage, a nonlinear static analysis is conducted, and the gravity loads are applied. The permanent loads are assumed to have a value of 8 kN/m^2 whilst the variable actions are equal to 2.5 kN/m^2, in accordance with EN 1991 [31], and all permanent and variable actions are applied. In the second step, the earthquake is simulated through a nonlinear implicit dynamic analysis. The acceleration time history is applied at the base of the structure while the static loads remain constant. The time history is processed, filtering for window frequencies matching the system modes and the natural frequency of the structure during an earthquake with a PGA of 0.35 g. In the third analysis stage, the thermal loads are applied to the deformed structure in the form of a time–temperature curve. The load combination in this stage is considered to be 100% of the permanent loads acting together with 60% of the variable actions [31]. The overall analysis is performed in a sequence to carry forward the deformations, stresses and damage caused to the structure during one stage to the next stage of the analysis. The key objective of the current study is to compare between the structural behaviour of structures subjected to a multi-hazard event with the behaviour of those exposed to a fire-only scenario. Thus, to compare with and examine the consequences of an earthquake directly preceding and possibly causing a fire, a fire-only event is also studied.

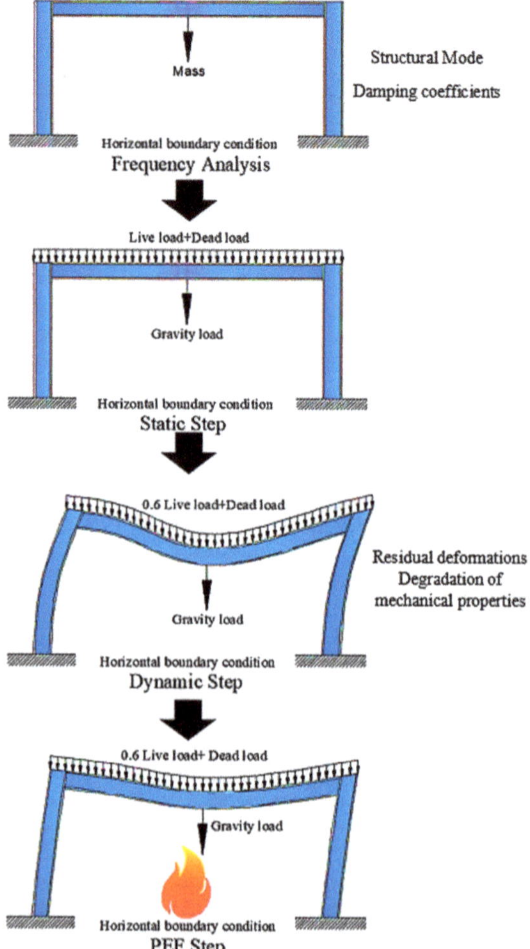

Figure 5. Methodology of the sequential analysis.

5. Results and Analysis

In this section, the results of the finite element analysis are presented and discussed. The first results presented are for the frequency analysis, in which a linear perturbation-frequency analysis is developed as a sub-step of analysis, followed by the results from the PEF structural simulations.

5.1. Frequency Analysis

The natural period of vibration of a dynamic system is an essential factor for the force-base design methodology ([33–35]). In this method, the base shear is the expected ultimate lateral load applied at the base of the structure during seismic activity. The natural period of vibration is a critical parameter in defining the design response spectrum and consequently in controlling the value of the base shear force. Hysteretic damping is applied in the restoring force, and viscous damping is considered by Rayleigh (proportional) damping, as provided in Equation (2):

$$[C] = \alpha_M \, [M] + \beta_K \, [K] \tag{2}$$

where α_M and β_K are the mass and stiffness proportional damping coefficients, and [M], [K], and [C] are the mass, stiffness, and damping $n \times n$ matrices, respectively. The damping ratio of the system for different natural frequencies (ξ_i) can be determined using Equation (3):

$$\xi_i = \frac{1}{2}\left[\frac{\alpha_c}{\omega_i} + \beta_c \omega_i\right] \tag{3}$$

In this expression, ω_i is the system-mode frequency. Owing to the orthogonality between the system mode and damping matrix, as well as the assumption of 5% damping for the system modes, the corresponding mass and stiffness coefficients of Rayleigh damping are calculated using Equations (4) and (5), respectively:

$$\alpha_M = \frac{2\omega_i\omega_j}{\omega_j^2 - \omega_i^2}\left(\xi_i\omega_j - \xi_j\omega_i\right) \tag{4}$$

$$\beta_K = \frac{2\left(\xi_j\omega_j - \xi_i\omega_i\right)}{\omega_j^2 - \omega_i^2} \tag{5}$$

where ω_i and ω_j are any two system-mode frequencies and ξ_i and ξ_j are the damping ratio at ω_i and ω_j, respectively. International design codes provide empirical formulae to estimate the fundamental period of vibration T of the structure. Eurocode 8 Part 1 [6] recommends using the Rayleigh method, as presented in Equation (6):

$$T = 2\pi\sqrt{\frac{\sum_{i=1}^{n}\left(m_i \cdot S_i^2\right)}{\sum_{i=1}^{n}\left(f_i \cdot S_i\right)}} \tag{6}$$

in which m_i represents storey mass, f_i represents horizontal forces, and S_i is the displacement of masses caused by horizontal forces. The first six natural vibration periods, the damping coefficients, and the natural vibration period of the system have been computed based on a linear perturbation-frequency analysis in accordance with EN 1998-1, and the findings are shown in Table 1. Figure 6 presents the corresponding mode shapes. In addition, Table 1 presents each of these natural vibration periods together with the value determined using EN 1998-1. The data presented in Table 1, together with the mode shapes in Figure 6, indicate that the first natural period, computed according to Eurocode 8 provisions, (0.16 sec) is between the second (0.296 sec) and third (0.106 sec) mode of the simulated values. It is also evident that the estimated natural period values decrease significantly for the first two modes after which the reduction changes more gradually for the remaining modes. Due to this, it has been concluded that it is important to consider more modes than just the first mode of the system in the seismic analysis, as has traditionally been the case. Accordingly, three input motions are considered in this paper, with natural vibration periods of 0.24 sec, 0.36 sec and the Eurocode 8 value of 0.16 sec, respectively.

Table 1. First six natural vibration periods and factors of Rayleigh damping.

Model	Natural Vibration Period (sec), T						Code	Damping Coefficients	
	FE Model						EN 1998-1		
	Modes								
	1	2	3	4	5	6		α_m	β_k
Value	0.36	0.296	0.106	0.101	0.09	0.081	0.16	0.959	0.0026

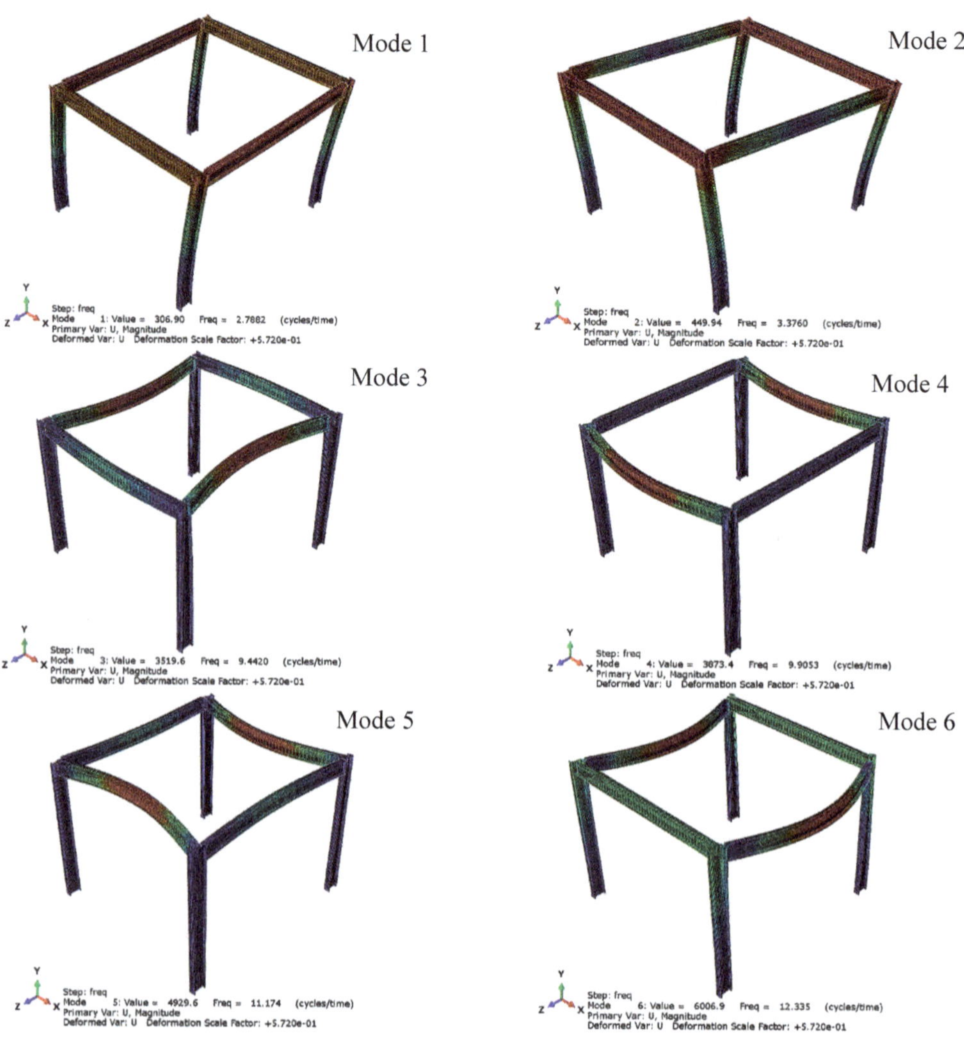

Figure 6. First six mode shapes for the steel framed structure following the frequency analysis.

5.2. Validation Study

Owing to a dearth of physical test data on a complete 3D structure, the numerical model is validated through a previously verified modelling approach, using the OpenSees FE software [14]. OpenSees (Open System for Earthquake Engineering Simulation) was initially developed at the University of California, Berkeley for seismic loading analysis [36] and was later extended to perform structural fire analyses at the University of Edinburgh [37]. Usmani et al. [37] found that OpenSees is capable of providing an accurate depiction of structural performance during fires. In this study, an identical steel frame has been modelled using OpenSees, and the results are presented in Figure 7, including (a) the time-displacement response for the fire-only scenario, (b) the temperature-displacement response for the fire-only scenario, (c) the time-displacement response for the PEF scenario and (d) the temperature-displacement response for the PEF scenario. All of the presented results are obtained from the mid-span location and the results from both the Abaqus and OpenSees models are presented. It is clear that both models and approaches provide almost identical results.

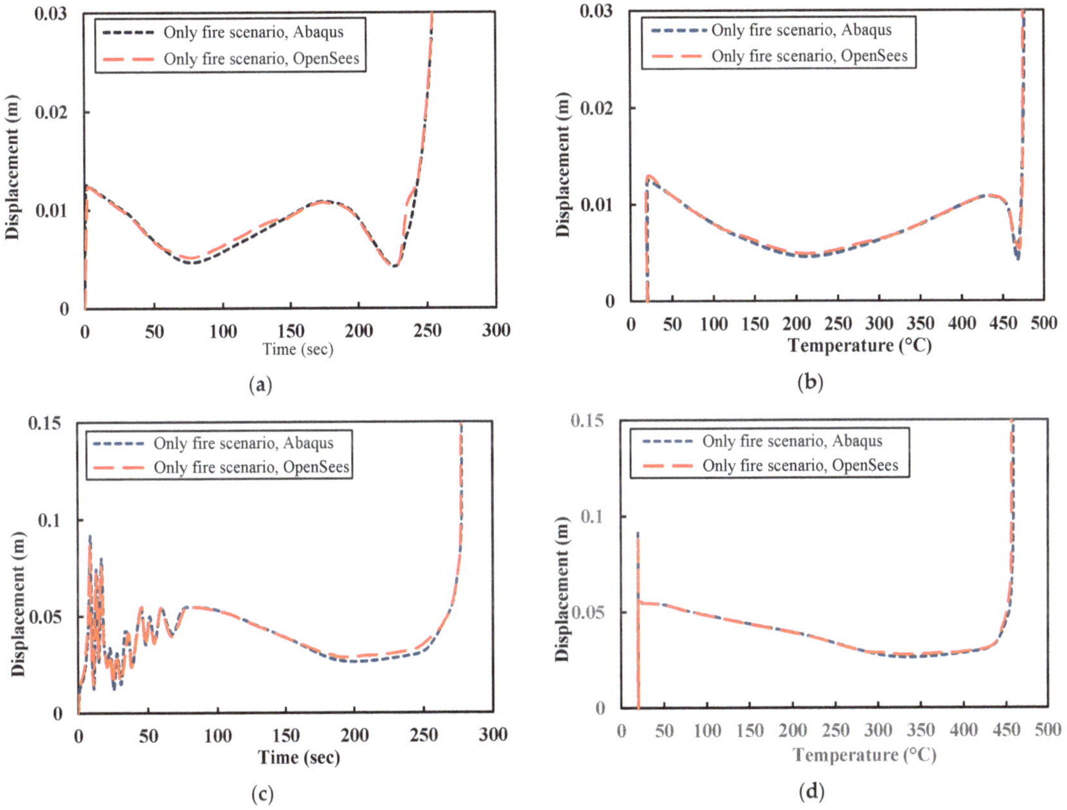

Figure 7. Comparison of Abaqus with OpenSees simulations, including (**a**) the time-displacement record for the fire-only scenario, (**b**) the temperature-displacement record for the fire-only scenario, (**c**) the time-displacement record for the PEF scenario, and (**d**) the temperature-displacement record for the PEF scenario.

5.3. Post-Earthquake Fire

In this section, the FE model developed in Abaqus that has been previously described is employed to assess and understand the post-earthquake fire (PEF) behaviour of steel framed structures. As stated before, the nonlinear sequential analysis [5] comprises a static stage, followed by the time history earthquake analysis after which the fire is applied. In the seismic analysis, the structure is subjected to two different time-history motions (referred to a Case I and Case II, respectively) which are matched to a particular predominant natural vibration period in accordance with the time period window resulting from a frequency analysis, as well as the natural period computed according to Eurocode 8 guidance. In addition, to replicate a real earthquake situation as accurately as possible, two types of excitation are applied, including unidirectional and bidirectional excitations for the different natural periods. Eurocode 8 requires that structures remain operational following relatively frequent earthquake events without incurring significant damage and incurring no structural damage. As such, the code defines an acceptable degree of reliability and validity for acceptable damage which must be reviewed during the design stage. The storey drift criterion is one of the primary stability criteria used in seismic codes and the Eurocode 8 limit is specified as 1% of the storey height under the ultimate design earthquake, which is 0.03 m in the present study.

In order to understand how an earthquake impacts upon a structure's fire resistance, a series of fire-only analyses are first presented. Figure 8 illustrates the collapse mechanism for a steel frame following a fire whilst Figure 9 shows the time-displacement and temperature-displacement curves, respectively, for the fire-only scenario. It is observed that local failure occurs concurrently for the two opposing beams in a symmetrical manner. The failure occurred around 260 sec after the fire began and at a temperature of approximately 480 °C.

The results from the PEF analysis for Case I, which involved an artificial earthquake, with a PGA of 0.35 g and a predominant natural vibration period of 0.16 sec, exposed to excitation in the Z direction, are presented in Figures 10 and 11. Figure 10 presents (a) the residual deformation that the steel frame experiences due to the earthquake excitation as well as (b) the shape and mechanism of failure of the structure (in the beam) after the PEF event for Case I. Whereas, Figure 11 presents the time-displacement results in the (a) z-direction, (b) y-direction and (c) the total displacement value respectively, as caused by PEF loading, as well as (d) temperature total displacement results due to PEF, for the case I scenario. The data from the corresponding fire-only analysis is also provided in these images.

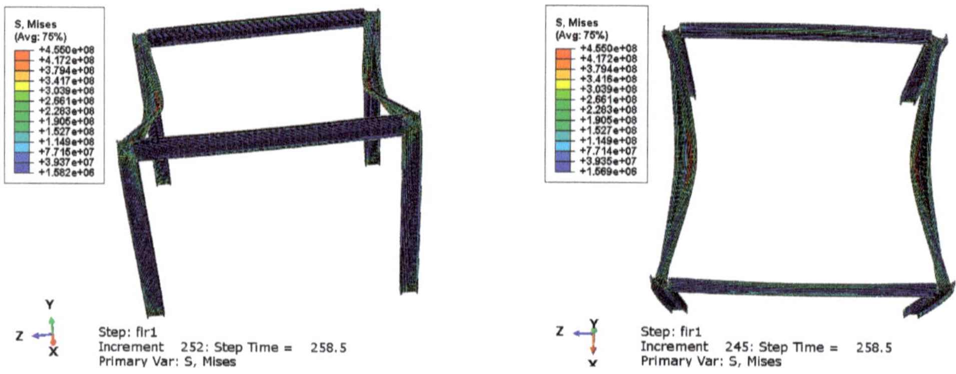

Figure 8. Failure mechanism (Fire-only scenario).

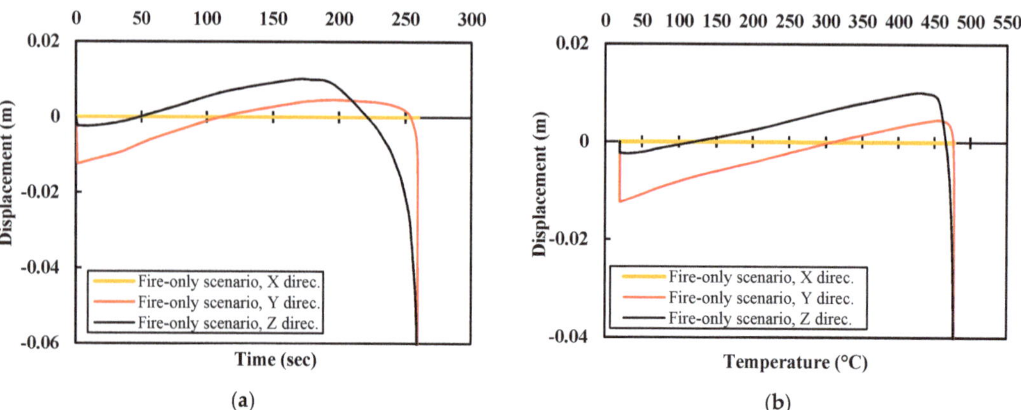

Figure 9. *Cont.*

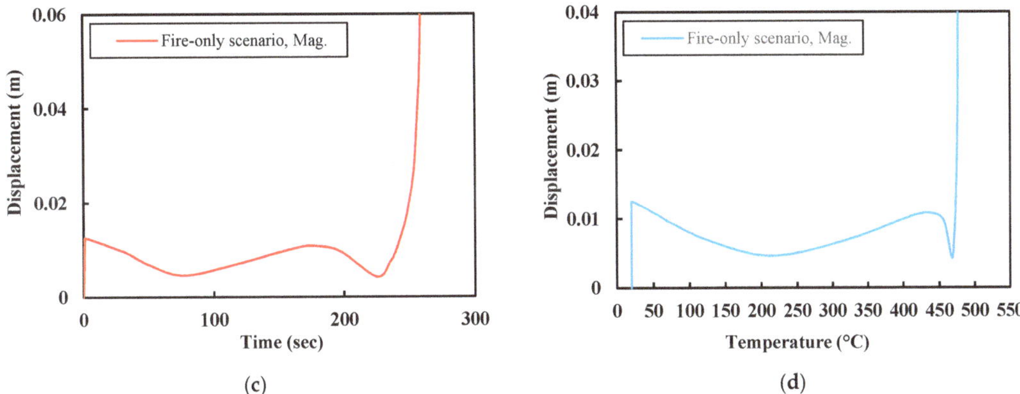

Figure 9. Results from the fire-only analysis of the steel framed structure including (**a**) the time-mid-span displacement; (**b**) the temperature-mid-span displacement; (**c**) the time-mid-span displacement data for the total displacement, and (**d**) the temperature-mid-span displacement record for the total displacement.

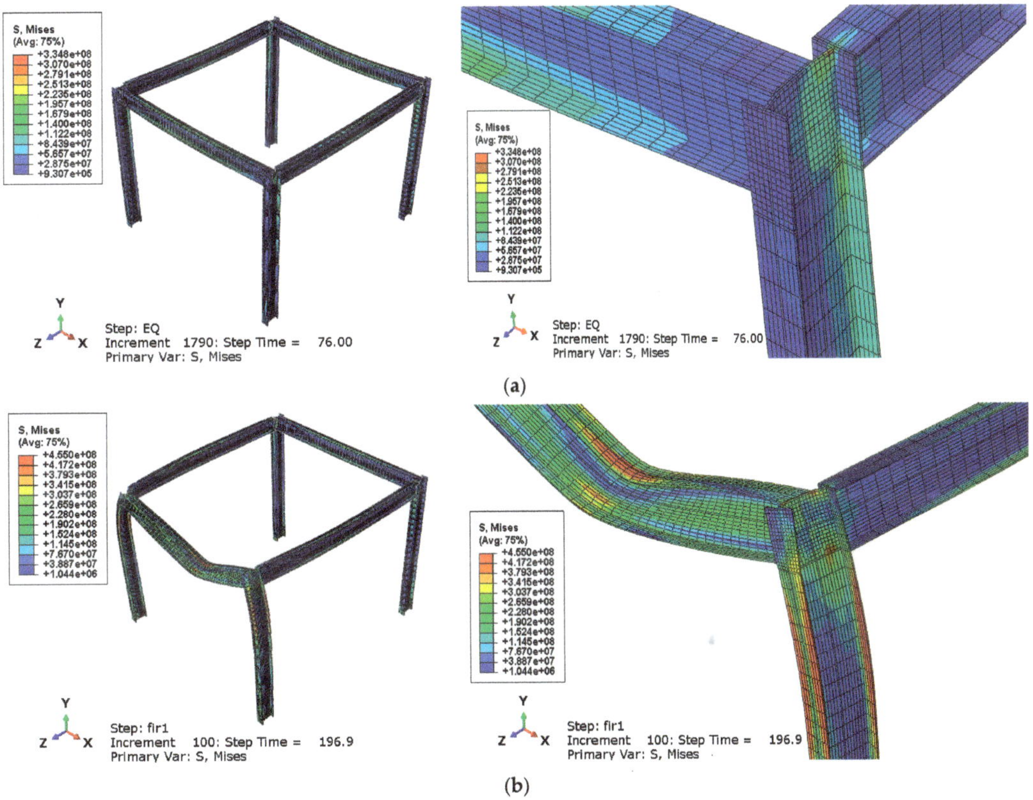

Figure 10. Images from a Case I PEF analysis with an artificial earthquake (PGA = 0.35 g, natural period = 0.16 sec), one-directional excitation in the z-direction including (**a**) the residual deformation of the structure at the end of earthquake event and (**b**) the shape and mechanism of failure of the structure after the PEF event.

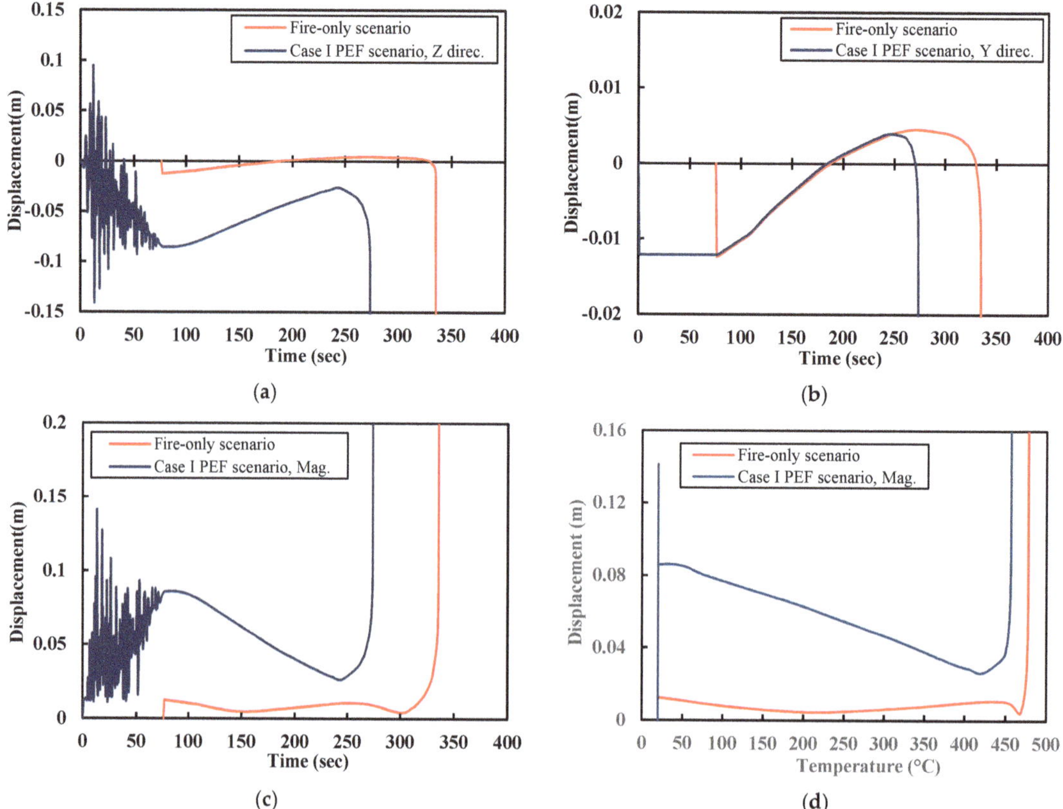

Figure 11. Comparison of the fire-only analysis versus the PEF analysis for Case I including (**a**) the time-mid-span displacement record in the z-direction, (**b**) the time-mid-span displacement in the y-direction, (**c**) the time-mid-span displacement record for the total displacement and (**d**) the temperature-mid-span displacement record for the total displacement.

The results indicate that the structure maintains the earthquake force successfully, experiencing geometrical and mechanical damage within the acceptable range of Eurocode 8. However, in comparison with the images for the fire-only scenario provided in Figure 10, it is clear that the failure shape in the PEF case is no longer symmetrical. In addition, the collapse occurs after just 272 sec, which is a 19% reduction from the fire-only case, and at a temperature of 455 °C. The storey drift value at collapse is 0.024 m and therefore remains within the 0.03 m limit stipulated by Eurocode 8. The corresponding results for the Case II scenario (PGA of 0.35 g and a natural period of 0.36 sec) are presented in Figures 12 and 13, respectively. It is clear that the failure mechanisms are unsymmetrical, and in this case, collapse occurs after 278 sec and at a temperature of 458 °C, which is almost identical to Case I.

The data presented for both Case I and Case II reflect the effect of an earthquake on the fire strength of the structure during unidirectional excitation. This kind of the excitation does not represent the situation of earthquake excitation in reality, which is also typically unidirectional. Due to this, more observations are obtained by examining the structural response to bidirectional excitation for a further two real and artificial motions (Cases III and IV, respectively).

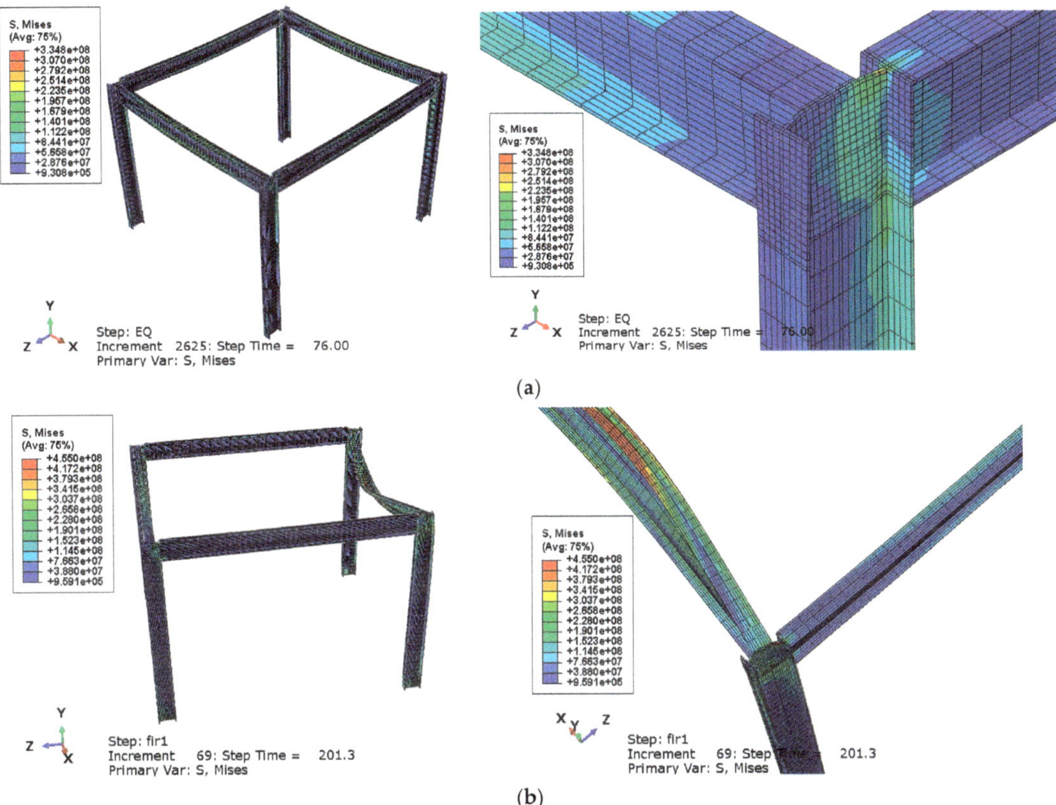

Figure 12. Images from a Case II PEF analysis with a real earthquake (PGA = 0.35 g, natural period = 0.36 sec) one-directional excitation in the z-direction, including (**a**) the residual deformation of the structure at the end of earthquake event and (**b**) the shape and mechanism of failure of the structure after the PEF event.

Figures 14 and 15 present the results from the analysis of a Case I earthquake with bidirectional excitation in both the x- and z-directions (referred to as Case III); these figures are presented in a similar format as before, for the purpose of comparison. It is clear that the global failure mechanism is dominant due to the combined effects of bidirectional excitation and the PEF event. The columns of one side of the structure completely collapsed in this scenario. The displacement records at a level of 1.4 m along the column length, for both the fire-only and PEF events are compared in Figure 16, which presents the time-mid-span displacement results at this position in (a) the x-direction, (b) the y-direction and (c) of the total displacement, respectively. Figure 16d presents the temperature-displacement response at the same point, 1.4 m from the column base. For this case, with bidirectional excitation, failure occurred after just 185 sec and at a temperature of 306 °C, representing a reduction of 45% compared with the fire-only analysis. The storey drift was 0.118 m, exceeding the allowable Eurocode 8 value. Similar behaviour and results are observed for Case IV, which has an identical input motion as Case II except with bidirectional excitation in both the x- and z-directions. The corresponding results are provided in Figures 17–19, in a similar format as before. It is clear that there is a significant reduction in the failure time for the PEF situation in Case IV of approximately 45% (to 185 sec) as well as a storey drift of 0.115 m, exceeding the allowable Eurocode 8 limit value by 85%. Significant local and global failure occurs in this case, preventing the structure from withstanding the applied loads.

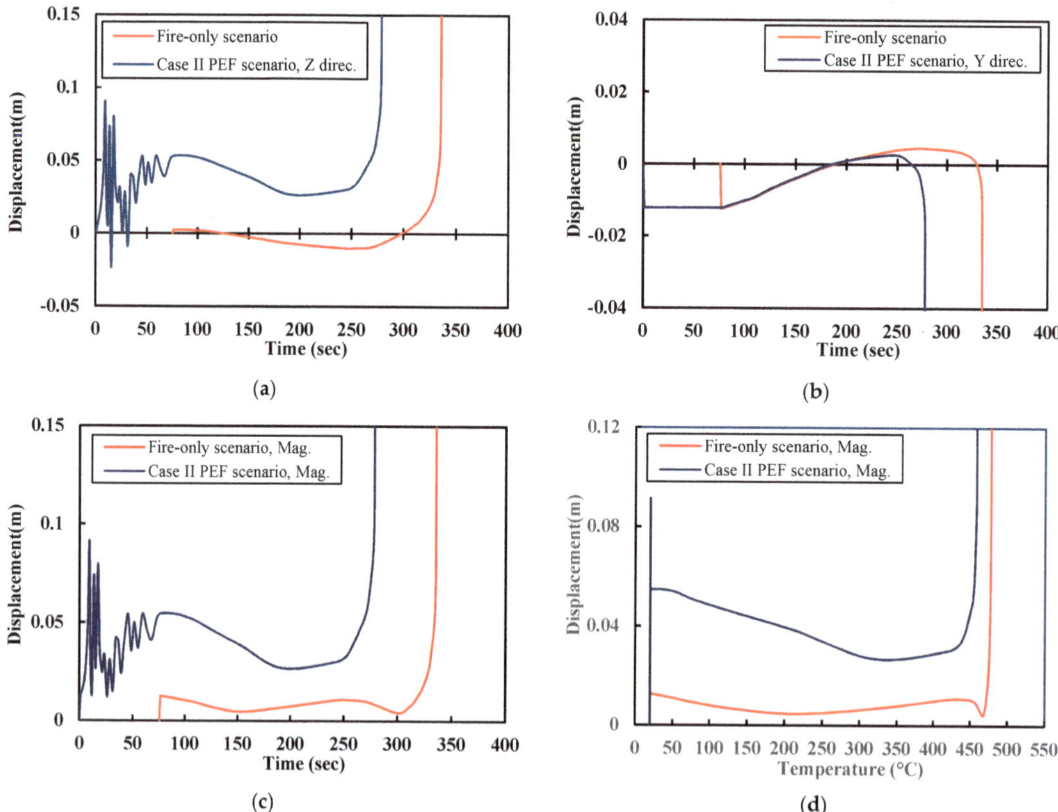

Figure 13. Comparison of the fire-only analysis versus the PEF analysis for Case II including (**a**) the time-mid-span displacement record in the z-direction, (**b**) the time-mid-span displacement in the y-direction, (**c**) the time-mid-span displacement record for the total displacement and (**d**) the temperature-mid-span displacement record for the total displacement.

In this section, a detailed numerical investigation into the behaviour of a steel-framed subject to a PEF event is presented. Structural damage, residual deformation, and stress degradation as result of earthquake excitation are considered and included in the multi-hazard analysis. Two different types of structural failure due to the effect of the combined hazards are observed, namely local and global failure. The failure times for all of the analysed cases are compared to the corresponding values from a fire-only analysis in Figures 9, 11, 13, 15, 16 and 18. In addition, Figure 20 presents a comparison of the fire-only analysis versus the PEF analysis for each of the four analysed cases (I–IV). It is shown that the geometrical and mechanical damage induced by an earthquake event can substantially decrease the fire resistance of the structure, specifically in the occurrence of bidirectional excitation (see Table 2). This observation has a significant consequence on the design aspects of the system for multi-hazard analysis. The design load combination, the number of structural modes incorporated in the seismic design as part of the multi-hazard investigation and the structural element section type are very influential parameters. Although the current study has not included a detailed investigation of the effects of different cross-section shapes, specifically tubular members, the results presented provide a valuable insight into the significant effects of a PEF event on a steel framed structure, and also on the importance of choosing a suitable column section in earthquake-prone zones. Furthermore, based on these results, it is proposed that using tubular sections is essential in earthquake zones to provide extra resistance in a PEF scenario, even though other sections may satisfy the seismic design requirements (that do not consider PEF). This is clearly an area that requires further research. Further, the load combinations provided in international codes do not currently include provisions for post-earthquake fire and each event is considered completely independently. The results presented herein do not support such an approach.

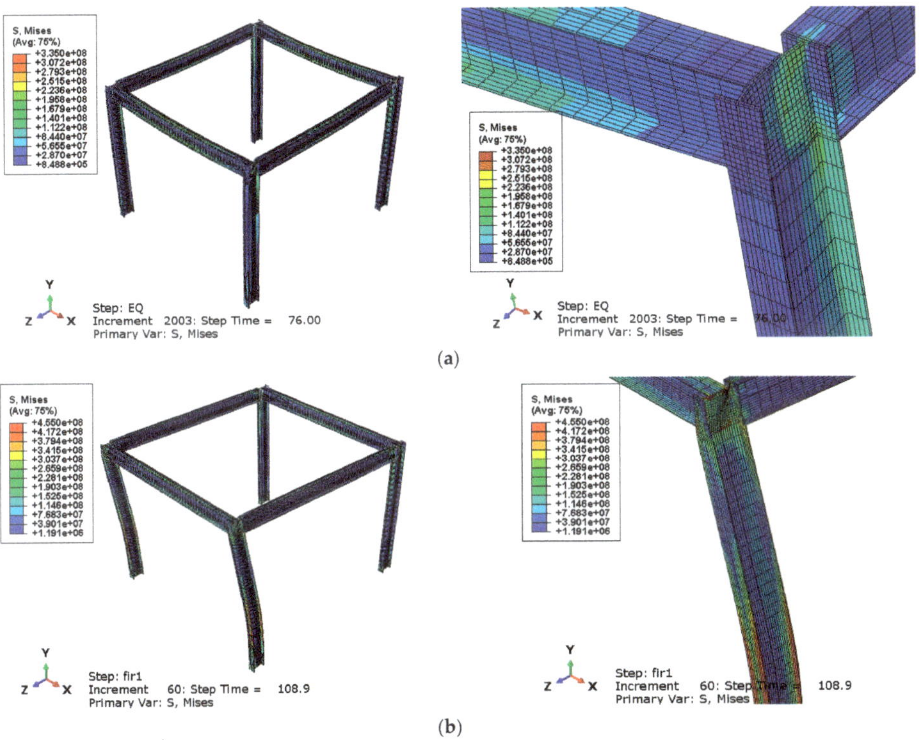

Figure 14. Images from a Case III PEF analysis with a real earthquake (PGA = 0.35 g, natural period = 0.24 sec), bi-directional excitation in the x- and z-direction including (**a**) the residual deformation of the structure at the end of earthquake event and (**b**) the shape and mechanism of failure of the structure after the PEF event.

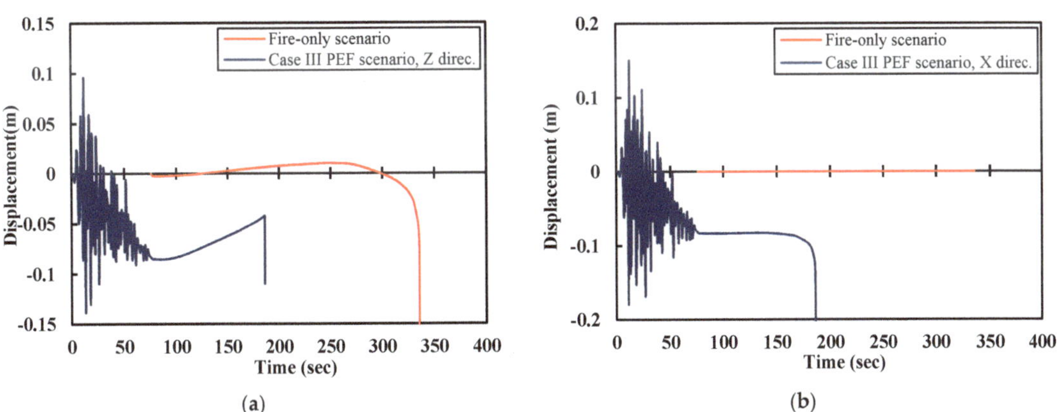

Figure 15. *Cont.*

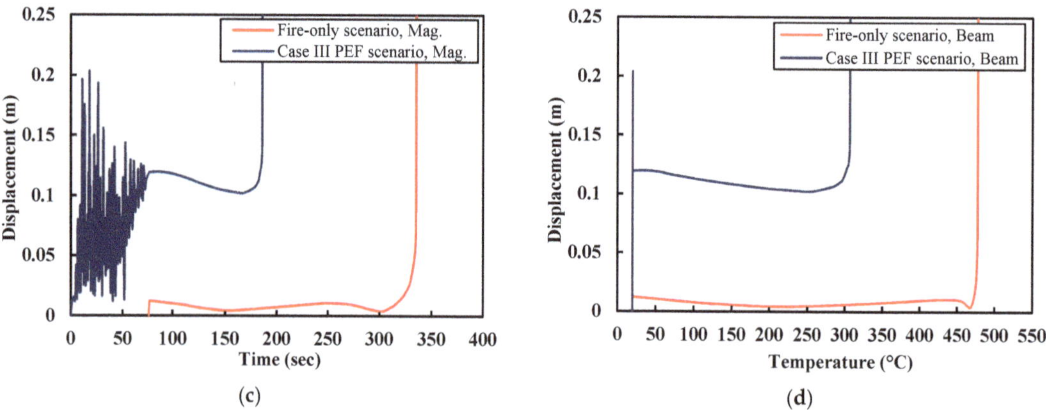

Figure 15. Comparison of the fire-only analysis versus the PEF analysis for Case III including (**a**) the time-mid-span displacement record in the z-direction, (**b**) the time-mid-span displacement record in the x-direction, (**c**) the time-mid-span displacement record for the total displacement and (**d**) the temperature-mid-span displacement record for the total displacement.

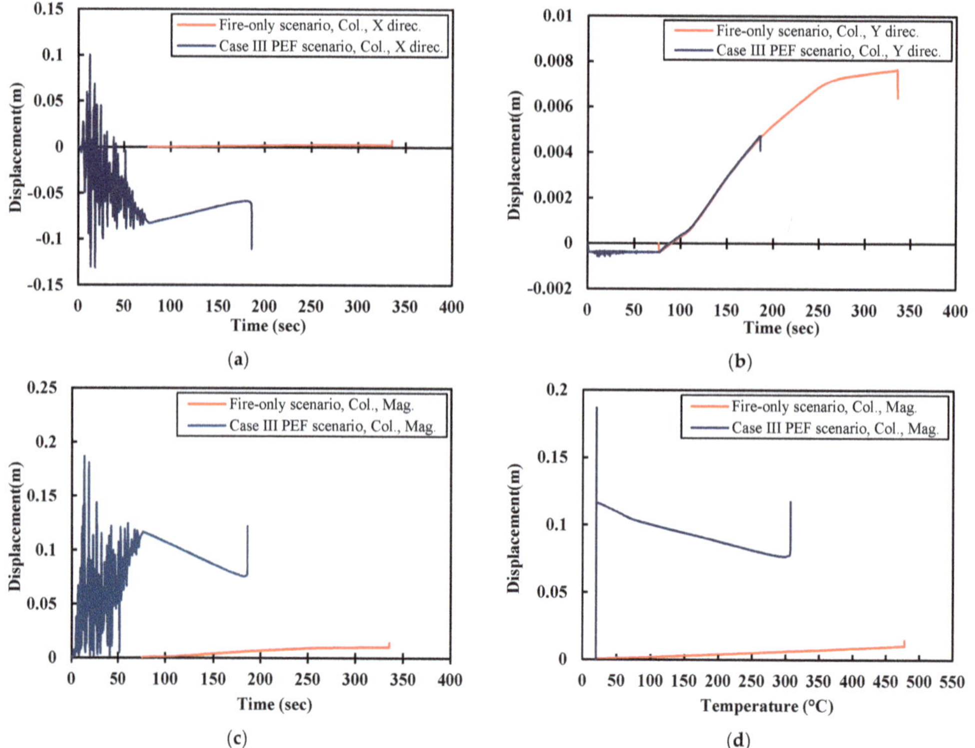

Figure 16. Comparison of the displacement values at a point which is 1.4 m along the column length for both the fire-only and PEF events for Case III including (**a**) the time-displacement record in the x-direction, (**b**) the time-displacement record in the y-direction, (**c**) the time-displacement record for total displacement value and (**d**) the temperature-displacement record for the total displacement value.

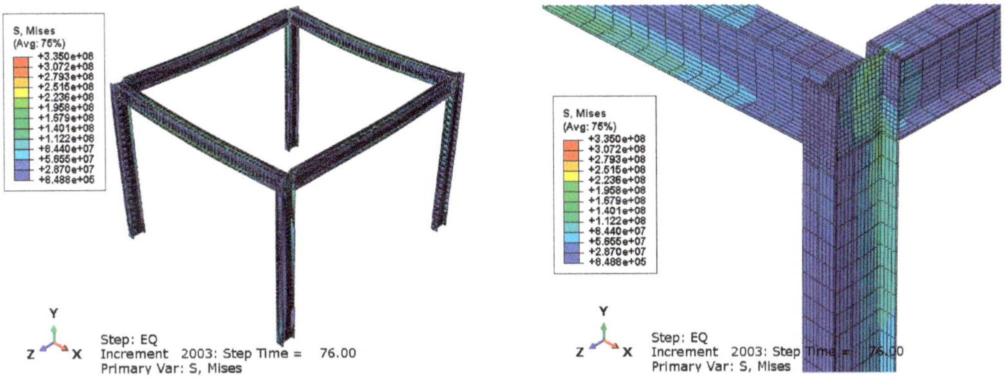

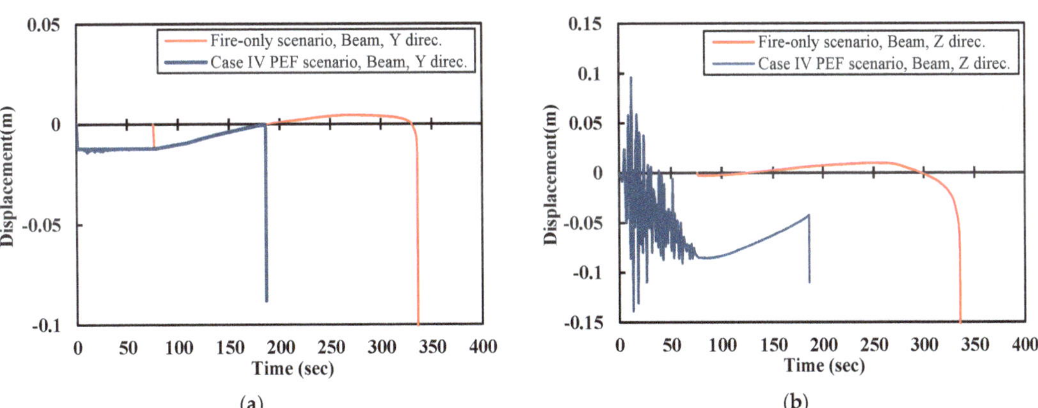

Figure 17. Images from a Case IV PEF analysis with an artificial earthquake (PGA = 0.35 g, natural period = 0.16 sec), bi-directional excitation in the x- and z-direction including (**a**) the residual deformation of the structure at the end of earthquake event and (**b**) the shape and mechanism of failure of the structure after the PEF event.

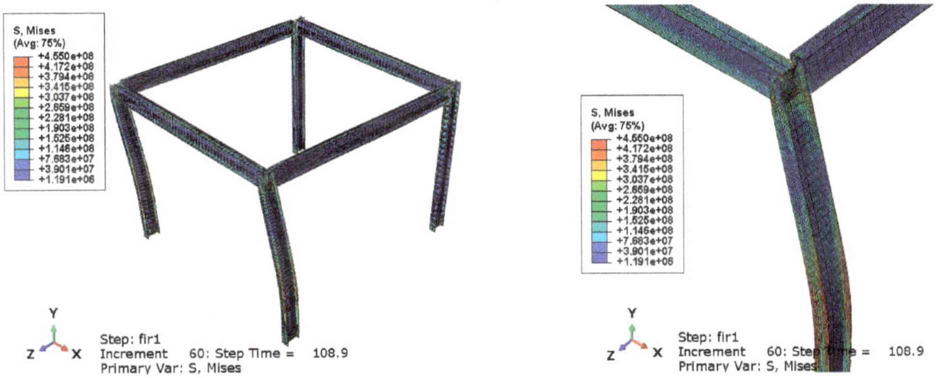

Figure 18. *Cont.*

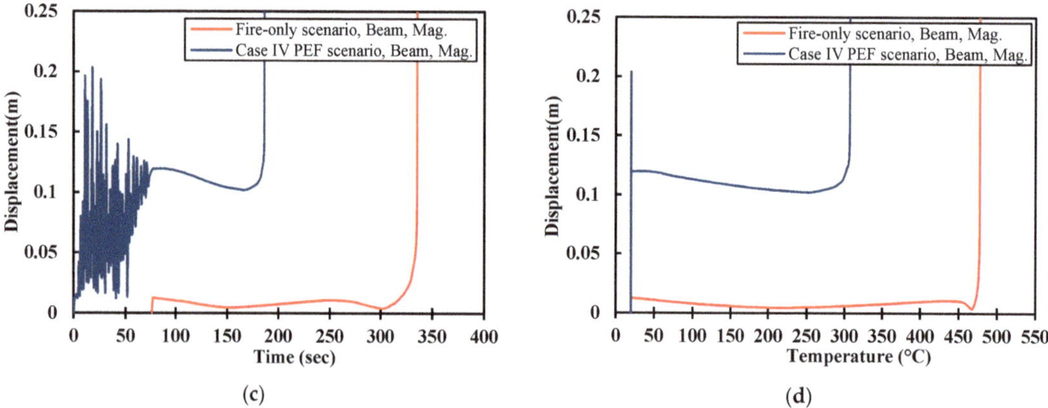

Figure 18. Comparison of the fire-only analysis versus the PEF analysis for Case IV including (**a**) the time-mid-span displacement record in the y-direction, (**b**) the time-mid-span displacement record in the z-direction, (**c**) the time-mid-span displacement record for the total displacement and (**d**) the temperature-mid-span displacement record for the total displacement.

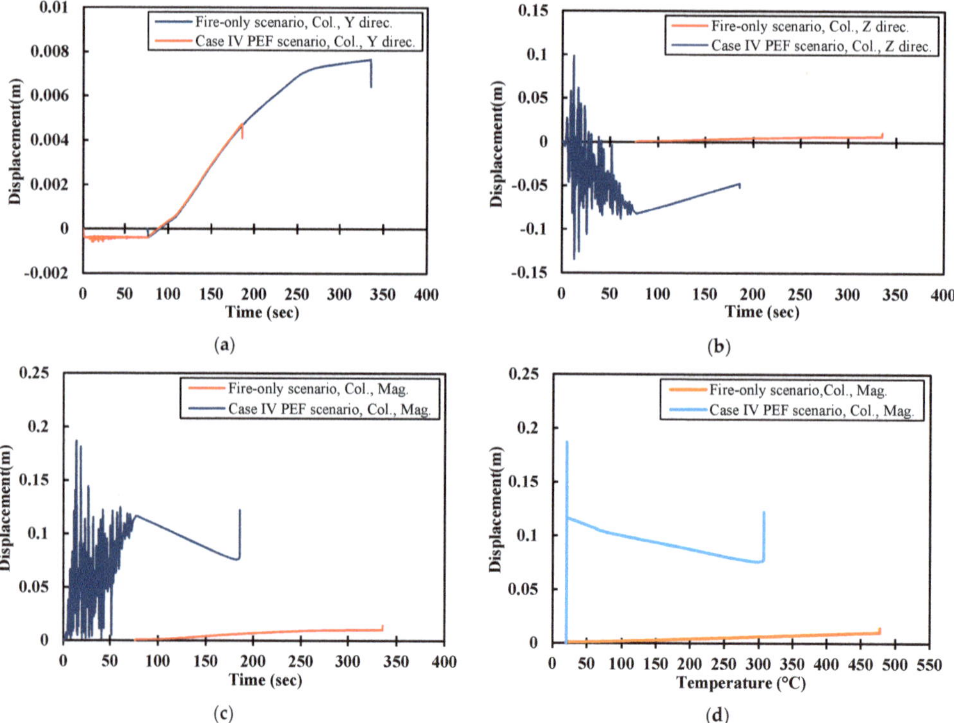

Figure 19. Comparison of the displacement values at a point which is 1.4 m along the column length for both the fire-only and PEF events for Case IV including (**a**) the time-displacement record in the y-direction, (**b**) the time-displacement record in the z-direction, (**c**) the time-displacement record for total displacement value and (**d**) the temperature-displacement record for the total displacement value.

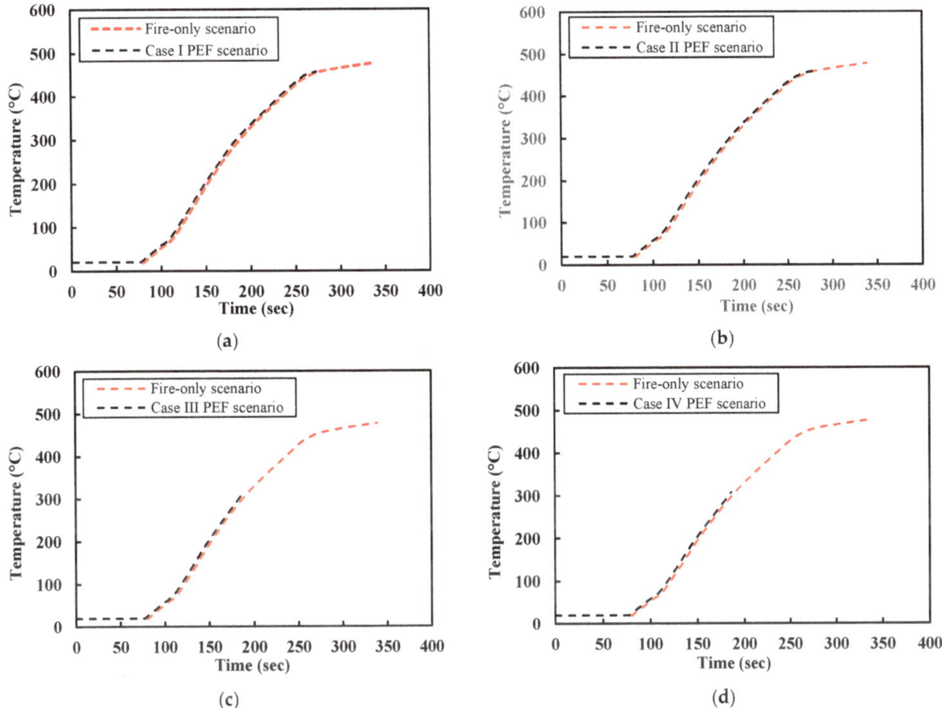

Figure 20. Comparison of the fire-only analysis versus the PEF analysis for the time -temperature response including (**a**) case I, (**b**) case II, (**c**) case III and (**d**) case IV.

Table 2. Results comparison for all analysed circumstances.

Case No.	Type of Analysis	Type of Excitation	Failure Time (sec)	Failure Tem. (°C)	Time Failure, Compared to Fire-Only Results	Type of Failure
Fire-Only	Fire-Only	No excitation	336	480	-	Local/Symmetrical
Case I	PEF	Unidirectional	272	455	−19%	Local/Asymmetrical
Case II	PEF	Unidirectional	277	455	−18%	Local/Asymmetrical
Case III	PEF	Bidirectional	185	306	−45%	Global/Asymmetrical
Case IV	PEF	Bidirectional	185	306	−45%	Global/Asymmetrical

6. Conclusions

This paper presents a detailed analysis of the influence of a post-earthquake fire on the behaviour of a steel framed building. It is clear that there are grave consequences in terms of occupant and structural safety during this type of multi-hazard scenario. Therefore, an accurate analysis of the response of structures exposed to such an event is required at the design stage, especially for very important buildings. The likelihood of a fire occurring following an earthquake is reasonably high, despite PEF being the subject of relatively little research in the available literature. In most design codes, structures exposed to multiple hazards such as earthquakes and then fires are analysed and designed separately. Structures subjected to an earthquake experience partial damage, and the subsequent occurrence of a fire may lead to structural collapse. Most available analysis procedures and design codes do not address the association of the two hazards. Thus, the design of structures based on existing standards may present a high risk of structural failure.

A suitable method of analysis has been developed in this paper to investigate the behaviour of structures that are exposed to such sequential hazards. Investigating the effects of PEF on structures classified as "ordinary" in the design codes (such as educational and residential buildings, for example) is necessary as these types of building are very common in urban and well-populated environments. A performance-based design consideration requires structures to remain within the 'life safety' level of response under the design for the occurrence of an earthquake and fire, separately. In the current paper, two types of failure mechanisms are detected for steel framed buildings subjected to PEF—global and local failure. Local failure happens in the beams, whereas global failure is evidenced by significant lateral movement in the columns due to bidirectional excitation. Interestingly, the majority of the fire-only analyses discussed herein resulted only in a local collapse, while all of the PEF analyses with bidirectional excitation resulted in a global collapse. Therefore, it is clear that the failure mode for a PEF can be quite different compared to a single hazard event. Consequently, it is suggested that columns with greater bi-directional stiffness (e.g., tubular sections) are likely to offer the greatest ultimate resistance in earthquake hazard zones under the combined effects of bidirectional earthquake excitation and subsequent fire. Despite the investigations in this paper being performed in relation to a particular class of structures, the results confirm the need to incorporate PEF as a load case during both the analysis and design stages. Further studies need to be performed either numerically or experimentally, using complete a seismic soil-structure interaction analysis, to develop a better understanding of the issue.

Author Contributions: Conceptualization, A.T.A. and P.E.F.C.; Methodology, A.T.A. and K.A.C.; Software, A.T.A.; Validation, A.T.A., P.E.F.C. and K.A.C.; Formal analysis, A.T.A.; Investigation, A.T.A.; Data curation, A.T.A., P.E.F.C. and K.A.C.; Writing—original draft preparation, A.T.A.; Writing—review and editing, P.E.F.C. and K.A.C.; Visualization, A.T.A.; supervision, P.E.F.C. and K.A.C.; project administration, P.E.F.C. All authors have read and agreed to the published version of the manuscript.

Funding: This research received no external funding.

Institutional Review Board Statement: Not applicable.

Informed Consent Statement: Not applicable.

Data Availability Statement: The study did not report any data.

Conflicts of Interest: The authors declare no conflict of interest.

References

1. Chicchi, R.; Varma, A.H. Research review: Post-earthquake fire assessment of steel buildings in the United States. *Adv. Struct. Eng.* **2018**, *21*, 138–154. [CrossRef]
2. Li, G.Q.; Zhang, C.; Jiang, J. A review on fire safety engineering: Key issues for high-rise buildings. *Int. J. High-Rise Build.* **2018**, *7*, 265–285. [CrossRef]
3. Yassin, H.; Iqbal, F.; Bagchi, A.; Kodur, V.K.R. Assessment of Post-Earthquake Fire Performance of Steel-Frame Buildings. In Proceedings of the 14th World Conference on Earthquake Engineering, Beijing, China, 12–17 October 2008.
4. Ronagh, H.R.; Behnam, B. Investigating the Effect of Prior Damage on the Post-Earthquake Fire Resistance of Reinforced Concrete Portal Frames. *Int. J. Concr. Struct. Mater.* **2012**, *6*, 209–220. [CrossRef]
5. Kamath, P.; Sharma, U.K.; Kumar, V.; Bhargava, P.; Usmani, A.; Singh, B.; Singh, Y.; Torero, J.; Gillie, M.; Pankaj, P. Full-scale fire test on an earthquake-damaged reinforced concrete frame. *Fire Saf. J.* **2015**, *73*, 1–19. [CrossRef]
6. British Standards Institution. *Eurocode 8: Design of Structures for Earthquake Resistance—Part 1: General Rules, Seismic Actions and Rules for Buildings Eurocode*; National Standards Authority of Ireland: Dublin, Ireland, 2004; Volume BS EN 1998, p. 231.
7. British Standards Institution. *Eurocode 3: Design of Steel Structures—Part 1–2: General Rules—Structural Fire Design*; European Committee for Standardization: Brussels, Belgium, 1993; Volume BS EN, p. 227.
8. Douglas, J.; Gkimprixis, A. Using targeted risk in seismic design codes: A summary of the state of the art and outstanding issues. In Proceedings of the 6th National Conference on Earthquake Engineering and 2nd National Conference on Earthquake Engineering and Seismology, Bucharest, Romania, 14–17 June 2017.
9. Solomos, G.; Pinto, A.; Dimova, S. A Review of the seismic Hazard Zonation in National Building Codes in the Context of Eurocode 8. *JRC Sci. Tech. Rep.* **2008**. Available online: https://publications.jrc.ec.europa.eu/repository/handle/JRC48352 (accessed on 1 July 2021).
10. Smith, M. *ABAQUS/Standard User's Manual*; Version 2018; Dassault Systèmes Simulia Corp: Providence, RI, USA, 2018; p. 1146.

11. Kobes, M.; Helsloot, I.; de Vries, B.; Post, J.G. Building safety and human behaviour in fire: A literature review. *Fire Saf. J.* **2010**, *45*, 1–11. [CrossRef]

12. Della Corte, G.; Landolfo, R.; Mazzolani, F.M. Post-earthquake fire resistance of moment resisting steel frames. *Fire Saf. J.* **2003**, *38*, 593–612. [CrossRef]

13. Ali, H.M.; Senseny, P.E.; Alpert, R.L. Lateral displacement and collapse of single-story steel frames in uncontrolled fires. *Eng. Struct.* **2004**, *26*, 593–607. [CrossRef]

14. Mousavi, S.; Bagchi, A.; Kodur, V.K.R. Review of post-earthquake fire hazard to building structures. *Can. J. Civ. Eng.* **2008**, *698*, 689–698. [CrossRef]

15. Zaharia, R.; Pintea, D. Fire after Earthquake Analysis of Steel Moment Resisting Frames. *Int. J. Steel Struct.* **2009**, *9*, 275–284. [CrossRef]

16. Ghoreishi, M.; Bagchi, A.; Sultan, M.A. Estimating the response of structural systems to fire exposure: State-of-the-art review. In Proceedings of the Fire and Materials 2009, 11th International Conference, San Francisco, CA, USA, 26–28 January 2009; pp. 475–483.

17. Memari, M.; Mahmoud, H.; Ellingwood, B. Post-earthquake fire performance of moment resisting frames with reduced beam section connections. *J. Constr. Steel Res.* **2014**, *103*, 215–229. [CrossRef]

18. Zhou, M.; Jiang, L.; Chen, S.; Cardoso, R.P.R.; Usmani, A. Remaining fire resistance of steel frames following a moderate earthquake—A case study. *J. Constr. Steel Res.* **2020**, *164*, 105754. [CrossRef]

19. Jelinek, T.; Zania, V.; Giuliani, L. Post-earthquake fire resistance of steel buildings. *J. Constr. Steel Res.* **2017**, *138*, 774–782. [CrossRef]

20. Alasiri, M.R.; Chicchi, R.; Varma, A.H. Post-earthquake fire behavior and performance-based fire design of steel moment frame buildings. *J. Constr. Steel Res.* **2021**, *177*, 106442. [CrossRef]

21. Fajfar, P. Structural Analysis in Earthquake Engineering—A Breakthrough of Simplified Non-Linear Methods. In Proceedings of the 12th European Conference on Earthquake Engineering, London, UK, 9–13 September 2002; p. 843.

22. Suwondo, R.; Gillie, M.; Cunningham, L.; Bailey, C. Effect of earthquake damage on the behaviour of composite steel frames in fire. *Adv. Struct. Eng.* **2018**, *21*, 2589–2604. [CrossRef]

23. Behnam, B.; Abolghasemi, S. Post-earthquake Fire Performance of a Generic Fireproofed Steel Moment Resisting Structure. *J. Earthq. Eng.* **2019**, *25*, 1–26. [CrossRef]

24. British Standards Institution. *Eurocode2: Design of Concrete Structures: Part 1–1: General Rules and Rules for Buildings*; Committee for Standardization: Brussels, Belgium, 2004; Volume BS EN 1991-1-1, p. 240.

25. SeismoSoft. Manual and Program Description of the Program SeismoStruct. Available online: http://www.seismosoft.com (accessed on 20 May 2020).

26. U.S. Geological Survey. PEER Ground Motion Database. Pacific Earthquake Engineering Research Centre. 2018. Available online: https://ngawest2.berkeley.edu/ (accessed on 12 March 2020).

27. Mondal, D.R. High Risk of Post-Earthquake Fire Hazard in Dhaka, Bangladesh. *Fire* **2019**, *2*, 24. [CrossRef]

28. Fischer, E.C.; Maddalozzo, W. Post-Earthquake Fire Performance of Industrial Facilities PEER Transportation Systems Research Program. Available online: https://peer.berkeley.edu/sites/default/files/walker_maddalozzo_peer2020_poster.pdf (accessed on 1 July 2021).

29. ISO 834-10. In *Fire Resistance Tests—Elements of Building Construction—Part 10: Specific Requirements to Determine the Contribution of Applied Fire Protection Materials to Structural Steel Elements*; International Organization for Standardization: Geneva, Switzerland, 2014.

30. British Standards Institution. *Eurocode: Basis of Structural Design*; Committee for Standardization: Brussels, Belgium, 2011; Volume BS EN 1990:2002+A1, p. 112.

31. British Standards Institution. *Eurocode 1: Actions on Structures Part 1–6: General Actions during Execution*; Committee for Standardization: Brussels, Belgium, 2011; Volume BS EN 1991-1, p. 31.

32. Carrera, E.; Fazzolari, A.F.; Cinefra, M. *Chapter 9—Static Response of Uncoupled Thermoelastic Problems*; Academic Press: Oxford, UK, 2017; pp. 311–326. Available online: https://www.sciencedirect.com/science/article/pii/B9780124200661000124 (accessed on 10 July 2020).

33. Khose, V.N.; Singh, Y.; Lang, D.H. A comparative study of design base shear for RC buildings in selected seismic design codes. *Earthq. Spectra* **2012**, *28*, 1047–1070. [CrossRef]

34. Crowley, H.; Pinho, R. Revisiting Eurocode 8 formulae for periods of vibration and their employment in linear seismic analysis. *Earthq. Eng. Struct. Dyn.* **2010**, *39*, 223–235. [CrossRef]

35. Zembaty, Z.; Kokot, S.; Kuś, J. Mitigating Rockburst Effects for Civil Engineering Infrastructure and Buildings. In *Rockburst: Mechanisms, Monitoring, Warning, and Mitigation*; Elsevier: Amsterdam, The Netherlands, 2018; pp. 541–548.

36. McKenna, F.T. *Object-Oriented Finite Element Programming: Frameworks for Analysis, Algorithms and Parallel Computing*; University of California: Berkeley, CA, USA, 1997.

37. Usmani, A.; Zhang, J.; Jiang, J.; Jiang, Y.; May, I. Using OpenSees for structures in fire. *J. Struct. Fire Eng.* **2010**, *3*, 919–926. [CrossRef]

Article

Parametric Analysis of a Steel Frame under Fire Loading Using Monte Carlo Simulation

Ragad Almadani and Feng Fu *

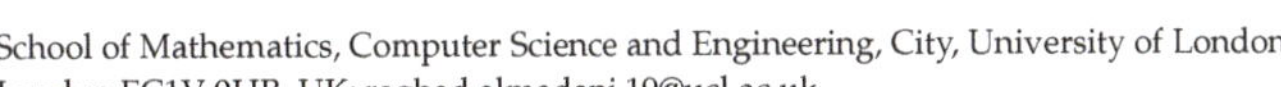

School of Mathematics, Computer Science and Engineering, City, University of London,
London EC1V 0HB, UK; raghad.almadani.19@ucl.ac.uk
* Correspondence: feng.fu.1@city.ac.uk

Abstract: In this paper, the parametric analysis of the thermal and structural response of a two-storey, single-zone steel frame building on fire is made considering different parameters Monte Carlo simulation is used to generate random variables for the opening factor, fire compartment area and finally the beam flange thickness. Using the random parameter generated, a sequential thermal and mechanical analysis was conducted using the finite element software ABAQUS. The first step was a heat transfer analysis, followed by mechanical analysis. The effect of different parameters on the thermal and mechanical response of the structure was studied.

Keywords: City University; fire temperature; opening factor; compartment area; thermal analysis

Citation: Almadani, R.; Fu, F. Parametric Analysis of a Steel Frame under Fire Loading Using Monte Carlo Simulation. *Fire* **2022**, *5*, 25. https://doi.org/10.3390/fire5010025

Academic Editor: Maged A. Youssef

Received: 28 December 2021
Accepted: 9 February 2022
Published: 14 February 2022

Publisher's Note: MDPI stays neutral with regard to jurisdictional claims in published maps and institutional affiliations.

1. Introduction

Steel has been the forefront of efficient construction in the last few years, where it has been used widely in the construction of high-rise buildings, industrial structures and residential structures. What makes steel one of the most appealing materials in the construction industry is its engineering properties. The most appealing properties of steel are its strength to weight ratio, ductility and flexibility. Such properties allow designers to build structures such as skyscrapers, which certainly would have not been possible with any other material. Steel can also be prefabricated and shipped to construction sites easily, which is quite beneficial when it comes to meeting the ever-increasing demands of new buildings. Nevertheless, there is a huge downside to using steel as a construction material because of its low resistance to fire when compared to other construction materials such as concrete. Steel loses almost half of its strength when subjected to temperature which is equal to or greater than 590 °C, which will eventually lead the structure to fail. The losses that follow structural failures caused by fire are colossal and can take different forms, such as loss of human lives, environmental loss and economical loss. Hence, the insurance of structural stability of a building under fire loading has been one of the most important and challenging aspects when it comes to designing a new structure [1]. It is important that in the event of a fire, structures are able to withstand the minimum level of life safety not only for the occupants but also fire fighters and the public that are in proximity of the building. The minimum level of fire safety design must ensure a reduction of the risk of deaths and injuries, protect the contents of a building, and ensure that the building continue to function after the fire with the least amount of repair possible.

In order to ensure that the structure meets the fire safety design objectives, designers have to follow one of two methods; the first is the prescriptive method, where a detailed description of the types and shapes of materials used in the design, the thickness of protection layer on structural elements and even the details of construction are given. However, this method relies solely on previous experience of the standard structural fire design tests. While this approach is very useful when it comes to static situations, it sometimes fails to meet the fire safety requirements of a building, raising concerns about the limitations of this method. One of its limitations is that the standard structural fire design tests assume that the structural elements of a building work independently, which is not the case in reality [2,3]. This approach is usually used for quick solutions and for junior designers, because it does not require an in-depth knowledge of the field.

The second method is the performance-based approach, which indicates how a structure will perform when subjected to different load conditions. Designers who use this method need to develop an accurate numerical simulation for fire loading to assess the fire safety resistance of the structure [4–7]. There are three main components to this approach: fire modelling, thermal analysis and structural analysis [2,3]. Keeping this in mind, this method allows designers to come up with solutions to build complex structures that were never possible with a standard prescriptive approach. This approach is usually adopted for more optimum solutions that require computational skills and deep understanding of the field.

There are a number of different modelling techniques commonly used today to predict the different fire scenarios instead of carrying out experimental tests. The techniques used range from simple hand calculations to more advanced computational techniques. Deciding which technique to use depends on the level of accuracy needed for the project, time restrictions and computational resources. With the development of software packages, designers can use a number of methods to predict possible fire scenarios in a building in order to further control the risk of a fire event. Some of these methods include zone models, computational fluid dynamics (CFD) [8] models and finite element (FE) models [9]. Zone models are simple computational models that operate on the basis of dividing a compartment area into separate zones, with the assumption that the temperature condition is uniform throughout each zone. CFD models are more sophisticated than zone models, because they analyze the fluid flow and heat transfer by solving the fundamental equations of fluid dynamics. Finally, FE models operate by dividing a large geometry into several hundred smaller parts that interact with each other. Given the different methods designers can adopt to predict the different fire scenarios, the uncertain nature of the different factors affecting a fire event the structural stability of a building one of the most challenging responsibilities for structural engineers. The fuel, load density and ventilation areas are all factors that contribute to a fire event and its duration. Moreover, the unpredictable nature of the response of structural elements to fire makes the whole fire process stochastic. As a result, the engineering design of structural fire is either based on empirical studies of the behavior of fire or on reliability analysis [10].

The FE modelling technique is one of the most straightforward methods used to predict the thermal and structural behavior of the structural members, according to [4–7]. The simulation works by breaking up large geometry to hundreds of smaller and simpler parts that interact together. In order to run the simulation using the finite element modelling technique, the thermal and mechanical properties of the material used in construction have to be calculated according to design codes. Design codes such as Eurocode provide formulas to calculate the thermal properties of a material and curves to obtain gas temperatures. The values obtained from the formulas provided can then be applied directly to the model using software packages.

However, most variables used during the fire analysis are either estimated or assumed, which makes the efficiency of the design in doubt and highlights the importance of the concept of intensive parametric study in structural fire design. As in reality, the parameters vary due to different fire scenarios; to effectively study the influence of different parameters, one of the promising methods is the Monte Carlo simulation [11]. This method generates hundreds of random variables for the different stochastic parameters that relate to fire, which allow designers to analyze the buildings under different fire scenarios, and thus to maximize the reliability and safety of the design.

Therefore, in this paper, the parametric analysis of the thermal and structural response of a two-story, single-zone steel frame building to fire is made considering different parameters; Monte Carlo simulation is used to generate random variables for the opening factor, fire compartment area and the beam flange thickness. Using the random parameters generated, a sequential thermal and mechanical analysis was conducted using the finite element software ABAQUS. The first step was a non-linear heat transfer analysis, followed by a non-linear mechanical analysis. The effect of different parameters on the thermal and mechanical response of the structure was studied.

2. Monte Carlo Simulation

The Monte Carlo Simulation is used here to generate the random parameters which will affect the fire scenarios and the thermal response of the structural steel members in a steel framed building. The key factors affecting the fire scenarios and thermal response of a structure under fire conditions are explained here

2.1. Time-Temperature Curves for a Compartment

The room temperature of a building in fire can be calculated using the formulas from Eurocode, BS EN 1993-1-2: Eurocode 1 part 1.2 [12] which gives the parametric time temperature of a compartment in fire:

$$\theta_g = 20 + 1325 \left(1 - 0.324 \, e^{-0.2t^*} - 0.204e^{-1.7t^*} - 0.472e^{-19t^*}\right) \tag{1}$$

With

$$t^* = t \times \Gamma \tag{2}$$

$$\Gamma = [O/b]^2 / (0.04/1160)^2 \tag{3}$$

$$O = A_v \sqrt{h_{eq}} / A_t \tag{4}$$

$$b = \sqrt{(\rho c \lambda)} \tag{5}$$

where θ_g is the gas temperature in the fire compartment (°C), t is the time (m), O is the opening factor ($m^{1/2}$), b is the thermal diffusivity ($J/m^2 \, s^{1/2} \, K$), ρ is the density (kg/m^3), C is the specific heat (J/kgK), λ is the thermal conductivity (W/mK), A_t is the total internal surface area of the compartment (m^2), A_v is the area of ventilation (m^2) and h_{eq} is the height of openings (m) [12,13].

2.2. Thermal Response of Structural Members

For unprotected steel sections, the increase of temperature in a small time interval is given by BS EN 1993-1-2: Eurocode 3 [13] as follows:

$$\Delta\theta_{a,t} = k_{sh} \frac{A_m/V}{c_a\rho_a} \dot{h}_{net}\Delta t \tag{6}$$

where, $\Delta\theta_{a,t}$ is the increase of temperature. A_m/V is the section factor for unprotected steel member. A_m is the exposed surface area of the member per unit length. V is the volume if the member per unit length. c_a is the specific heat of steel. ρ_a is density of the steel.

2.3. Monte Carlo Simulation

Based on above formulas from the design code, three main parameters dominate the room temperature and thermal response of the structural member. They are the opening factor (O), the compartment area (A_{com}) and the cross-sectional area of the structural members (A_m). Therefore, in the Monte Carlo simulation, the opening factor (O), the compartment area (A_{com}) and the thickness of the flange for the steel I-section beam were selected as the key parameters for the random simulation. These three parameters were selected based on Equations (1) and (3), since these three parameters play important roles in determining the thermal response of the structural members. The Monte Carlo simulation code was developed using MATLAB [11]. The code developed uses an inbuilt command called 'normrnd' to generate normal random numbers between the specified range to form a set of inputs that generate the different fire scenarios. The specified limits for the opening factor (O) and the compartment area (A_{com}) were set according to the British Standard Institutes [12], and the limits for the flange thickness of the I-beam section were chosen according to the British Standard Institutes [12,13], as shown in Table 1. The number of simulations chosen was 500, which is believed to be sufficent to produce reasonable parameters in a real fire scenario.

Table 1. Range of the parameters used for Monte Carlo simulation.

Opening Factor, O ($m^{1/2}$)	$0.02 \leq O \leq 0.2$ [12]
Compartment Area, A_{com} (m^2)	$A_{com} \leq 500$ [12]
Beam Flange Thickness, T (mm)	$7.5 < T < 55$ [13]

The MATLAB code generated 500 random numbers for each parameter. Amongst the 500, only five numbers were chosen in this paper. These values are tabulated in Table 2.

Table 2. Chosen values from the Monte Carlo simulation.

Opening Factor, O ($m^{1/2}$)	O = 0.0251 O = 0.0554 O = 0.1129 O = 0.1689 O = 0.1961
Compartment Area, A_{com} (m^2)	Acom = 32.9 Acom = 42 Acom = 56 Acom = 70 Acom = 79.8
Beam Flange Thickness, T (mm)	T = 7.9066 T = 9.3202 T = 16.7989 T = 19.3693 T = 25.8854

The values noted above were used to run the heat transfer analysis in Section 3. The corresponding unique nodal temperatures were extracted for all cases and used to run the mechanical analysis explained in Section 3.

3. Finite Element Analysis

3.1. Finite Element Model

The finite element analysis procedure was split into two stages: (1) Heat Transfer Analysis and (2) Mechanical Analysis. The FE model is a two-story, one-bay by one-bay steel frame structure. Each story is 3 m high, making the whole structure 6 m high in total and 7 m by 5.5 m in width. This is a typical steel frame office dimension, according to (Tagawa et al., 2015). Different element types have been tried in order to choose the suitable element to simulate the behavior of the composite connections. 3D continuum elements were used. C3D8R element with reduced integration (1 Gauss point) has been chosen for the simulation of all the components in the model. The model simulates a corner fire with a three hour duration. Due to time constrictions, only the steel frame was modelled without the slab. ABAQUS was used for both simulations.

3.2. Heat Transfer Analysis

In order to construct the two-story steel frame structure on ABAQUS, three main parts were created: a 6 m column, 5.5 m and 7 m beams, and a partitioned midway. The values of the steel's conductivity (λa), steel's specific heat (Ca) and steel's density (ρa) were obtained from Eurocode. All three parts were assigned the same steel material. Furthermore, an instance was created and the three different parts were added in order to assemble the elements as one whole structure. This was done with the use of the offset, rotate and translate commands.

After that, a heat transfer step was created with a time period of 10,800 s, and the maximum increment per step was set to 10. Before simulating the heat transfer analysis, an amplitude of the gas temperature was obtained from Equation (1) for simulating the different fire scenarios using the parameters generated from the Monte Carlo simulations.

ABAQUS will fail to run any heat transfer analyses if the absolute zero temperature and Stefan-Boltzmann constant (σ) were not inputted. Hence, from model attributes, a value of -273.5 was given to the former and a value of 5.67×10^{-8} to the latter. Before the analysis was submitted for results, the entire structure was meshed using hexagonal element shapes and assigned an element type of heat transfer. After the analysis was submitted, and the heat analysis results were completed, the unique nodal temperatures at three different locations were extracted and the average value was taken in order to apply them at the heat load analysis stage.

For the convergence criterion, the default solution control parameters defined in ABAQUS/Standard are designed to provide reasonably optimal solutions of complex problems involving combinations of nonlinearities as well as efficient solutions of simpler nonlinear cases. However, the most important consideration in the choice of the control parameters is that any solution accepted as "converged" is a close approximation to the exact solution of the nonlinear equations. In this context "close approximation" is interpreted rather strictly by engineering standards when the default value is used, as described below.

3.3. Mechanical Analysis

The mechanical analysis was subsequently carried out using ABAQUS. The true stress-strain curve was used for the material model of the steel. The most relevant mechanical properties for this model are the yield stress (σy), plastic strain (εp), expansion Co-efficient (α), young's modulus (E) and Poisson's ratio (ν). Thereafter, a solid homogeneous section was created with the new material, which was then assigned to the three different parts.

Following that, a step from the general static type was created, with a time period of 10,800 s, and the maximum number of increments was set to 1000, initial of 10, minimum of 0.108 and maximum of 60. The next step was to create a boundary condition for the fixed supports of the steel frame and apply the gravity load (dead and live loads) to the entire structure; this was done directly from the model tree.

Before the load was applied, the unique nodal temperature obtained from the heat transfer analysis was used to create an amplitude. The temperature on each node from the heat transferring analysis step was added through a predefined field; hence, a predefined field of the type 'temperature' was then created, and the relevant parts of the geometry were selected in order to simulate the same corner fire from the previous step. This was to simulate the corner fire test from the Cardington fire test, which is believed to be the worst-case scenario for a building on fire. The magnitude of the predefined field was set to 1 and the amplitude to the one created using the unique nodal temperatures from the heat transfer analysis.

For this type of analysis, the software for the normal hexagonal elements type was first used for meshing the structure; however, due to the complexity of the geometry of the structure, the assembly could not be meshed properly. Instead, the structure was meshed using the tetrahedral elements type. About 10 different mesh sizes were tried before a final mesh size was determined. This was done through assigning different seeds in the mesh module in ABAQUS, and ABAQUS generated the mesh automatically. Finally, a job was created and submitted; the energy results were then viewed, and the values for the displacement of the beams in all x, y and z directions were extracted.

4. Parametric Analysis

4.1. Effect of Opening Factor

The first set of results were obtained from both finite element analyses: (1) Heat transfer analysis, and (2) heat load analysis corresponded to the different opening factors obtained from the Monte Carlo simulation (Table 2). The different opening factors were applied to a steel frame with fixed dimensions (5.5 × 7 m) in terms of temperature amplitudes The results of the heat transfer load are shown in the form of nodal temperatures (NT11), whereas the results obtained from the mechanical analysis are shown in the form of displacement (U).

- **Scenario 1**: An opening factor with a value of 0.0251 was used, and the corresponding results are shown in Figure 1a,b.

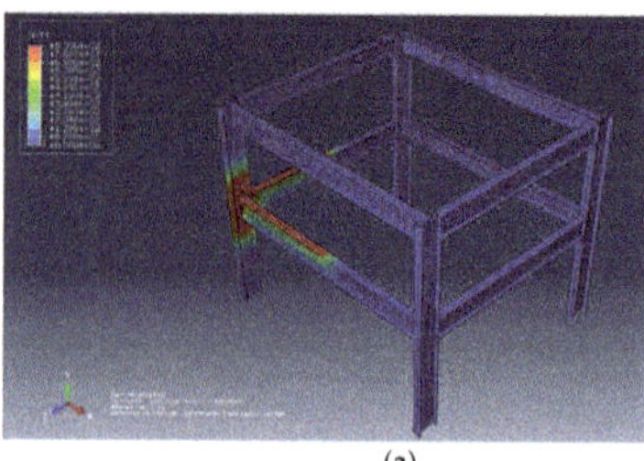
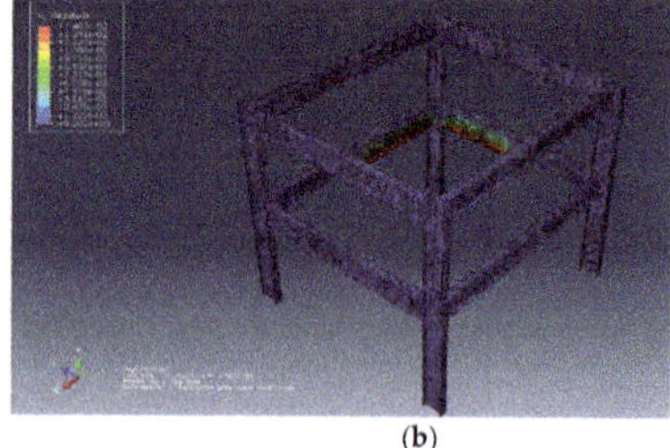

(a)　　　　　　　　　　　　　　　　　　(b)

Figure 1. Effect of opening factors. (**a**) The nodal temperature corresponding to the opening factor of 0.0251. (**b**) The displacement corresponding to the opening factor of 0.0251.

- **Scenario 2**: An opening factor with a value of 0.0554 was used, and the corresponding results are shown in Figure 2a,b.

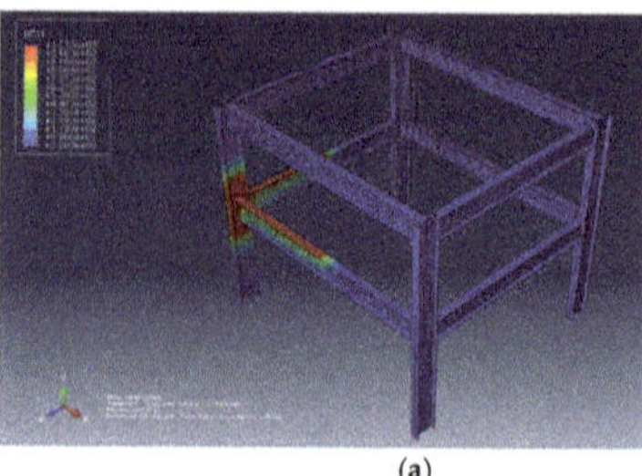
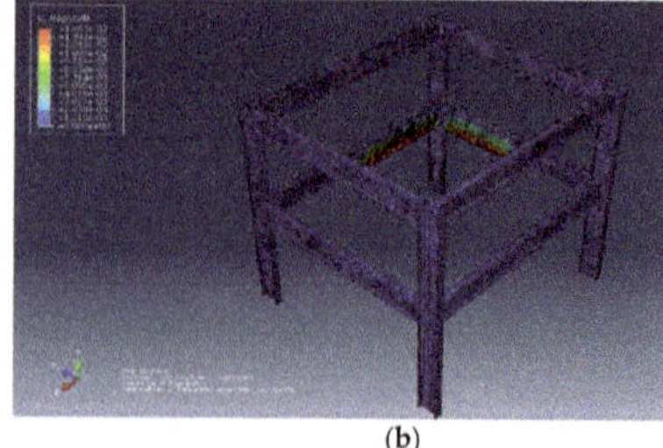

(a)　　　　　　　　　　　　　　　　　　(b)

Figure 2. Effect of opening factors. (**a**) The nodal temperature corresponding to the opening factor of 0.0554. (**b**) The displacement corresponding to the opening factor of 0.0554.

- **Scenario 3**: An opening factor with a value of 0.1129 was used, and the corresponding results are shown in Figure 3a,b.

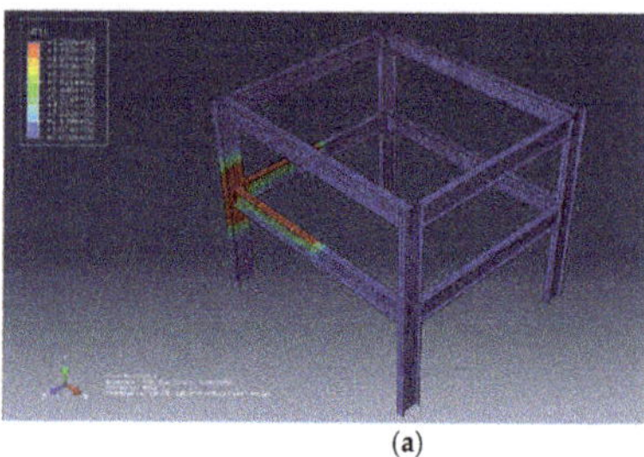
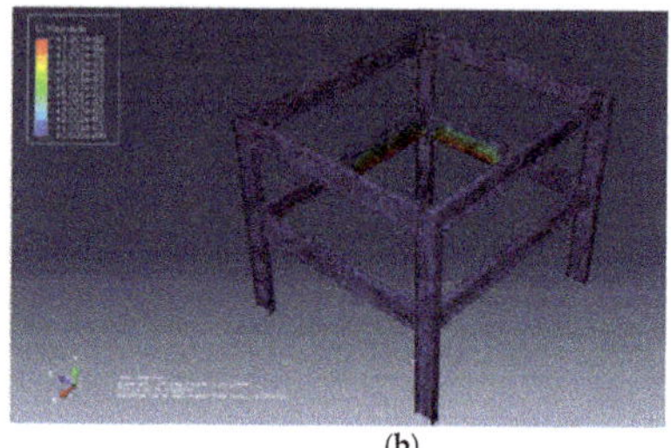

(a)　　　　　　　　　　　　　　　　　　(b)

Figure 3. Effect of opening factors. (**a**) The nodal temperature corresponding to the opening factor of 0.1129. (**b**) The displacement corresponding to the opening factor of 0.1129.

- **Scenario 4:** An opening factor with a value of 0.1689 was used, and the corresponding results are shown in Figure 4a,b.

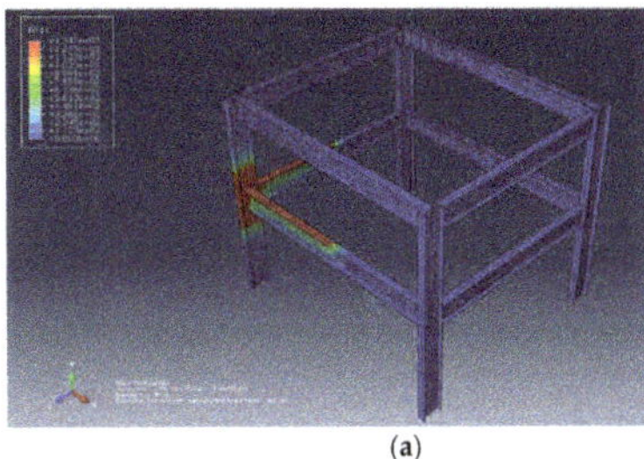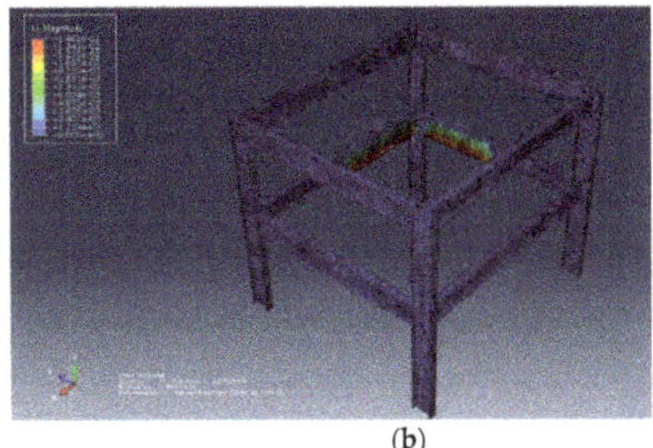

(a)   (b)

Figure 4. Effect of opening factors. (**a**) The nodal temperature corresponding to the opening factor of 0.1961. (**b**) The displacement corresponding to the opening factor of 0.1961.

- **Scenario 5**: An opening factor with a value of 0.1961 was used, and the corresponding results are shown in Figure 5a,b.

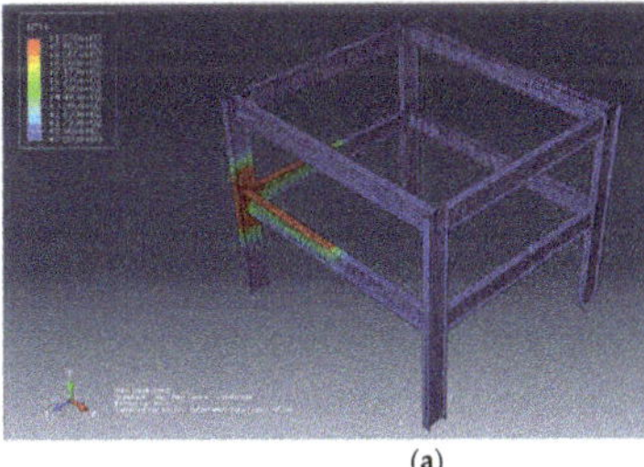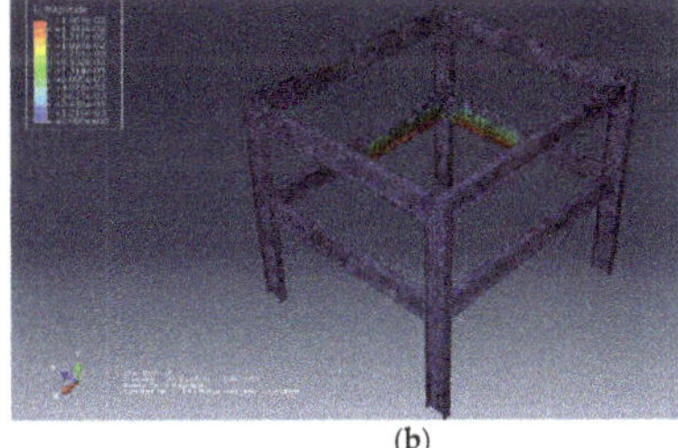

(a)   (b)

Figure 5. Effect of opening factors. (**a**) The nodal temperature corresponding to the opening factor of 0.0554. (**b**) The displacement corresponding to the opening factor of 0.0554.

4.2. Effect of Compartment Area

The second set of results shown in this section represent the finite element analyses that correspond to the different compartment area dimensions (Table 2). While the dimension of the plan of the steel frame changed, the temperature amplitude and heat load remained constant. In order to achieve the desired compartment area size, one of the beams was given a fixed length of 7 m, while the other beam changes with every scenario. The results are represented below in the form of NT11 and U for each scenario generated by the Monte Carlo simulation.

- **Scenario 1**: A compartment area with a size of 32.9 m^2 was modelled, and the beam lengths were 7 and 4.7 m in length. The corresponding results are shown in Figure 6a,b.

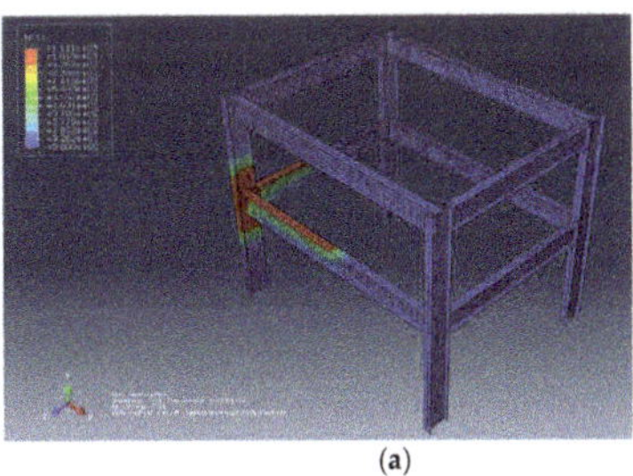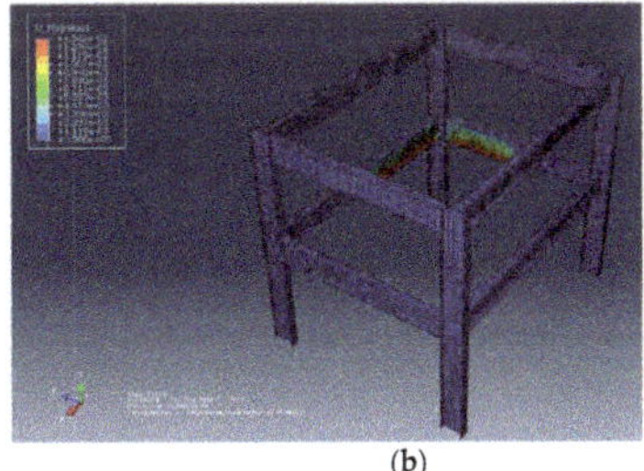

(a)   (b)

Figure 6. Effect of Compartment Area. (**a**) The nodal temperature corresponding to the compartment area of 32.9 m^2. (**b**) The displacement corresponding to compartment area of 32.9 m^2.

- **Scenario 2**: A compartment area with a size of 42 m^2 was modelled, and the beams were 7 and 6 m long. The corresponding results are shown in Figure 7a,b.

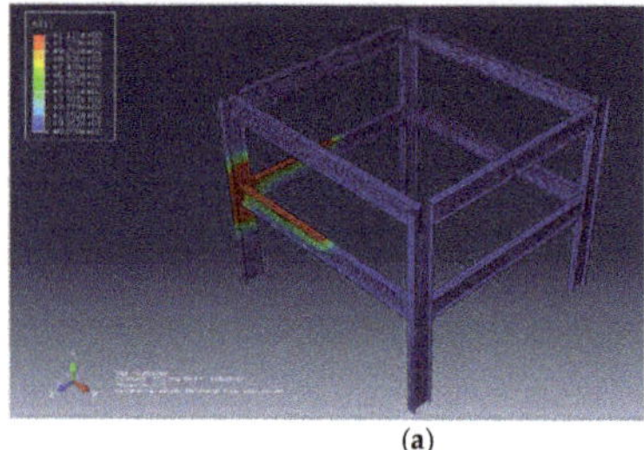
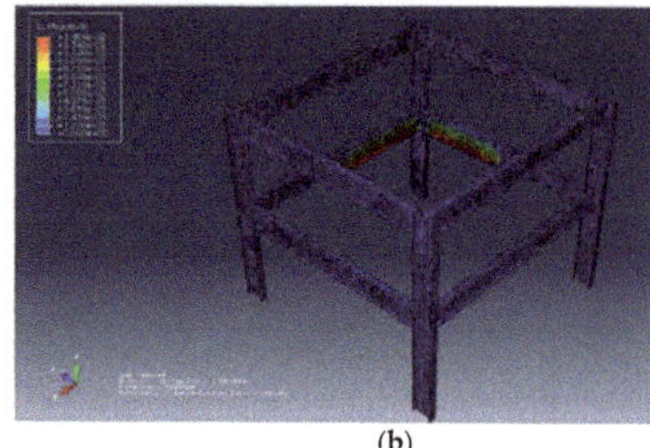

(a)   (b)

Figure 7. Effect of Compartment Area. (**a**)The nodal temperature corresponding to the compartment area of 42 m^2. (**b**) The displacement corresponding to the compartment area of 42 m^2.

- **Scenario 3**: A compartment area with a size of 56 m^2 was modelled, and the beams were 7 and 8 m long. The corresponding results are shown in Figure 8a,b.

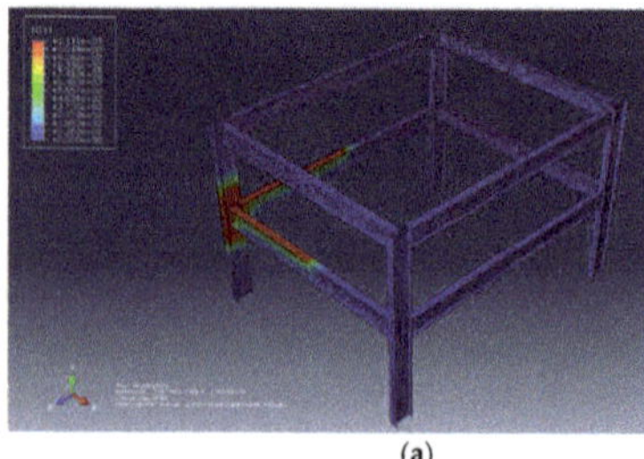
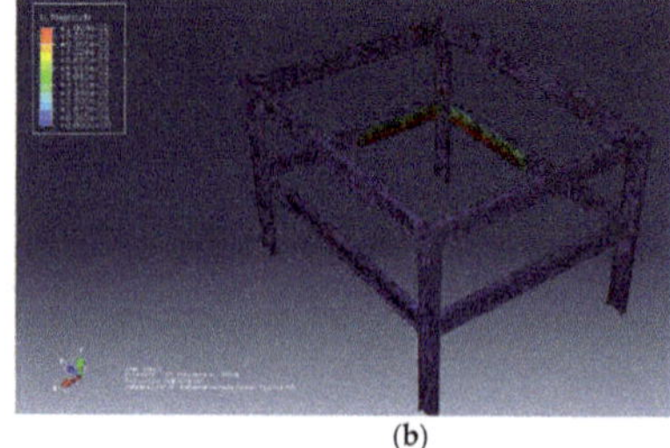

(a)   (b)

Figure 8. Effect of Compartment Area. (**a**) The nodal temperature corresponding to the compartment area of 56 m^2. (**b**) The displacement corresponding to the compartment area of 56 m^2.

- **Scenario 4:** A compartment area with a size of 70 m^2 was modelled, and the beams were 7 and 10 m long. The corresponding results are shown in Figure 9a,b.

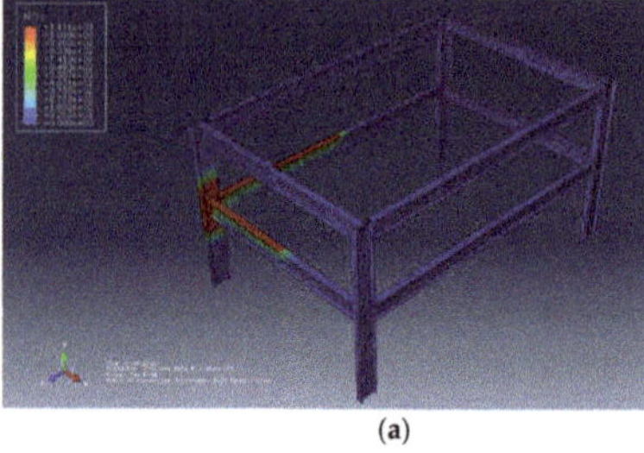
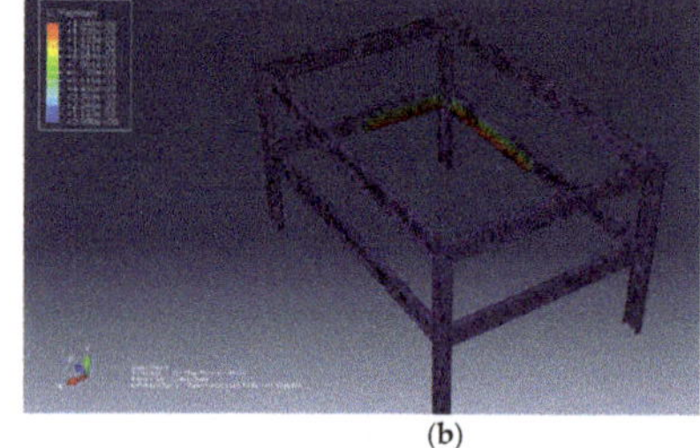

(a)   (b)

Figure 9. Effect of Compartment Area. (**a**) The nodal temperature corresponding to compartment area of 70 m^2. (**b**) The displacement corresponding to to compartment area of 70 m^2.

- **Scenario 5:** A compartment area with a size of 79.8 m^2 was modelled, and the beams were 7 and 11.4 m long. The corresponding results are shown in Figure 10a,b.

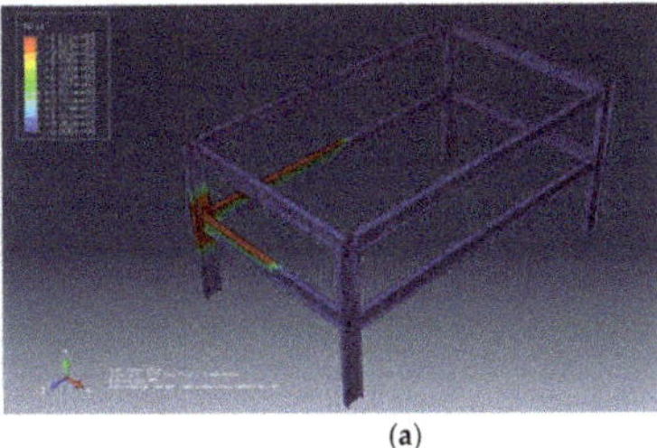 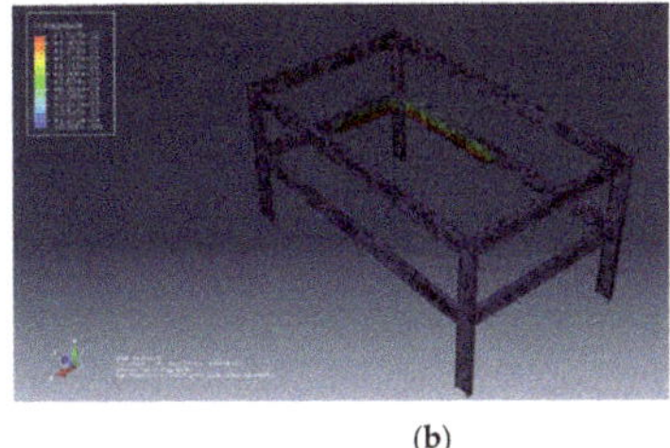

(a) (b)

Figure 10. Effect of Compartment Area. (**a**) The nodal temperature corresponding to the compartment area of 79.8 m^2. (**b**) The displacement corresponding to the compartment area of 79.8 m^2.

4.3. Effect of Flange Thickness

The third and last set of results represented in this section are related to the beam flange thickness. The thickness of the flange varied from 7.9 to 25.8 mm while the dimensions of the steel frame including the beams webs were fixed. The temperature amplitude and heat load applied to the structures were also fixed. The results represented are in the form of NT11 and U.

- **Scenario 1**: The thickness of the flange was set to 7.9 mm, and the corresponding results are shown in Figure 11a,b.

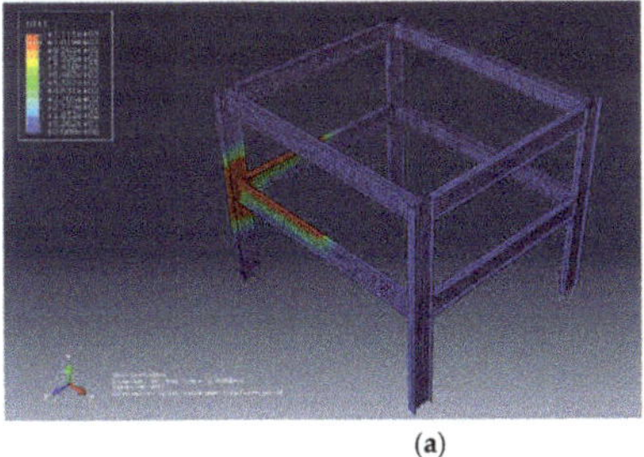 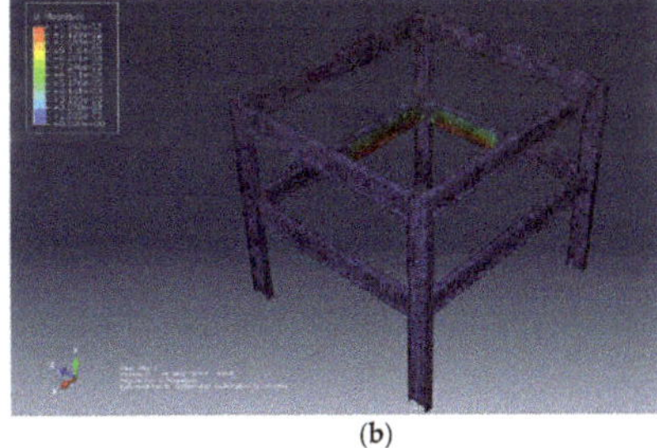

(a) (b)

Figure 11. Effect of flange thickness. (**a**) The nodal temperature corresponding to the flange thickness of 7.9 mm. (**b**) The displacement corresponding to the flange thickness of 7.9 mm.

- **Scenario 2**: The thickness of the flange was set to 9.3 mm, and the corresponding results are shown in Figure 12a,b.

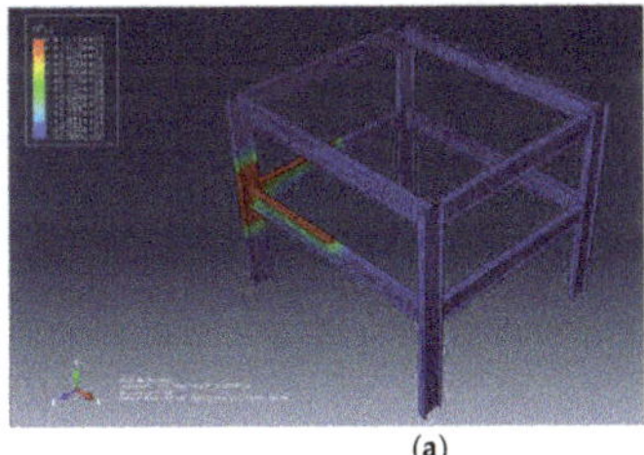 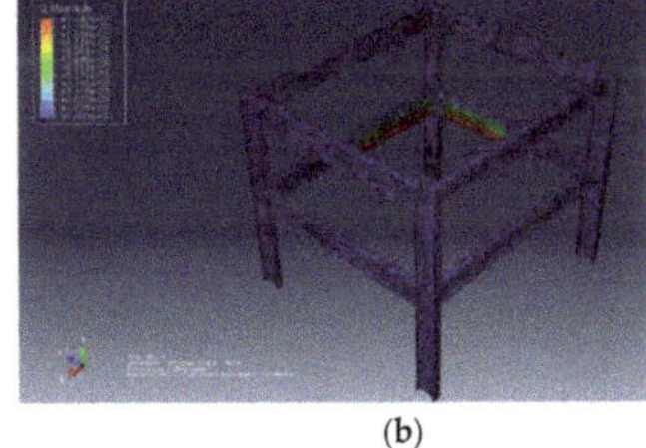

(a) (b)

Figure 12. Effect of flange thickness. (**a**) The nodal temperature corresponding to flange thickness of 9.3 mm. (**b**) The displacement corresponding to flange thickness of 9.3 mm.

- **Scenario 3**: The thickness of the flange was set to 16.7 mm, and the corresponding results are shown in Figure 13a,b.

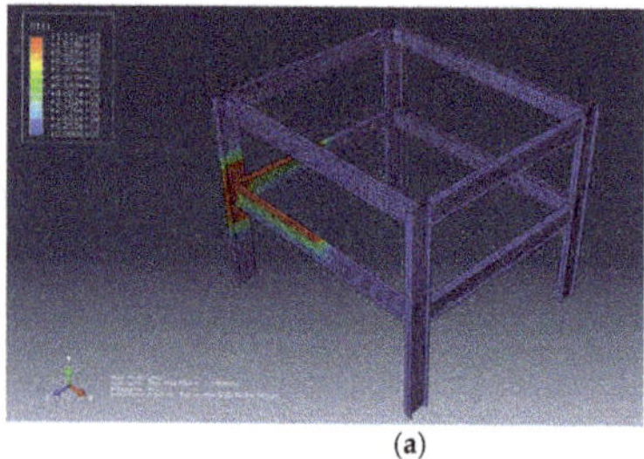 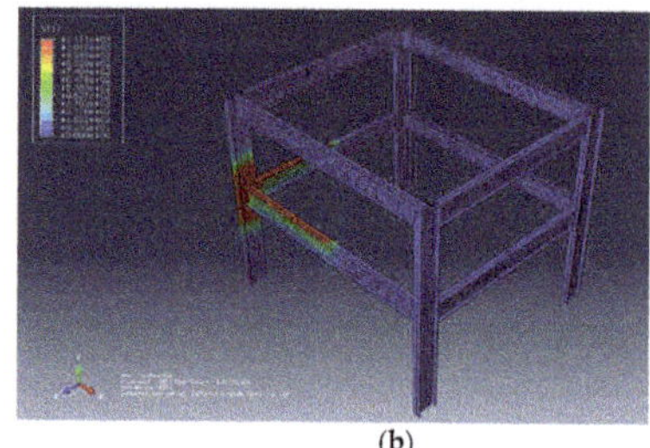

(a) (b)

Figure 13. Effect of flange thickness. (**a**) The nodal temperature corresponding to flange thickness of 16.7 mm. (**b**) The displacement corresponding to flange thickness of 16.7 mm.

- **Scenario 4:** The beam flange thickness was set to 19.3 mm, and the corresponding results are shown in Figure 14a,b.

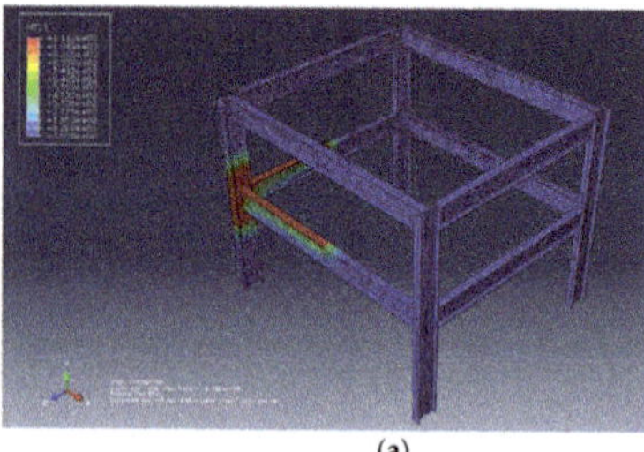 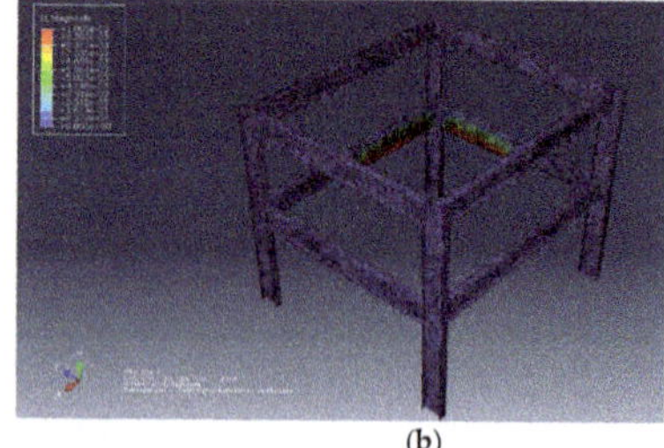

(a) (b)

Figure 14. Effect of flange thickness. (**a**) The nodal temperature corresponding to flange thickness of 19.3 mm. (**b**) The displacement corresponding to flange thickness of 19.3 mm.

- **Scenario 5:** The beam flange thickness was set to 25.8 mm, and the corresponding results are shown in Figure 15a,b.

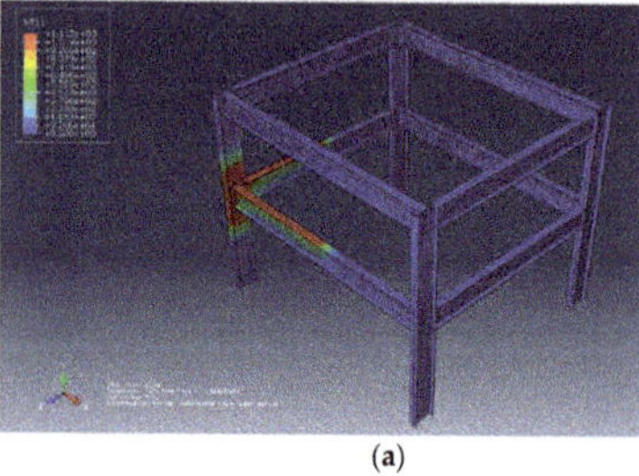 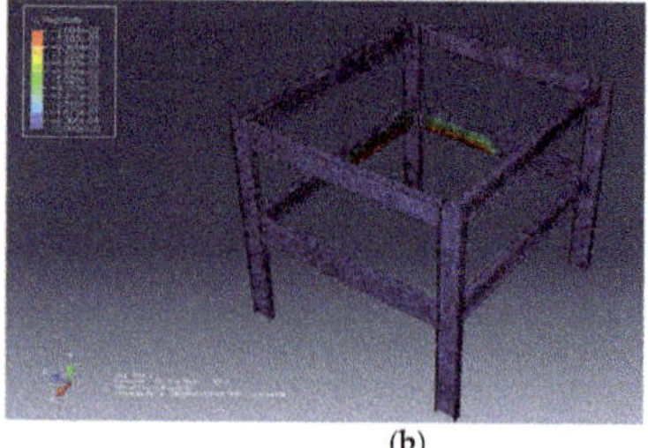

(a) (b)

Figure 15. Effect of flange thickness. (**a**) The nodal temperature corresponding to flange thickness of 25.8 mm. (**b**) The displacement corresponding to flange thickness of 25.8 mm.

4.4. Summary

From the parametric temperature-time curve Equation (1), it is clear to see that the opening factor plays an important role when it comes to determining the spread time of the fire along with the gas temperature increase in the fire compartment. Therefore, when looking at the gas temperature curves obtained from the different opening factors, it can be observed that the larger the opening factor is, the higher the gas temperature will be. The highest opening factor reaches the highest temperature in the shortest time. In fact, the highest gas temperature reached was by the largest opening factor, around 1300 °C. This is almost twice as much as the maximum gas temperature reached by the smallest opening factor, which was around 800 °C. However, the maximum specified limit of opening factor by Eurocode is 0.2.

From the ABAQUS model, it can be also seen that with different opening factors, the regions with the highest nodal temperatures were the upper flanges of both the beams and the inner web of the column, which was expected due to the fact that the load in both analyses was applied in those regions. The highest observed nodal temperature was found to be 1347 °C, corresponding to the opening factor of 0.1961; similar results were obtained for the opening factor of 0.1689, where it was around 1343 °C. On the other hand, the lowest observed temperature was found when the opening factor was 0.0251, where the temperature only reached 838.6 °C. When compared to the recorded temperatures from the Cardington tests [5], the temperatures from the numerical analysis are higher. In fact, the maximum nodal temperatures observed here are almost the same as the gas temperatures reached. This is mainly due to the fact that the steel was designed to be unprotected and there were no slabs modelled. Thus, the upper flange of the beam has a hotter temperature because it is in direct contact with the gas temperature.

The maximum displacement in the y-direction was plotted for the different opening factors as shown in Figure 16. In general, all beams started to deflect at around the same time; however, the highest deflection recorded was associated with the one of the highest opening factors (O = 0.1689), where it reached to around 0.014 m.

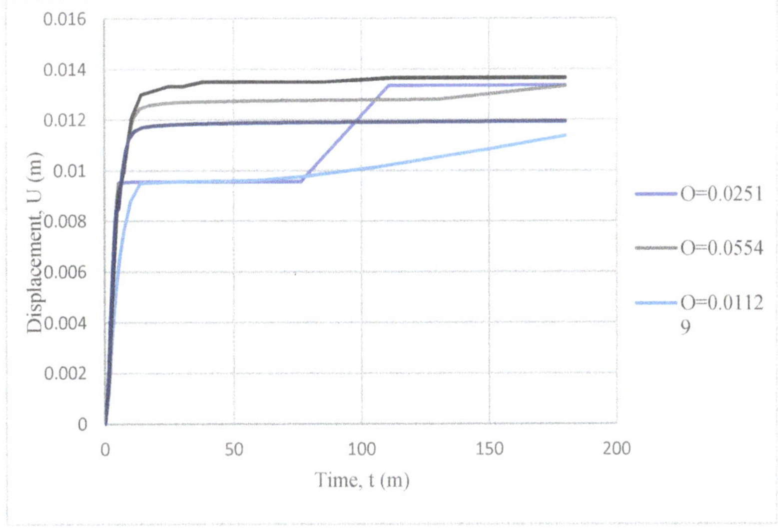

Figure 16. Beam displacement in the y-direction corresponding to the different opening factors.

The results of the displacement in the y-direction for the different compartment areas are plotted in Figure 17. The plot shows that the highest deflection recorded was for the area of 70 m^2, and it was around 0.014 m. It was also noted that in general there were around two general trends followed: one was followed by the plots corresponding to both 32.9 and 70 m^2, and the other one followed by the remaining three. However, the latter curves started to deflect quicker than the former. The reason for this could be the fact that the results extracted from the model were chosen at random, even though an average of value was taken from three different nodes to minimize errors.

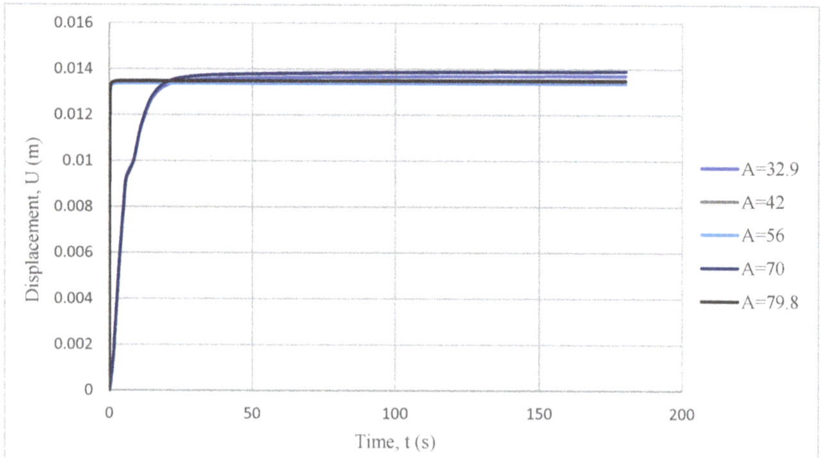

Figure 17. Beam displacement in the y-direction corresponding to the different compartment areas.

The displacement in the y-direction corresponding to the different flange thickness are plotted in Figure 18. From the figure, it can be said that the beams with the thinner flange thickness deflected quicker than the thicker flanges, which can be explained by the amount of time it takes for the heat to radiate from the top flange to the bottom one.

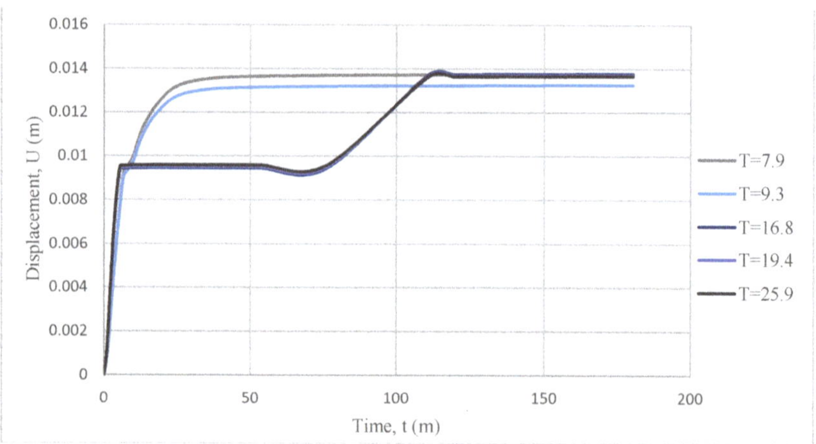

Figure 18. Beam displacement in the y-direction corresponding to the different beam flange thickness.

From all scenarios, it can be observed that the general behavior of the beam is similar, despite the different factors affecting the final result. This is not surprising because in some cases the heat load applied or the compartment area was fixed.. From the general displaced shapes for all cases, shown in Section 4, it is safe to say that the expansion co-efficient of steel had the most influence when it comes to heat load, even though it was also observed that some of the results are exaggerated in some cases, such as the nodal temperatures; this is mainly due to the absence of the slab.

As explained in [14], the tortional restraint is very important for the bare steel beams. It should be modeled in the ABAQUS model, but in reality, the beams are restrained directly by the slabs, which actually help to resist the lateral tortional buckling of the beam. We did not model the lateral restraint so as to provide a worst-case scenario for the study. The fire protection is an important factor affecting the global and local response of a structure on fire, but in this study, non-fire protection is assumed; this also provided a worst-case scenario for the study.

As explained in [15,16], the joints play an important role in the local and global behavior of the structure, as the moment and rotation capacity of the joints will affect the response of the structure. Therefore, the most accurate way to assess the joints is to physically model the bolts and endplate. Attempts have been made by the authors, but for a 3D frame system, the modeling of the bolted connection causes a convergence problem in the model, due to the complexity of the frame. Therefore, the beams are directly tied to the columns to replicate a perfect rigid connection, to reduce the computational cost while remaining reasonably accurate.

A further study will be performed by the authors to address the above limitations.

5. Conclusions

The aim of this project was to apply a parametric study of a steel frame in fire through sequential thermal and mechanical analyses. A two-story steel frame structure was simulated to investigate the influence of the opening factor, compartment area, and beam flange thickness using a Monte Carlo simulation to generate the random parameters. The following conclusions can be made:

1. The opening factor was found to be the parameter with the most influence on the rate of spread and temperature increase in the fire compartment.
2. The flange thickness also has significant influence on the response of structural members, due to its influence on the section factors.
3. The compartment area has less of an effect on the response of the structural member

Author Contributions: Conceptualization, F.F. Investigation, R.A. and F.F.; Methodology, F.F.; Supervision, F.F.; original draft, R.A.; Writing—review & editing, F.F. All authors have read and agreed to the published version of the manuscript.

Funding: This research received no external funding.

Data Availability Statement: Not applicable.

Conflicts of Interest: The authors declare no conflict of interest.

Nomenclature

A_{com}	Area of Compartment
A_t	Area of total internal surface of fire compartment
A_v	Area of ventilation
B	Width of Section
b	Thermal Diffusivity
C	Specific Heat
Ca	Specific Heat of Steel
D	Depth of Section
heq	Height of Openings
O	Opening Factor
T	Thickness of Flange
t	Time
U	Displacement
α	Expansion Co-efficient
ε_p	Plastic Strain
θ_a	Steel's Temperature
θg	Gas Temperature in the Fire Compartment
λ	Thermal Conductivity
λ_a	Thermal Conductivity of Steel
ρ	Density
ρ_a	Density of Steel

σ	Stephan-Boltzmann Constant
σ_y	Yield Stress
ν	Poisson's Ratio

References

1. Fitzgerald, R.W. *Structural Integrity During Fire*; National Fire Protection Association: Quincy, MA, USA, 1997.
2. Bailey, C. Prescriptive Design Methods in Fire. 2004. Available online: http://www.mace.manchester.ac.uk/project/research/structures/strucfire/Design/prescriptive/default.htm (accessed on 5 January 2018).
3. Bailey, C. Zone Models. 2004. Available online: http://www.mace.manchester.ac.uk/project/research/structures/strucfire/Design/performance/fireModelling/zoneModels/default.htm?p=28\T1\textbar{}#28 (accessed on 5 January 2018).
4. Fu, F. *Advanced Modelling Techniques in Structural Design*; John Wiley & Sons, Ltd.: Chichester, UK, 2015.
5. Fu, F. *Structural Analysis and Design to Prevent Disproportionate Collapse*; CRC Press: Boca Raton, FL, USA, 2016.
6. Fu, F. *Fire Safety Design for Tall Buildings*; Taylor Francis: Boca Raton, FL, USA, 2021; ISBN 978-0-367-44452-5.
7. Fu, F. *Design and Analysis of Tall and Complex Structures*; Butterworth-Heinemann: Oxford, UK, 2018; ISBN 978-0-08-101121-8.
8. Tavelli, S.; Rota, R.; Derudi, M. A critical comparison between CFD and zone models for the consequence analysis of fires in congested environments. *AIDIC* **2014**, *36*. [CrossRef]
9. Tagawa, H.; Miyamura, T.; Yamashita, T.; Kohiyama, M.; Ohsaki, M. Detailed finite element analysis of full-scale four-storey steel frame structure subjected to consecutive ground motions. *Jpn. Int. J. High-Rise Build.* **2015**, *4*, 65–73.
10. Wong, J. *Reliability of Structural Fire Design*; University of Canterbury: Christchurch, New Zealand, 1999.
11. MathWorks Perform Sensitivity Analysis through Random Parameter. 2018. Variation. Available online: https://www.mathworks.com/discovery/monte-carlo-simulation.html (accessed on 28 March 2018).
12. *EN 1991-1-2 (2005b)*; Eurocode 1. Actions on Structures, Part 1-2: General Actions—Actions on Structures Exposed to Fire. Commission of the European Communities: Brussels, Belgium, 2005.
13. *EN 1993-1-2 (2005a)*; Eurocode 3. Design of Steel Structures, Part 1-2; General Rules. Structural Fire Design. Commission of the European Communities: Brussels, Belgium, 2005.
14. Tartaglia, R.; D'Aniello, M.; Wald, F. Behaviour of seismically damaged extended stiffened end-plate joints at elevated temperature. *Eng. Struct.* **2021**, *247*, 113193. [CrossRef]
15. Qiang, X.; Bijlaard, F.S.K.; Kolstein, H.; Jiang, X. Behaviour of beam-to-column high strength steel endplate connections under fire conditions—Part 2: Numerical study. *Eng. Struct.* **2014**, *64*, 39–51. [CrossRef]
16. Shakil, S.; Lu, W.; Puttonen, J. Response of high-strength steel beam and single-storey frame in fire: Numerical simulation. *J. Constr. Steel Res.* **2018**, *148*, 551–561. [CrossRef]

fire

Numerical and Experimental Analysis of Fire Resistance for Steel Structures of Ships and Offshore Platforms

Marina Gravit and Daria Shabunina *

Peter the Great St. Petersburg Polytechnic University, 195251 St. Petersburg, Russia; marina.gravit@mail.ru
* Correspondence: shabunina.de@edu.spbstu.ru

Abstract: The requirements for the fire resistance of steel structures of oil and gas facilities for transportation and production of hydrocarbons are considered (structures of tankers and offshore platforms). It is found that the requirements for the values of fire resistance of structures under hydrocarbon rather than standard fire conditions are given only for offshore stationary platforms. Experimental studies on the loss of integrity (E) and thermal insulating capacity (I) of steel bulkheads and deck with mineral wool under standard and hydrocarbon fire regimes are presented. Simulation of structure heating was performed, which showed a good correlation with the experimental results (convective heat transfer coefficients for bulkheads of class H: 50 W/m²·K; for bulkheads of class A: 25 W/m²·K). The consumption of mineral slabs and endothermic mat for the H-0 bulkhead is predicted. It is calculated that under a standard fire regime, mineral wool with a density of 80–100 kg/m² and a thickness of 40 to 85 mm should be used; under a hydrocarbon fire regime, mineral wool with a density above 100 kg/m² and a thickness of 60–150 mm is required. It is shown that to protect the structures of decks and bulkheads in a hydrocarbon fire regime, it is necessary to use 30–40% more thermal insulation and apply the highest density of fire-retardant material compared to the standard fire regime. Parameters of thermal conductivity and heat capacity of the applied flame retardant in the temperature range from 0 to 1000 °C were clarified.

Keywords: oil and gas facility; offshore platform; tanker; steel structure; bulkhead; deck; hydrocarbon fire mode; fire-resistance limit; fire protection

Citation: Gravit, M.; Shabunina, D. Numerical and Experimental Analysis of Fire Resistance for Steel Structures of Ships and Offshore Platforms. *Fire* **2022**, *5*, 9. https://doi.org/10.3390/fire5010009

Academic Editor: Maged A. Youssef

Received: 19 December 2021
Accepted: 14 January 2022
Published: 16 January 2022

Publisher's Note: MDPI stays neutral with regard to jurisdictional claims in published maps and institutional affiliations.

1. Introduction

Building structures of reservoirs, equipment and structures in an accident, accompanied by fire and explosion, are subjected to high-temperature impact due to the large number and type of fire load [1,2]. In Europe and the USA, combustion of hydrocarbons (oil, oil products) and the development of fire are considered on the hydrocarbon fire curve, at which, in the first minutes of the fire, the temperature reaches 1000 °C and higher [3,4]. In the design of structures of the oil and gas complex (O&G) in Russia, the condition of fire development on the standard ("cellulose") curve according to ISO 834 [5] is used.

Tankers are in second place in the total transportation volume of oil and petroleum products (after oil pipes). The highest risk of formation of explosive mixtures inside the tanker occurs during tanker unloading. When the liquid level drops, the air is exhausted into the tank and mixed with petroleum product vapors [6]. As petroleum vapors are heavier than air, they can spread through tanker rooms and ignite over large areas. Ships and offshore platforms consist of decks, compartments and interior spaces that contain several systems, subsystems and components necessary for operation. Explosion, fire or flooding of compartments can damage equipment and cause a critical risk to operations [7–9]. In [10], an empirical method was used to calculate the compressive strength limit in the center of the deck, according to the results of which, the maximum compressive stress on the deck was 175.53 MPa; the deflection value in the middle part of the deck did not exceed the acceptable value. In [11], the design of a working barge with a displacement of 5000 tons was demonstrated. Mechanical calculation showed sufficient strength under normal loading conditions and even in an emergency. In [12], the steel deck's behavior under different hydrocarbon ignition scenarios using ANSYS software was studied. Numerical studies of steel decks under the combined action of mechanical load and hydrocarbon fire regime are given, showing an increased deformation of the deck and reduced deck fire resistance under the considered fire scenarios. In [13,14], a fire was simulated using FDS structures of offshore platforms, and the fire risk was calculated. The authors investigated the behavior of steel structures of the upper part of an offshore platform under fire and hydrocarbon explosion and under wind load; the calculation was performed in ABAQUS software [15]. The thermophysical characteristics of the intumescent paints used as fire protection of steel structures were obtained in [16].

Steel structures in the ship's hull and structures of cargo tanks, decks and bulkheads that separate industrial rooms are designed with certain fire-resistance classes, depending on the parameters of the fire-resistance limits and temperature exposure modes: A, B, C and H (standard regime—A, B, C classes, and hydrocarbon—H class). The same fire-resistance classes are established for oil platforms [17]. In [18], a simulation of the thermal impact on the steel structure A-60 was presented, from the results of which the temperature distribution was calculated. The analysis results allow consideration of the design and safety planning aspects of an offshore living compartment.

According to SOLAS Regulation II-2/17 [19], decks and bulkheads shall be made of non-combustible materials and are classified as follows:

(1) "B" class divisions: B-15 and B-0;
(2) "A" class divisions: A-60, A-30, A-15 and A-0;
(3) "C" class divisions: divisions constructed of approved non-combustible materials.

Another classification of decks and bulkheads is also regulated in [17]:

(4) "H" class divisions: H-120, H-60 and H-0.

Figure 1 shows the location of the H-120 deck and A-60 and B-15 class bulkheads on the tanker.

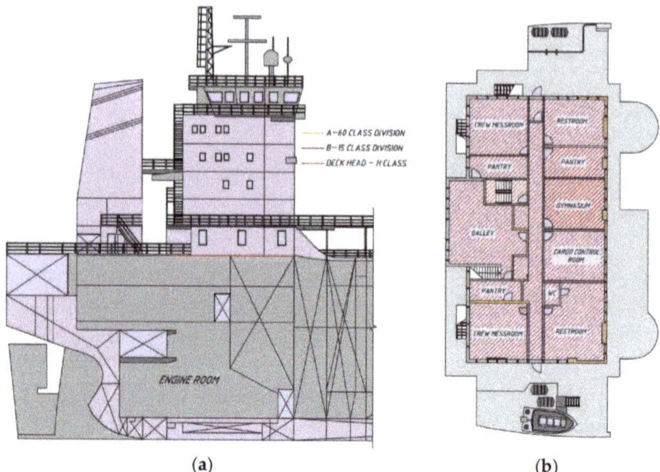

(a)   (b)

Figure 1. (**a**) Tanker with H-120 deck location. (**b**) Fragment of the section of the first deck with the arrangement of the bulkheads.

Figure 2 shows the location of the H-120 deck and A-60 and H-120 class bulkheads on an offshore platform.

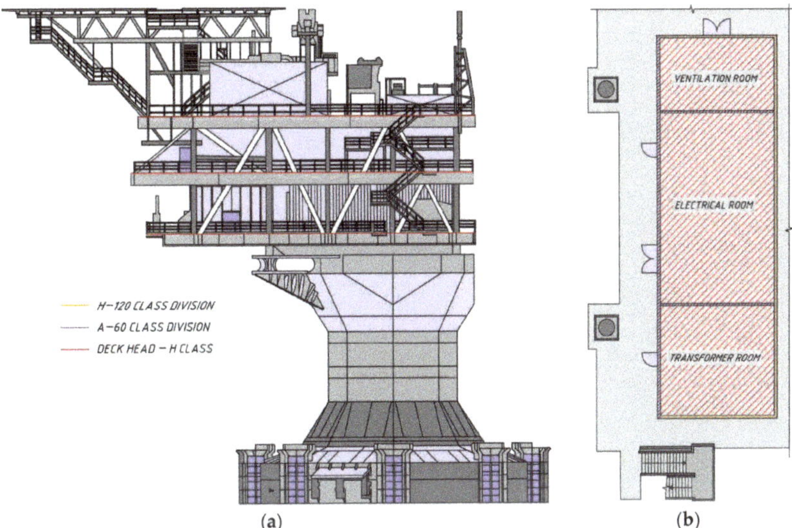

Figure 2. (**a**) Offshore platform with deck location. (**b**) Fragment of the section of the first deck with the arrangement of the bulkheads.

Fire-resistance tests of structures for ships and offshore structures are conducted following the requirements stated in SOLAS Regulation II-2/17 [19], International Maritime Organization (IMO) resolutions and guidelines of IMO member countries, for example, American Bureau of Shipping (ABS) [20] and Russian Maritime Register of Shipping (RS) [21]. Tests for fire resistance are carried out using both methods for determining the fire resistance of structures by the standard temperature regime (curves for A, B, C), which is similar to that established in ISO 834 [5], and by the hydrocarbon fire regime (curve H) for island structures and floating platforms. In the USA, the standard UL 1709 [22] is applied, which differs from the European EN 1363-2:1999 [23] in the development of a fire in the first minutes [24,25].

According to ISO 834-75 [5], IMO Res. A.754 [26] and the Russian State Standard GOST 30247.1 "Elements of building constructions. Fire-resistance test methods. Loadbearing and separating constructions" [27] harmonized with ISO 834 [5], the following limit conditions are distinguished for fire-resistance limits of enclosure structures, which include bulkheads and decks of tankers and platforms:

— Loss of integrity resulting from the formation of through cracks or openings in the structures through which combustion products or flame (E) penetrate to the unheated surface;
— Loss of thermal insulating capability (I) due to an average temperature rise of more than 140 °C at the unheated surface of the structure or at any point on that surface of more than 180 °C compared with the temperature of the structure before the test or more than 220 °C regardless of the temperature of the structure before the test (additional limit conditions of structures and the criteria of their occurrence are established, if necessary, in the standards for tests of particular structures). It should be noted that the same requirements for the quantitative values of the thermal insulating capacity (I) and qualitative features of the loss of integrity (E) are established for the enclosing structures.

Minimum requirements for fire resistance of bulkheads and decks are established in [17,20,28], for example, for bulkheads in [17] (Table 1).

Table 1. Fire integrity of bulkheads separating adjacent spaces/areas.

Spaces		(1)	(2)	(4)	(4)	(5)	(6)	(7)	(8)	(9)	(10)	(11)	(12)
Control stations including central process control rooms	(1)	A-0	A-0	A-60	A-0	A-15	A-60	A-15	H-60	A-60	A-60	*	A-0
Corridors	(2)		C	B-0	B-0 A-0	B-0	A-60	A-0	H-60	A-0	A-0	*	B-0
Accommodation spaces	(3)			C	B-0 A-0	B-0	A-60	A-0	H-60	A-0	A-0	*	C
Stairways	(4)				B-0 A-0	B-0 A-0	A-60	A-0	H-60	A-0	A-0	*	B-0 A-0
Service spaces (low risk)	(5)					C	A-60	A-0	H-60	A-0	A-0	*	B-0
Machinery spaces for category A	(6)						*	A-0	H-60	A-60	A-60	*	A-0
Other machinery spaces	(7)							A-0	H-0	A-0	A-0	*	A-0
Process areas, storage tank areas, wellhead/manifold areas	(8)				(Symmetrical)				–	H-60	H-60	*	H-60
Hazardous areas	(9)									–	A-0	*	A-0
Service spaces (high risk)	(10)										A-0	*	A-0
Open decks	(11)											–	*
Sanitary and similar spaces	(12)												C

Note: * The division is to be of steel or equivalent material, but is not required to be of an A-class standard.

Insulation materials should generally be non-combustible or show low combustion spreading to ensure structural fire resistance of ships and platforms [29–31]. Mineral wool of various densities is widely used in passive fire protection (PFP) [32,33] and less commonly used in epoxy-based fire-retardant intumescent paints [4]. Fire protection is applied (mounted) between thin metal walls as bulkhead panels on vertical structural elements of offshore structures. Studies related to the design, calculation and modeling of decks and bulkheads include either only calculations of the compressive strength and deflection values at the center of the structure [10,11] or only modeling of hydrocarbon fire and explosion scenarios [12–15,18]. In [34], two experiments of bulkheads under standard and hydrocarbon fire regimes are given, with their subsequent modeling confirming the correlation of the obtained temperatures, from which the conclusion about the possibility of prediction and justification of the fire-resistance limits by simulation is made.

The purpose of this article is to simulate experimental data for determining the fire-resistance limit of bulkheads of different classes and deck for an offshore platform to solve the following problems: calculation of the parameters of thermal insulation of bulkheads and deck; prediction of the fire-resistance limits of the structure on the example of the H-0 bulkhead depending on the thickness of mineral wool and its density for the H-0 bulkhead under a hydrocarbon fire regime with the variant to replace the used fire protection to endothermic mat based on ceramics and basalt fibers; calculation of the H-0 bulkhead on a deflection in the center of the considered structure under thermal load; and clarification of calculated coefficients of thermal conductivity and heat capacity for mineral wool in the temperature range from 0 to 1000 °C.

2. Materials and Methods

Experimental samples of H-class bulkheads and deck were tested to determine the time of reaching the limit state during fire exposure according to IMO FTP Code Part 3 IMO Res. A.754 (18) [26] under the condition of establishing a hydrocarbon temperature regime in the fire chamber of the furnace according to EN 1363-2: 1999 [23], characterized by dependence (1):

$$T - T_0 = 1080 \times \left(1 - 0.325 \times e^{-0.167t} - 0.675 \times e^{-2.5t}\right) \tag{1}$$

where T means the temperature inside the furnace in °C, corresponding to the relevant time t; T_0 is the temperature in °C inside the furnace prior to the start of heat impact; t is the time in minutes from the start of the test.

Experimental samples of A-class bulkheads were tested to determine the time of reaching the limit state during fire exposure according to IMO FTP Code Part 3 IMO Res. A.754 (18) [26] under the condition of creating in the fire chamber of the furnace a standard temperature regime according to ISO 834 [5], characterized by dependence (2):

$$T - T_0 = 345 \times \lg(8t + 1) \tag{2}$$

The furnace temperature was determined by means of twelve thermoelectric transducers with a switching head uniformly distributed at a distance of approximately 100 mm from the exposed side of the test sample according to IMO Res. A.754 (18) [26]. The temperature in the furnace during the fire tests was maintained according to the appropriate temperature regimes [23]. The temperature on the test samples was measured by cable thermoelectric chromel–alumel thermocouples. According to the test reports of the structure, each thermocouple is inserted through a steel pipe of standard weight, and the end of the pipe from which the welded junction protrudes is to be open. The thermocouple junction protrudes $\frac{1}{2}$ in (12.7 mm) from the open end of the pipe.

The ambient temperature during the tests was averaged according to the test reports and assumed 20 °C.

The software package (SP) ELCUT [35] was used to analyze the temperature distribution over the cross-section of the considered structures.

2.1. Experiments on Bulkhead and Deck Structures

The fire resistance of H-class bulkheads (H-0, H-60, H-120), A-class bulkheads (A-15, A-60) and deck (H-120) with mineral wool materials was investigated.

Fire tests of the H-0, H-60 and H-120 bulkheads were carried out at the Danish Institute of Fire and Security Technology (DIFT). The bulkheads were installed in a reinforced concrete frame and welded on four sides to the restraint frame. The dimensions of the structural core were following IMO Resolution A.754 (18). The test samples were tested with the insulation and the stiffeners toward the furnace.

The H-0 bulkhead has the following external dimensions: height 2480 mm, width 2420 mm, thickness 64.5/129.5 mm. The bulkhead consisted of a standard structural steel core insulated with Rockwool insulation (Hedehusene, Denmark), attached to the bulkhead with ø3 mm pins and ø28 mm washers. The pins on the level were located in 3 lines and a line on each of the stiffeners. The vertical center distance between the pins on the level was 400 mm along all lines. The vertical center distance between the pins on the stiffeners was 300 mm along all lines. The steel sheet thickness of 4.5 mm with the pins on the stiffeners at a distance of 600 mm was insulated with two layers of 30 mm Rockwool HC Firebatt (Hedehusene, Denmark) mineral wool with a density of 150 kg/m^3.

The H-60 bulkhead has the following external dimensions: height 2480 mm, width 2420 mm, thickness 75/115 mm. The steel sheet thickness of 5 mm with the pins on the stiffeners at a distance of 600 mm was insulated with two layers of mineral wool: 40 mm Rockwool HC Wired Matt (Hedehusene, Denmark) with a density of 150 kg/m^3 and 30 mm Rockwool HC Firebatt with a density of 150 kg/m^3. Installation of the insulation to the bulkhead is similar to the H-0 bulkhead.

The H-120 bulkhead has the following external dimensions: height 2480 mm, width 2420 mm, thickness 95/155 mm. The steel sheet thickness of 5 mm with the pins on the stiffeners at a distance of 600 mm was insulated with two layers of mineral wool: 40 mm Rockwool HC Wired Matt with density of 150 kg/m^3 and 50 mm Rockwool HC Firebatt with a density of 150 kg/m^3. Installation of the insulation to the bulkhead is similar to the H-0 bulkhead.

The temperature on the test samples was determined by cable thermoelectric chromel–alumel thermocouples designed as described in IMO Resolution A.754 (18) [26] and mounted on the unheated surfaces of the sample (Figure 3). The location of the thermocouples for the H-60 bulkhead is the same as on the H-120 bulkhead.

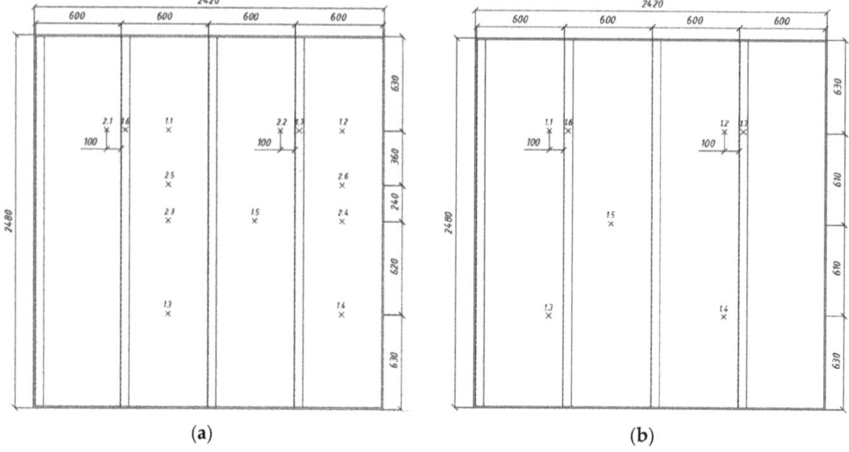

(a)　　　　　　　　　(b)

Figure 3. (a) Location of thermocouples on bulkhead H-120. (b) Location of thermocouples on bulkhead H-0.

The A-15 and A-60 (sample No. 1 and sample No. 2) bulkheads were built following IMO Resolution A.754 (18) [26] and insulated on the stiffened side not exposed to the fire. The mineral wool panels are secured to the bulkhead plate through steel pins and washers welded with a pitch of 300 mm.

Fire tests of A-15 and A-60 (sample No. 1) class bulkheads were performed at RINA Services Spa (Genoa, Italy); fire test of A-60 (sample No. 2) class bulkhead was performed at FGBU VNIIPO EMERCOM of Russia (Balashikha, Moscow region, Russia). The bulkheads were tested in the vertical position exposing to the fire the uninsulated bulkhead side, mounted within a steel restraint frame having a refractory concrete lining 50 mm thick. The temperature on the test samples was measured by cable thermoelectric chromel–alumel thermocouples, installed in the amount of 7 pieces on the unheated surfaces of the sample.

The A-15 bulkhead has the following external dimensions: height 3020 mm, width 3020 mm, thickness 44.5/69.5 mm. The steel sheet thickness of 4.5 mm with the pins on the stiffeners at a distance of 600 mm was insulated with one layer of mineral wool: 40 mm PAROC Marine Fire Slab (Helsinki, Finland) with a density of 80 kg/m^3.

The A-60 bulkhead (sample No. 1) has the following external dimensions: height 3020 mm, width 2420 mm, thickness 65/90 mm. The steel sheet thickness of 5 mm with the pins on the stiffeners at a distance of 600 mm was insulated with two layers of mineral wool: 60 mm and 25 mm PAROC Fire Slab (Helsinki, Finland) with a density of 100 kg/m^3.

The A-60 bulkhead (sample No. 2) has the following external dimensions: height 2480 mm, width 2420 mm, thickness 54.5/79.5 mm. The steel sheet thickness of 4.5 mm with the pins on the stiffeners at a distance of 600 mm was insulated with two layers of 25 mm TIZOL-FLOT Fire (Yekaterinburg, Russia) with a density of 100 kg/m^3.

Fire tests of the steel deck H-120 were carried out at the Fire Safety Scientific and Test Center of the FGBU VNIIPO EMERCOM of Russia (Balashikha, Moscow region, Russia). The temperature on the sample was measured by thermocouples, installed in the amount of 7 pieces on the unheated surface of the sample.

The H-120 deck has the following external dimensions: height 2440 mm, width 3040 mm, thickness 126/246 mm. The steel sheet thickness of 6 mm with the pins on the stiffeners at a distance of 600 mm was insulated with two layers of 60 mm Rockwool mineral wool panels with a density of 100 kg/m^3.

2.2. Simulation of Bulkhead and Deck Section Heating

SP ELCUT allows solving tasks related to the heating of structures [36]. All calculations of the structures are performed by the finite element method based on the two-dimensional finite element model in the ELCUT software. To solve the task, it is necessary to specify the geometry, describe the properties of the medium and define the boundary conditions. The input of the task parameters consists of marks divided into three groups [35]:

Block marks that describe the material properties in the model;

Rib marks describing the boundary conditions on the outer and inner surfaces of the model;

Vertex marks that describe the anchoring conditions (boundary conditions) applied to certain points in the model.

In the simulation of heating, the thermal conductivity equation is used in the flat case (3) [37]:

$$\frac{\partial}{\partial x}\left(\lambda_x \frac{\partial T}{\partial x}\right) + \frac{\partial}{\partial y}\left(\lambda_y \frac{\partial T}{\partial y}\right) = -q - c\rho \times \frac{\partial T}{\partial t} \tag{3}$$

where T is the temperature in °C; t is the time in seconds; λ means the components of the thermal conductivity tensor in W/(m·K); q is the specific power of heat source in W/m^3; c is the specific heat capacity in J/(kg·K); and ρ is the density in kg/m^3.

A number of boundary conditions, such as temperature, heat flow, convection and radiation, are set at the outer and inner boundaries of the computational domain. The value of T_0 is given as a linear function of coordinates. The heat flow is described by the following Relations (4) and (5) [35]:

$$F_n = -q_s \text{—on the outer borders} \tag{4}$$

$$F_n^+ - F_n^- = -q_s \text{—on the inner borders} \tag{5}$$

where F_n is the normal component of the density vector of heat flow, where "+" and "−" mean "left of the border" and "right of the border," respectively, in W/m^2; q_s is the power surface of the source for the inner border, for the outer, the known value of heat flow through the border in W/m^2.

Convective heat transfer is determined according to (6) [38]:

$$F_n = \alpha \times (T - T_0) \tag{6}$$

where α is the convective heat transfer coefficient in W/m^2·K; T_0 is the ambient temperature in K.

The radiation condition is set at the outer border of the model; the radiation heat transfer is determined according to (7) [35]:

$$F_n = k_{SB} \times \beta \times \left(T^4 - T_0^4\right) \tag{7}$$

where k_{SB} is the Stefan–Boltzmann constant in W/(m$^2 \cdot$ K^4); β is the surface absorption coefficient; and T_0 is the temperature of an absorbing medium in K.

Simulations were performed for bulkheads and deck under hydrocarbon and standard fire regimes.

Initial steel characteristics: steel grade D36 [39]; density 7800 kg/m^3; thermal conductivity and heat capacity are variable depending on temperature (values are taken from the program reference book). The boundary conditions are presented in Table 2.

Table 2. Boundary conditions set in the SP ELCUT.

Name of the Value	Value	Information Source
Degree of blackness of ship's alloy steel	0.35	[40]
Degree of blackness of mineral wool	0.92	[40]
Degree of blackness of endothermic mat	0.96	[40]
Convection heat transfer coefficient at standard temperature regime, W/(m$^2 \cdot$ K)	25	[38]
Convection heat transfer coefficient at hydrocarbon temperature regime, W/(m$^2 \cdot$ K)	50	[38]
Surface absorption coefficient	0.5	[37]
The emissivity of steel	0.8	[37]
The emissivity of mineral wool	0.7	[37]
Initial ambient temperature, °C *	20	-
Time step for calculating the temperature gradient of the structure, second	60	-

Note: * According to the test reports, the temperature measured by thermocouples in the furnace was determined as an absolute value, and the temperature on the unheated surface was recorded and displayed as the difference between the ambient temperature and the temperature on the unheated surface.

The characteristics of the mineral wool for the different bulkheads and deck are shown in Table 3. It is assumed that the density value does not change during heating. The value of heat capacity is assumed to be averaged for all types of mineral wool according to manufacturer's website and [41]; the trend of heat capacity change with temperature is assumed according to [42]. Moreover, the main influence on the heat transfer in the solid material layer has thermal conductivity [37].

Table 3. The main characteristics of mineral wool for structures.

Structure/ Manufacturer	ρ, KT/M^3	λ, W/(m·K)			C_p, J/(kg·K)			φ, %	Organic Substances, %	Thickness of Plates, mm
		10 °C	100 °C	300 °C	10 °C	100 °C	300 °C			
H-0 (Rockwool)	150	0.034	0.045	0.078	840	860	900	0.24	1.30	60/125
H-60 (Rockwool)	150	0.034	0.045	0.078	840	860	900	0.20	0.40	70/110
H-120 (Rockwool)	150	0.034	0.045	0.078	840	860	900	0.20	0.40	90/150
A-15 (PAROC)	80	0.037	0.047	0.095	840	860	900	0.34	1.50	40/65
A-60 (PAROC) *	100	0.037	0.047	0.095	840	860	900	0.28	3.09	60/85
A-60 (TIZOL) **	100	0.035	0.046	0.085	840	860	900	0.25	2.00	50/75
H-120 (Rockwool)	100	0.034	0.045	0.078	840	860	900	0.20	0.40	120/240

Note: * Sample No. 1; ** Sample No. 2.

3. Results and Discussion

3.1. Experimental and Simulation Results

For the H-0 bulkhead, the fire test was stopped at 122 min according to the requirements for this class of bulkheads (the limit condition for H-0 bulkheads is loss of integrity (E)). When the required time was reached, no smoke and flame penetration to the unheated side was observed, the integrity of the sample was preserved, and the deflection of the sample in the center of the bulkhead (60 mm) and the changing of the color of the open surface to yellow were recorded. According to the test results, it was found that the H-0 bulkhead with a steel sheet thickness of 4.5 mm, insulated with mineral wool with a thickness of 60/125 mm and a density of 150 kg/m^3, has fire resistance under the action of a hydrocarbon fire regime for at least 30 min before reaching the parameter of thermal insulating capacity (I) and at least 120 min before reaching the parameter of loss of integrity due to the temperature increase on the unheated surface of the structure on average more than 140 °C. According to the DIFT report, the H-0 bulkhead may also be classified as H-30 regarding the experimental data obtained.

For the H-60 bulkhead, the fire test was stopped at 123 min when the critical temperature on the unheated surface of the structure reached an average of more than 140 °C (the technical customer's request extended the test), no smoke and flame penetration on the unheated side was observed, the integrity of the sample was preserved, and the deflection of the sample in the center of the bulkhead (42 mm) and the changing of the color of the open surface to yellow were recorded. According to the test results, it was found that the H-60 bulkhead with a steel sheet thickness of 5 mm, insulated with mineral wool with a thickness of 70/110 mm and a density of 150 kg/m^3, has fire resistance under the action of a hydrocarbon fire regime for at least 120 min.

For the H-120 bulkhead, the fire test was stopped at 125 min when the critical temperature on the unheated surface of the structure reached an average of more than 140 °C (the technical customer's request extended the test), no smoke and flame penetration on the unheated side was observed, the integrity of the sample was preserved, and the deflection of the sample in the center of the bulkhead (24 mm) and the changing of the color of the open surface to yellow were recorded. According to the test results, it was found that the H-120 bulkhead with a steel sheet thickness of 5 mm, insulated with mineral wool with a thickness of 90/150 mm and a density of 150 kg/m^3, has fire resistance under the action of a hydrocarbon fire regime for at least 120 min.

The H-class bulkheads' appearance before and after the fire test did not change, and the deflection at the center of the bulkheads did not reach the limit value of 1/20 in each test [27]. For example, the heated and unheated sides before and after the fire test of H-120 bulkheads (Figures 4 and 5) and mineral wool after the fire test of H-120 and H-0 bulkheads (Figure 6) are shown.

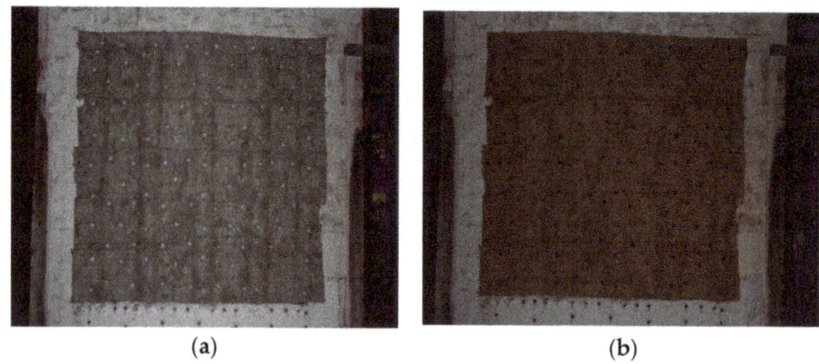

(a)   (b)

Figure 4. (**a**) Heated side of H-120 bulkhead before fire test. (**b**) Heated side of H-120 bulkhead after fire test.

(a)   (b)

Figure 5. (**a**) H-120 bulkhead at the beginning of fire test. (**b**) H-120 bulkhead at the end of fire test.

(a)   (b)

Figure 6. (**a**) Mineral wool after fire test of H-120 bulkhead. (**b**) Mineral wool after fire test of H-0 bulkhead.

For the A-15 bulkhead, the fire test was stopped at 30 min according to the requirements. No smoke and flame penetration on the unheated side was observed, the sample's integrity was preserved, and the deflection of the sample in the center of the bulkhead (105 mm) was recorded. No cracks and holes in the sample were found. According to the test results, it was found that the A-15 bulkhead with a steel sheet thickness of 4.5 mm, insulated with mineral wool with a thickness of 40 mm and a density of 80 kg/m^3, has fire resistance under the action of a standard fire regime for at least 15 min.

For the A-60 bulkhead (sample No. 1), the fire test was stopped at 60 min when the critical temperature on the unheated surface of the structure reached an average of more than 140 °C. No smoke and flame penetration on the unheated side was observed, the sample's integrity was preserved, and the deflection of the sample in the center of the bulkhead (70 mm) was recorded. No cracks and holes in the sample were found. According to the test results, it was found that the A-60 bulkhead (sample No. 1) with a steel sheet thickness of 5 mm, insulated with mineral wool with a thickness of 60/85 mm and a density of 100 kg/m^3, has fire resistance under the action of a standard fire regime for at least 60 min.

For the A-60 bulkhead (sample No. 2), the fire test was stopped at 60 min according to the customer's requirements. No smoke and flame penetration on the unheated side was observed, and the integrity of the sample was preserved. No cracks, holes or other visible changes on the sample were found, and the deflection value was not measured. According to the test results, it was found that the A-60 bulkhead (sample No. 2) with a steel sheet thickness of 4.5 mm, insulated with mineral wool with a thickness of 50/75 mm and with a density of 100 kg/m^3, has fire resistance under the action of a standard fire regime for at least 60 min.

Considered A-class bulkheads did not change their appearance before and after the fire test, and the deflection at the center of the bulkheads did not reach the limit value of 1/20 in each test [27]. For example, the heated and unheated sides after the fire test of the A-15 bulkhead are shown (Figure 7).

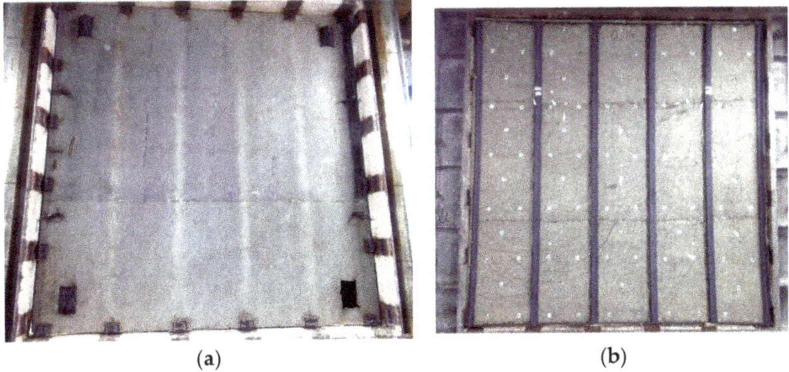

(a) (b)

Figure 7. (**a**) Heated side of A-15 bulkhead at the end of the fire test. (**b**) Unheated side of A-15 bulkhead at the end of the fire test.

For the deck H-120, the fire test was stopped at 125 min when the critical temperature on the unheated surface of the structure reached an average of more than 140 °C. No smoke and flame penetration on the unheated side was observed, and the integrity of the sample was preserved. No cracks, holes or other visible changes on the sample were found, and the deflection value was not measured (Figure 8). According to the test results, it was found that deck H-120, with a steel sheet thickness of 6 mm, insulated with mineral wool with a thickness of 120/240 mm and a density of 100 kg/m^3, has fire resistance under the action of a hydrocarbon fire regime for at least 120 min.

Figure 9 shows the time–temperature curves of the bulkheads and deck during the fire test. The graph shows the averaged values of the difference between the values of thermocouples located directly on the unheated surface of the sample and the initial ambient temperature (20 °C, Table 3).

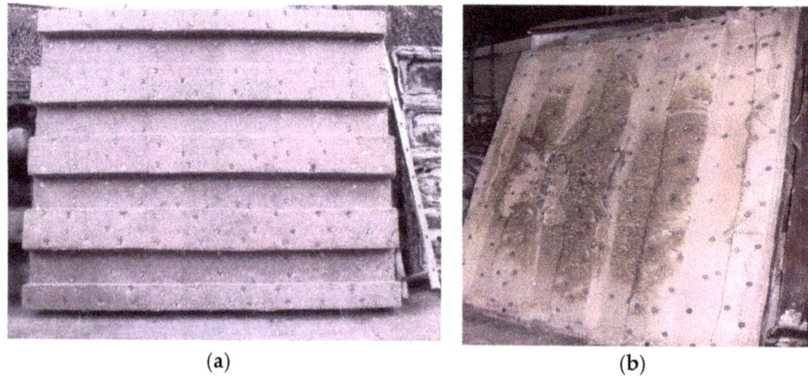

Figure 8. (**a**) Deck H-120 before the fire test. (**b**) Deck H-120 after the fire test.

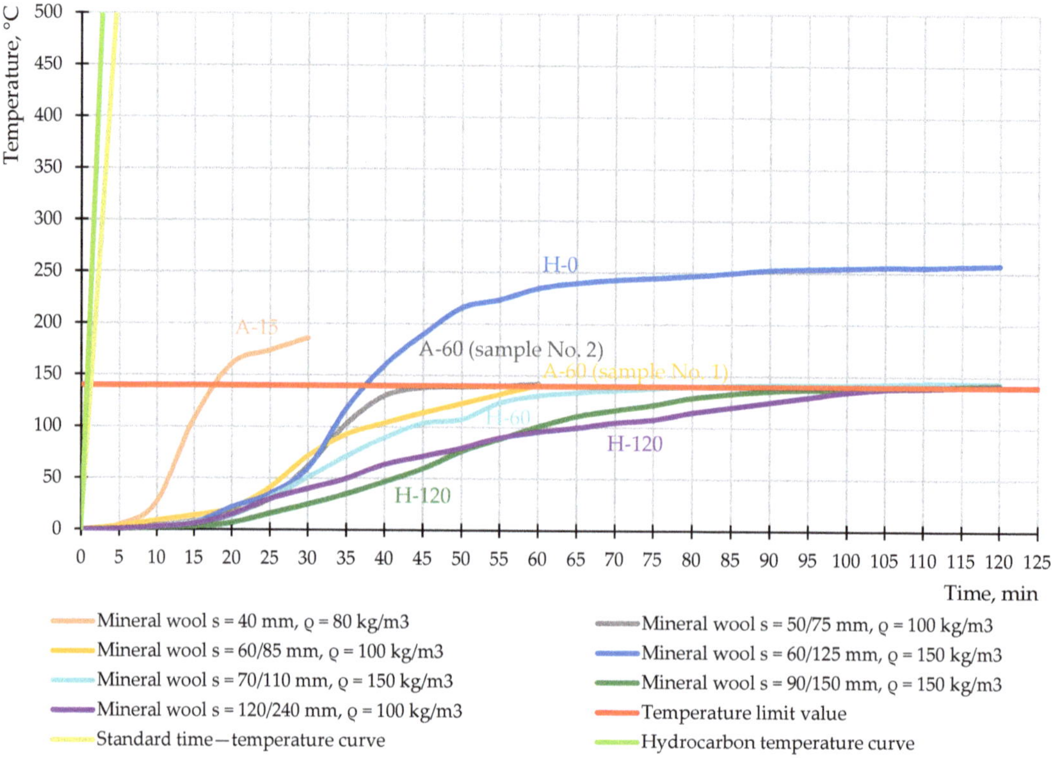

Figure 9. Temperature curves of experimental samples during fire tests.

As a result of the simulation, visualizations of the heating of the experimental bulkheads and deck were obtained (Figure 10). The location of the thermocouples on the structures is shown in the analytical model. Each analytical model represents $\frac{1}{4}$ of the structure since it consists of similar and repeating fragments. For H-class bulkheads and H-120 deck, the fire exposure was from the mineral wool side, and for A-class bulkheads, from the steel plate side.

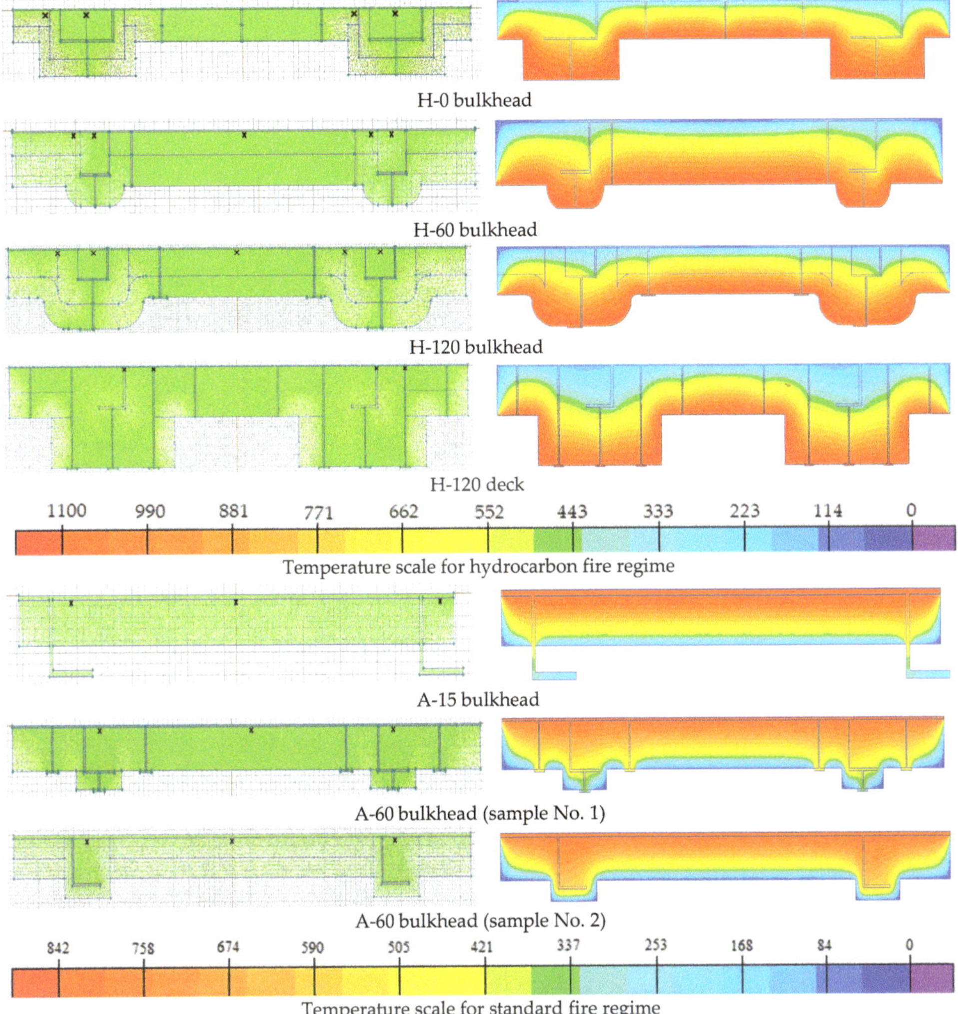

Figure 10. Analytical models of structures and thermocouple locations and visualization of the heating of bulkhead and deck structures.

The temperature–time dependences at the thermocouple location on the unheated surface were obtained for H-class bulkheads and deck H-120 (Figure 11).

The graph shows the averaged values of the difference between the values of thermocouples located directly on the unheated surface of the sample and the initial ambient temperature (20 °C, Table 3). The different location of the thermocouples over the cross-section of the samples is shown in Figure 10. Heat and mass transfer processes were not considered in the modeling.

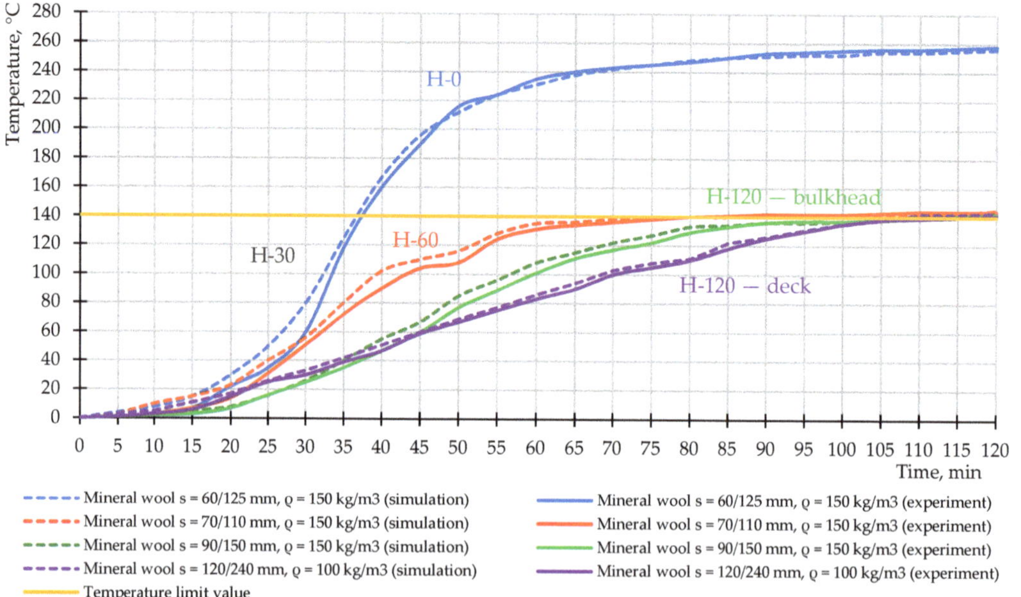

Figure 11. Experimental and simulated temperature curves of samples during the fire test under the hydrocarbon fire regime.

The results of the simulation show excellent correlation of the results (difference in values not more than 5%), except for the results for the H-0 bulkhead (25% in the range of 20 to 30 min), which has a smaller plate thickness (60 mm) compared to the other samples. Mineral wool is a dry fire retardant; however, it contains organic substances and water (Table 3). At a sharp temperature effect in the hydrocarbon fire regime in the range from 30 to 100 °C, the processes of heat and mass transfer are intensified, which may explain the excess of simulation results compared with the experimental values of temperatures.

Analysis of Figure 11 shows that the graph for the H-0 bulkhead grows more rapidly because the bulkhead warms up faster due to the fact that it has a smaller plate thickness (60/125 mm) at the same density (150 kg/m^3). The graph for the deck H-120 with lower density (100 kg/m^3) during the first 30 min shows higher temperature values compared to the H-120 bulkhead (150 kg/m^3); after 45 min, due to the higher insulation thickness (120/240 mm), the graph for the deck H-120 shows a smoother temperature increase until the limit value is reached.

The temperature–time dependences at the thermocouple location on the unheated surface were obtained for A-class bulkheads (Figure 12).

Analysis of Figure 12 shows that the graph for the A-15 bulkhead grows more rapidly because the bulkhead warms up faster due to the fact that it has a smaller plate thickness (40 mm) and a lower density (80 kg/m^3) while the A-60 bulkheads have higher values: density 100 kg/m^3 and plate thickness 60/85 mm for sample No. 1 and 50/75 mm for sample No. 2. Samples No. 1 and No. 2 for A-60 bulkheads with a density of 100 kg/m^3 have similar dynamics of temperature increase up to 30 min, then sample No. 1 continues to heat uniformly, while sample No. 2, which has a plate thickness of 10 mm less than the thickness of sample No. 1, increases sharply and reaches equilibrium state after 40 min of heating.

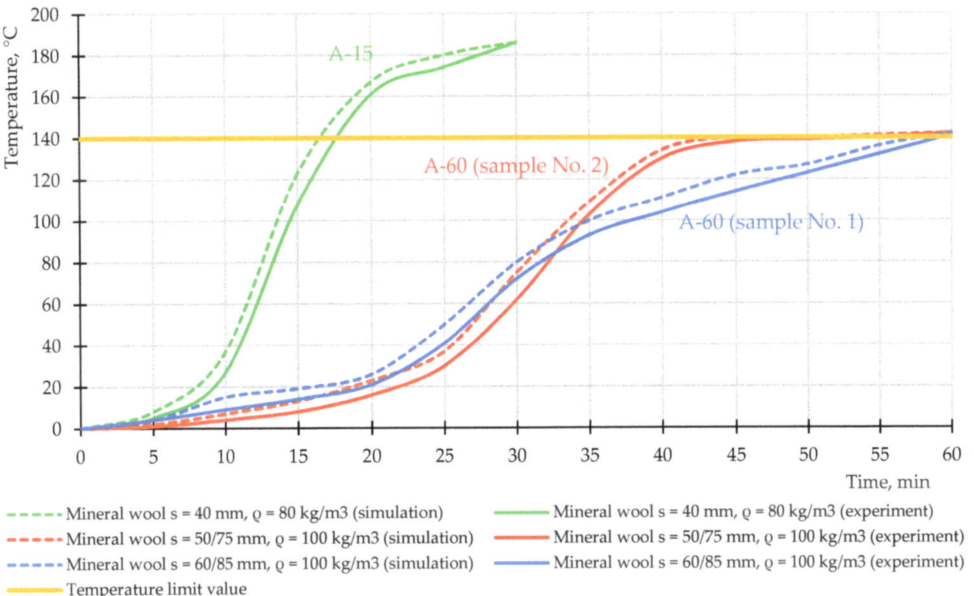

Figure 12. Experimental and simulated temperature curves of samples during the fire test under the standard fire regime.

In the example of the H-0 bulkhead, the deflection in the center of the considered structure under the thermal load was calculated in SP ELCUT using the connection of tasks of unsteady heat transfer and mechanical stresses and strains, which resulted in a deformation diagram shifted by 63 mm relative to the original position (Figure 13).

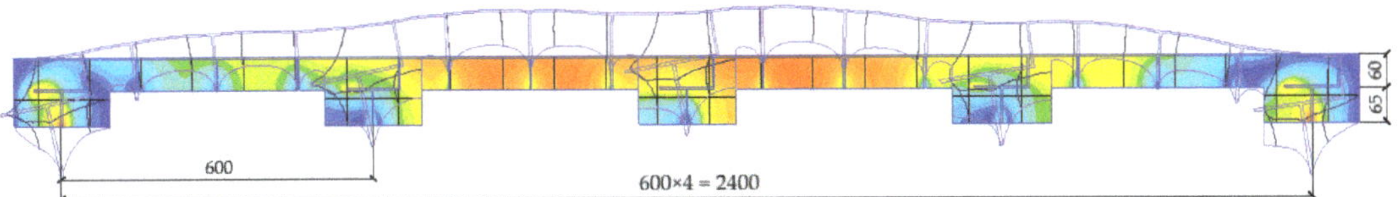

Figure 13. Deformation diagram of H-0 bulkhead under heat load.

According to [27], the limit value of deflection at the center of the bulkhead is determined according to (8):

$$\Delta = \frac{l}{20} = \frac{2480}{20} = 124\ \text{mm} \tag{8}$$

Thus, the deflection value obtained during the experiment (60 mm) and from the simulation (63 mm) does not exceed the acceptable value and confirms that the H-0 bulkhead subjected to high-temperature fire exposure maintains its integrity throughout the test.

3.2. Discussion

The calculated temperature values on the considered structures, obtained from the simulation results in the SP ELCUT, perfectly correlate with the experimentally obtained temperature values in any time period. In the example of the H-0 bulkhead, the thickness of used mineral wool was evaluated, and other variants of the rate of consumption of mineral wool under the hydrocarbon fire regime were presented (Figure 14). The choice of the bulkhead is justified by the maximum temperatures obtained from the experiment and simulation. The graph shows the averaged values of the difference between the values of thermocouples located directly on the unheated surface of the sample and the initial ambient temperature (20 °C, Table 3). The different locations of the thermocouples over the cross-section of the samples are shown in Figure 10. The H-0 bulkhead is also certified as H-30. According to the initial data, the H-0 bulkhead has two layers of mineral wool with a total thickness of 60 mm (2 × 30 mm) with a density of 150 kg/m^3. One of the ways to reduce the consumption of mineral plates is to increase their density. For example, when choosing the density of mineral wool as 240 kg/m^3 (PAROC mineral wool [43]), the temperature at 30 min is reduced by 40 °C, which shows the overconsumption of used fire protection. There are two variants to reduce the thickness of mineral wool with a density of 240 kg/m^3: using two layers of thickness of 25 mm and two layers of thickness of 20 mm. At a total thickness of 40 mm (2 × 20 mm), the temperature at 30 min reaches 138 °C, which shows optimal thickness use, providing the required fire protection efficiency in hydrocarbon fire mode.

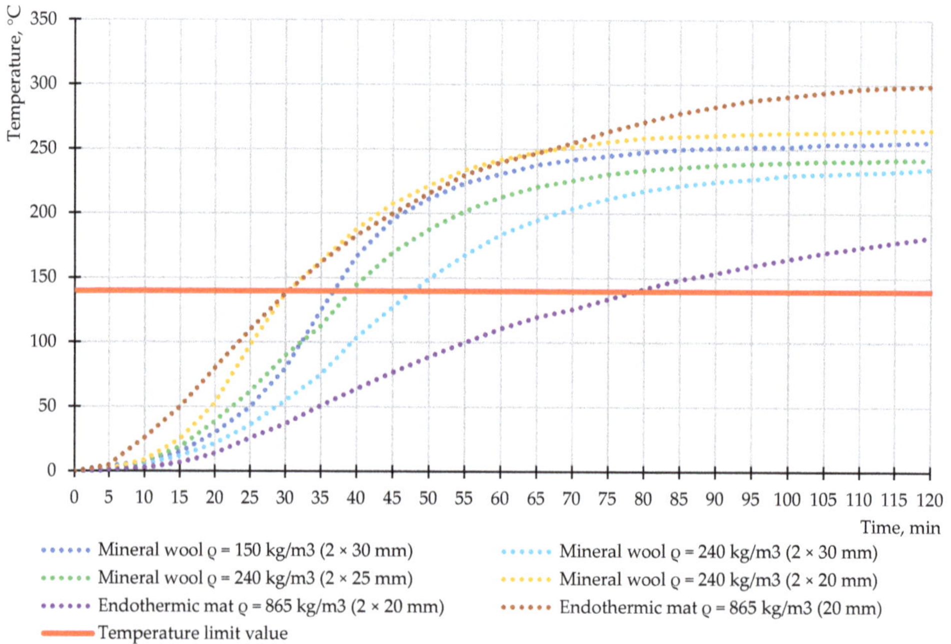

Figure 14. Variations of flame-retardant application for H-0 bulkhead.

Endothermic mats with high fire-resistance limits and high cost relative to mineral wool are also used as insulation systems in O&G [44]. Thermophysical characteristics were taken for flexible endothermic mat "3M Interam" with a thickness of 20 mm with basalt fiber and endothermic ingredients (Table 4).

Table 4. Coefficients of thermal conductivity and heat capacity of the endothermic mat "3M Interam" as a function of temperature [45].

T, °C	93	177	316	399	482
λ, W/K·m	0.151	0.175	0.100	0.118	0.140
C, J/K·m	1155	1155	1155	1155	1155
ρ, kg/m^3	865	865	865	865	865

When using an endothermic mat with a thickness of 40 mm, the temperature at 30 min reaches 37 °C, which proves the need to reduce the consumption of the used insulation system. When reducing the thickness of the endothermic mat to 20 mm, the temperature at 30 min reaches 139 °C, which proves the use of optimal thickness, at which the required fire protection efficiency in hydrocarbon fire mode is provided.

The values of coefficients of thermal conductivity and heat capacity of mineral wool manufacturers Rockwool (1), PAROC (2) and TIZOL (3) were clarified (Table 5). The calculated values of thermophysical characteristics obtained from the simulation correlate with the values in Table 3. For temperatures above 400 °C, the obtained values require experimental confirmation but can be used in solving thermal engineering tasks with unsteady thermal conductivity.

Table 5. Calculation coefficients of thermal conductivity and heat capacity of mineral wools.

T, °C	0	100	200	300	400	500	600	700	800	900	1000
λ (1), W/K·m	0.0340	0.0450	0.0595	0.0780	0.1021	0.1314	0.1669	0.2047	0.2432	0.2810	0.3195
λ (2), W/K·m	0.0370	0.0470	0.0652	0.0950	0.1384	0.1967	0.2589	0.3207	0.3826	0.4453	0.5086
λ (3), W/K·m	0.0350	0.0460	0.0729	0.0850	0.1196	0.1652	0.2222	0.2724	0.3229	0.3741	0.4269
C, J/K·kg	840	860	877	900	913	929	943	955	971	988	1004

Analysis of Table 5 shows that the best characteristics (the lowest thermal conductivity in the range from 0 to 1000 °C) to perform the functions of fire protection and thermal insulation have material 1 (Rockwool). As stated in Section 2, the heat capacity values were given the same and averaged for all types of mineral wool due to the lack of accurate data at elevated temperatures from manufacturers.

4. Conclusions

Test methods of the fire resistance of steel structures for hydrocarbon fuel transportation facilities are similar to the requirements of onshore structures of the oil and gas complex. The same parameters under different fire regimes are applied: loss of integrity and thermal insulating capacity. Based on experimental results, simulation of the fire resistance of bulkheads of different classes and decks for an offshore platform was carried out. It was found that to obtain the required fire-resistance limits of bulkheads in standard fire conditions (A-class), mineral wool with a density of 80–100 kg/m^2 and fire protection material consumption of 40 to 85 mm (plate thickness) should be used; hydrocarbon mode (H-class) requires the use of the densest mineral wool (from 100 kg/m^2) with a material consumption of 60–150 mm. Thus, to protect steel decks and bulkheads in a hydrocarbon fire with a structural steel thickness of 4.5–5 mm, it is necessary to use 30–40% more thermal insulation and apply the highest density of fire-retardant material compared to the standard fire.

Simulations have shown that constructing the H-0 bulkhead certified as H-30 (for loss of integrity and thermal insulating capacity) and H-0 (for loss of integrity) with mineral wool with a density of 150 kg/m^3 and a thickness of 60 mm is not optimal in relation to fire protection overage.

The simulated H-0 bulkhead with mineral wool with a density of 240 kg/m^3 and an insulation thickness of 40 mm and endothermic mat with a density of 865 kg/m^3 and a thickness of 20 mm can reduce the consumption of fire protection by 33% and 66%, respectively, providing the required fire resistance H-30. It is expected in the future to use fire protection and thermal insulation plates containing a combination of super-thin basalt fiber and ceramic fibers, designing a high fire-resistance rating and an adequate final product cost.

The obtained values of the coefficients of thermal conductivity and heat capacity for mineral wool can be used in the calculation of structures for fire resistance with the considered type of fire protection in the temperature range from 0 to 1000 °C.

Author Contributions: Conceptualization, M.G.; data curation, D.S. All authors have read and agreed to the published version of the manuscript.

Funding: The research is partially funded by the Ministry of Science and Higher Education of the Russian Federation under the strategic academic leadership program "Priority 2030" (Agreement 075-15-2021-1333 dated 30 September 2021).

Institutional Review Board Statement: Not applicable.

Informed Consent Statement: Not applicable.

Data Availability Statement: Testing laboratories of the FGBU VNIIPO EMERCOM of Russia, Danish Institute of Fire and Security Technology and RINA Services Spa.

Acknowledgments: The authors would like to thank Nikolai Ivanovich Vatin, Peter the Great St. Petersburg Polytechnic University, St. Petersburg, Russia, for valuable and profound comments.

Conflicts of Interest: The authors declare no conflict of interest.

References

1. Andryushkin, A.Y.; Kirshina, A.A.; Kadochnikova, E.N. The evaluation of the fire-retardant efficiency of intumescent coatings of steel structures exposed to high-temperature gas flows. *Pozharovzryvobezopasnost/Fire Explos. Saf.* **2021**, *30*, 14–26. [CrossRef]
2. Gravit, M.; Zimin, S.; Lazarev, Y.; Dmitriev, I.; Golub, E. Fire Simulation of Bearing Structures for Natural Gas Module Plant. *Adv. Intell. Syst. Comput.* **2020**, 365–376. [CrossRef]
3. Imran, M.; Liew, M.S.; Nasif, M.S.; Gracia, E.M.; Danyaro, K.U.; Niazi, M.U. Thermal and Mechanical Response of Partially Protected Steel I-Beam under Fire. *MATEC Web Conf.* **2018**, *203*, 06009. [CrossRef]
4. Gravit, M.; Gumerova, E.; Bardin, A.; Lukinov, V. Increase of Fire Resistance Limits of Building Structures of Oil-and-Gas Complex Under Hydrocarbon Fire. *Adv. Intell. Syst. Comput.* **2018**, *692*, 818–829. [CrossRef]
5. ISO 834-75. Elements of Building Constructions. Fire-Resistance Test Methods. General Requirements. Available online: https://docs.cntd.ru/document/9055248 (accessed on 2 November 2021).
6. Dmochowska, A. Threats related to accidental release of LPG in rail transport. *Zesz. Nauk. SGSP* **2020**, *73*, 37–54. [CrossRef]
7. Aksu, S. Assessing compartment-wide damage consequences in ships and offshore vessels using Fault Trees with System Location Attribution. *Ocean Eng.* **2019**, *192*, 106510. [CrossRef]
8. Zakhmatov, V.D.; Tursenev, S.A.; Mironchev, A.V.; Chernyshov, M.V.; Ozerov, A.V.; Dorozhkin, A.S. Analysis of Existing and Justification of Applying New Automatic System for Fire-and-Explosion Prevention at Vessels, Ships, Offshore Oil Platform. *Pozharovzryvobezopasnost/Fire Explos. Saf.* **2018**, *27*, 50–63. [CrossRef]
9. Kim, M.; Kim, G.; Oh, M. Optimized Fire Protection for Offshore Topside Structure with 3-Sides PFP Application. In Proceedings of the International Offshore and Polar Engineering Conference, San Francisco, CA, USA, 25–30 June 2017; pp. 740–746.
10. Salazar-Domínguez, C.M.; Hernández-Hernández, J.; Rosas-Huerta, E.D.; Iturbe-Rosas, G.E.; Herrera-May, A.L. Structural Analysis of a Barge Midship Section Considering the Still Water and Wave Load Effects. *J. Mar. Sci. Eng.* **2021**, *9*, 99. [CrossRef]
11. Nitonye, S.; Ezenwa, O. Calculation for Hull Strength Construction in Offshore Structures (A Case Study of 5000t Work Barge). *West Afr. J. Ind. Acad. Res.* **2013**, *8*, 3–12.
12. Chandrasekaran, S.; Nagavinothini, R. Behavior of stiffened deck plates under hydrocarbon fire. *Mar. Syst. Ocean Technol.* **2020**, *15*, 95–109. [CrossRef]
13. Niazi, U.M.; Nasif, M.S.; Bin Muhammad, M.; Imran, M. Integrated Consequence Modelling for Fire Radiation and Combustion Product Toxicity in offshore Petroleum Platform using Risk Based Approach. *MATEC Web Conf.* **2018**, *225*, 06013. [CrossRef]
14. Manco, M.R.; Vaz, M.A.; Cyrino, J.C.R.; Landesmann, A. Evaluation of localized pool fire models to predict the thermal field in offshore topside structures. *J. Braz. Soc. Mech. Sci. Eng.* **2020**, *42*, 1–15. [CrossRef]
15. Chandrasekaran, S.; Pachaiappan, S. Numerical analysis and preliminary design of topside of an offshore platform using FGM and X52 steel under special loads. *Innov. Infrastruct. Solut.* **2020**, *5*, 1–14. [CrossRef]
16. Cirpici, B.K.; Wang, Y.; Rogers, B. Assessment of the thermal conductivity of intumescent coatings in fire. *Fire Saf. J.* **2016**, *81*, 74–84. [CrossRef]
17. ABS Rules for Building and Classing Facilities on Offshore Installations. Rules for Building and Classing Facilities on Offshore Installations 2021. Available online: https://ww2.eagle.org/content/dam/eagle/rules-and-guides/current/offshore/63_rulesforbuildingandclassingfacilitiesonoffshoreinstallations_2021/fac-rules-july21.pdf (accessed on 8 November 2021).
18. Seo, J.K.; Lee, S.E.; Park, J.S. A method for determining fire accidental loads and its application to thermal response analysis for optimal design of offshore thin-walled structures. *Fire Saf. J.* **2017**, *92*, 107–121. [CrossRef]
19. A Study of SOLAS Regulation II-2/17 on Alternative Design and Arrangements for Fire Safety. Available online: https://core.ac.uk/download/pdf/38467591.pdf (accessed on 2 November 2021).
20. Bureau of Shipping, A. ABS Rules for Building and Classing Mobile Offshore Units. Available online: https://ww2.eagle.org/content/dam/eagle/rules-and-guides/current/offshore/3_mobileoffshoreunits_2021/mou-part-5-july21.pdf (accessed on 3 December 2021).

21. Russian Maritime Register of Shipping. Available online: https://rs-class.org/en/ (accessed on 22 November 2021).
22. UL 1709. Rapid Rise Fire Tests of Protection Materials for Structural Steel. Available online: https://nd.gostinfo.ru/print.aspx?control=27&id=4546871&print=yes (accessed on 2 November 2021).
23. EN 1363-2:1999. Fire Resistance Tests—Part 2: Alternative and Additional Procedures. Available online: https://nd.gostinfo.ru/document/6239985.aspx (accessed on 22 November 2021).
24. Gravit, M.V.; Golub, E.V.; Antonov, S.P. Fire Protective Dry Plaster Composition for Structures in Hydrocarbon Fire. *Mag. Civ. Eng.* **2018**, *3*, 79. [CrossRef]
25. Gravit, M.; Golub, E.; Klementev, B.; Dmitriev, I. Fire Protective Glass Fiber Reinforced Concrete Plates for Steel Structures under Different Types of Fire Exposure. *Buildings* **2021**, *11*, 187. [CrossRef]
26. IMO Resolution A.754 (18). Recommendation for Testing Fire Resistance of Class "A", "B" and "F" Slabs. Available online: http://www.rise.odessa.ua/texts/A754_18.php3 (accessed on 2 November 2021).
27. Russian Government Standard GOST 30247.1-94. Elements of Building Constructions. Fire-Resistance Test Methods. Loadbearing and Separating Constructions. Available online: https://docs.cntd.ru/document/9055247 (accessed on 20 November 2021).
28. Palfy, J. Guidance Notes on Alternative Design and Arrangements for Fire Safety. Available online: https://ww2.eagle.org/content/dam/eagle/rules-and-guides/current/design_and_analysis/122_altdesignandarrangforfiresafety/fire_safety_guidance_e-july10.pdf (accessed on 2 November 2021).
29. Imran, M.; Liew, M.S.; Nasif, M.S.; Niazi, U.M.; Yasreen, A. Hydrocarbon Fire and Explosion's Safety Aspects to Avoid Accident Escalation for Offshore Platform. *ICIPEG 2016* **2017**, 801–808. [CrossRef]
30. Imran, M.; Liew, M.S.; Nasif, M.S.; Niazi, U.M.; Yasreen, A. Hazard Assessment Studies on Hydrocarbon Fire and Blast: An Overview. *Adv. Sci. Lett.* **2017**, *23*, 1243–1247. [CrossRef]
31. Jafarov, E. Causes and Consequences of Fire Emergencies on Oil and Gas Platforms. *Revista Gestão Inovação e Tecnologias* **2021**, *11*, 1253–1258. [CrossRef]
32. Park, D.K.; Kim, J.H.; Park, J.S.; Ha, Y.C.; Seo, J.K. Effects of the structural strength of fire protection insulation systems in offshore installations. *Int. J. Nav. Arch. Ocean Eng.* **2021**, *13*, 493–510. [CrossRef]
33. Zhukov, A.; Konoval'Tseva, T.; Bobrova, E.; Zinovieva, E.; Ivanov, K. Thermal insulation: Operational properties and methods of research. *MATEC Web Conf.* **2018**, *251*, 01016. [CrossRef]
34. Gravit, M.; Dmitriev, I. Numerical Simulation of Fire Resistance of Steel Ship Bulkheads. *Transp. Res. Procedia* **2021**, *54*, 733–743. [CrossRef]
35. ELCUT. Modeling of Two-Dimensional Fields by the Finite Element Method. Available online: https://elcut.ru/free_doc_r.htm (accessed on 6 January 2022).
36. Dudin, M.O.; Vatin, N.I.; Barabanshchikov, Y. Modeling a set of concrete strength in the program ELCUT at warming of monolithic structures by wire. *Mag. Civ. Eng.* **2015**, *54*, 33–45. [CrossRef]
37. Markus, E.S.; Snegirev, A.Y.; Kuznetsov, E.A. *Numerical Simulation of a Fire Using Fire Dynamics*; St. Petersburg Polytech-Press: St. Petersburg, Russia, 2021; p. 175.
38. EN 1991-1-2: Eurocode 1: Actions on Structures—Part 1–2: General Actions—Actions on Structures Exposed to Fire. Available online: https://www.phd.eng.br/wp-content/uploads/2015/12/en.1991.1.2.2002.pdf (accessed on 25 October 2021).
39. Russian Government Standard GOST 52927-2015. Rolled of Normal, Increased-and High-Strength Steel for Shipbuilding. Specifications. Available online: https://docs.cntd.ru/document/1200122434 (accessed on 6 January 2022).
40. Recommendations for Optimizing the Operation of Fire Extinguishing, Smoke Removal and Ventilation Systems in Fires. Available online: https://fireman.club/literature/optimizatsiya-deystviy-sistem-pozharotusheniya-dyimoudaleniya-i-ventilyatsii-pri-pozharah-2005/ (accessed on 6 January 2022).
41. Code of Practice. SP 50.13330.2012. Thermal Performance of the Buildings. Available online: https://docs.cntd.ru/document/1200095525 (accessed on 11 January 2022).
42. Paudel, D.; Rinta-Paavola, A.; Mattila, H.-P.; Hostikka, S. Multiphysics Modelling of Stone Wool Fire Resistance. *Fire Technol.* **2021**, *57*, 1283–1312. [CrossRef]
43. PAROC Building Insulation. Available online: http://kosko.ru/sites/kosko.ru/files/files/opisanie-svojstv-paroc.pdf (accessed on 28 November 2021).
44. Gravit, M.; Shabunina, D. Structural Fire Protection of Steel Structures in Arctic Conditions. *Buildings* **2021**, *11*, 499. [CrossRef]
45. Product Data Sheet "3M Interam" Endothermic Mat E-5A-4. Available online: https://www.aircraftspruce.com/catalog/pdf/05-00948tech.pdf (accessed on 18 December 2021).

Article

Thermal–Mechanical Coupling Evaluation of the Panel Performance of a Prefabricated Cabin-Type Substation Based on Machine Learning

Xiangsheng Lei [1], Jinwu Ouyang [2], Yanfeng Wang [1], Xinghua Wang [1], Xiaofeng Zhang [3], Feng Chen [3], Chang Xia [2], Zhen Liu [2,*] and Cuiying Zhou [2,*]

[1]   Power Grid Planning Research Center of Guangdong Power Grid Limited Liability Company, Guangzhou 510699, China; leixiangsheng@gd.csg.cn (X.L.); wangyanfeng@gd.csg.cn (Y.W.); wangxinghua@gd.csg.cn (X.W.)

[2]   Guangdong Engineering Research Centre for Major Infrastructure Safety, Sun Yat-Sen University, Guangzhou 510275, China; ouyjw@mail2.sysu.edu.cn (J.O.); xiach9@mail2.sysu.edu.cn (C.X.)

[3]   China Energy Construction Group Guangdong Electric Power Design and Research Institute Limited Liability Company, Guangzhou 510060, China; zhangxiaofeng@gedi.com.cn (X.Z.); chenfeng@gedi.com.cn (F.C.)

*   Correspondence: liuzh8@mail.sysu.edu.cn (Z.L.); zhoucy@mail.sysu.edu.cn (C.Z.)

Citation: Lei, X.; Ouyang, J.; Wang, Y.; Wang, X.; Zhang, X.; Chen, F.; Xia, C.; Liu, Z.; Zhou, C. Thermal–Mechanical Coupling Evaluation of the Panel Performance of a Prefabricated Cabin-Type Substation Based on Machine Learning. *Fire* **2021**, *4*, 93. https://doi.org/10.3390/fire4040093

Academic Editor: Maged Youssef

Received: 15 November 2021
Accepted: 7 December 2021
Published: 9 December 2021

Publisher's Note: MDPI stays neutral with regard to jurisdictional claims in published maps and institutional affiliations.

Abstract: The panel performance of a prefabricated cabin-type substation under the impact of fires plays a vital role in the normal operation of the substation. However, current evaluations of the panel performance of substations under fire still focus on fire resistance tests, which seldom consider the relationship between fire behavior and the mechanical load of the panel under the impact of fires. Aiming at the complex and uncertain relationship between the thermal and mechanical performance of the substation panel under impact of fires, this paper proposes a machine learning method based on a BP neural network. First, the fire resistance test and the stress test of the panel is carried out, then a machine learning model is established based on the BP neural network. According to the collected data, the model parameters are obtained through a series of training and verification processes. Meanwhile, the correlation between the panel performance and fire resistance was obtained. Finally, related parameters are input into the thermal–mechanical coupling evaluation model for the substation panel performance to evaluate the fire resistance performance of the substation panel. To verify the correctness of the established model, numerical simulation of the fire test and stress test of the panel is conducted, and numerical simulation samples are predicted by the trained model. The results show that the prediction curve of neural network is closer to the real results compared with the numerical simulation, and the established model can accurately evaluate the thermal–mechanical coupling performance of the substation panel under fire.

Keywords: prefabricated cabin-type substation; panel; BP neural network; thermal–mechanical coupling; machine learning; fire behavior; impact of fires

1. Introduction

With the development of the national economy, the demand for electricity, from all walks of life, has increased. After a period of rapid development, large-scale centralized new energy power generation has gradually extended in the direction of decentralization and miniaturization. The requirements of new energy construction cannot be met by conventional transmission substations. Technological development and the improvement of prefabricated substations have become increasingly prominent. As a new type of prefabricated substation [1–3], the prefabricated cabin-type substation is becoming an important development direction benefiting from its high degree of integration and high level of intensiveness. Fire has an important effect on the safety of buildings and structures [4,5], thus the performance of the prefabricated substation panel under impact of fires is a guarantee of safety and plays a vital role in the normal operation of the substation. As a structural stress component of the substation panel, at the beginning of the design, the fire safety of the panel needs to be considered to ensure the safety of the overall structure of the substation. A high temperature causes the deterioration of the mechanical properties of the substation panel material, which will bring about different degrees of damage to the substation panel. Therefore, before the construction of the substation, it is necessary to carry out a fire resistance performance test under fire on the panel to ensure the fire resistance safety of the entire project in the event of a fire. Therefore, accurately describing the fire performance of substation panels has become an important issue for the stability of current substations.

Since the substation panels are mainly reinforced concrete structures, the fire performance of the substation panels can refer to the fire resistance test [6–10] and numerical simulation method to analyze fire behavior. Naser and Kodur [11] conducted an experimental study on the fire behavior of composite steel girders subjected to high shear loading. Hawileh et al. [12–14] predicted the performance of concrete beams using a finite element model. Aguado et al. [15] used a 3D finite element model for predicting the fire behavior of hollow-core slabs. However, the current research on the performance of substation panels rarely considers correlations, with little consideration of the nonlinear relationship between stress performance and fire resistance under impact of fire.

The neural network, a method of machine learning, is widely used in various fields [16–23]. Abuodeh et al. [24,25] used machine learning techniques to predict behavior of RC beams and compressive strength of ultra-high-performance concrete. Liu et al. [26] established machine-learning-based models to predict shear transfer strength of concrete joints. The neural network also has a precedent in the application of substation [27–31]. Da Silva et al. [32] proposed the use of artificial neural networks to solve the problem of fault location in substations; Wang et al. [33] used deep learning methods to identify the switch status of substations; Jiang Hongyu et al. [34] proposed an adaptive suppression method of transformer noise in substations based on genetic wavelet neural networks for the problem of transformer noise control; Oliveira et al. [35] carried out automatic monitoring on the construction site of substations based on deep learning. Neural networks [36–38] with self-learning, self-organization, and extremely strong linear fidelity capabilities can accurately reflect the nonlinear relationship between input and output variables to maintain high accuracy in short-term prediction. Therefore, machine learning is used to establish a non-linear relationship between panel stress and fire resistance from the perspective of thermal–mechanical coupling, which is a worthwhile means for evaluating the performance of substation panels under impact of fire.

To solve the above problem, this paper proposes a machine learning method based on the principle of BP (back propagation) neural networks to analyze the thermal–mechanical coupling performance of substation panels under fire. The evaluation factors are selected, such as the substation panel geometric data, mechanical performance parameters, and fire resistance performance data. After the model training ends, the relationship between panel mechanical performance and fire resistance is established. Finally, predictive samples are input into the model to evaluate the fire resistance performance of the panel. Then, fire resistance test and the stress test of the panel is carried out. A BP neural network model is trained and built through a series of training the samples. Then, numerical simulation of the fire test and stress test of the panel is conducted, and numerical simulation samples is predicted by the trained model and compared with the real results. The results show that predicted samples fit well with the actual output values and better than the result of numerical simulation. Thus, the established model can accurately evaluate the thermal–mechanical coupling performance of the panel under fire.

2. Research Methods and Contents

2.1. The Research Process for Thermal–Mechanical Coupling Evaluation of Prefabricated Cabin-Type Substation Panel Performance

The key to the thermal–mechanical coupling evaluation process of a prefabricated substation panel is to establish an evaluation model based on BP neural networks. By inputting the stress state data of the substation panel into the evaluation model, the corresponding fire resistance parameters can be obtained. The thermal–mechanical coupling performance of the prefabricated substation panel can then be evaluated. The research process of the thermal–mechanical coupling evaluation of prefabricated substation panel performance is shown in Figure 1.

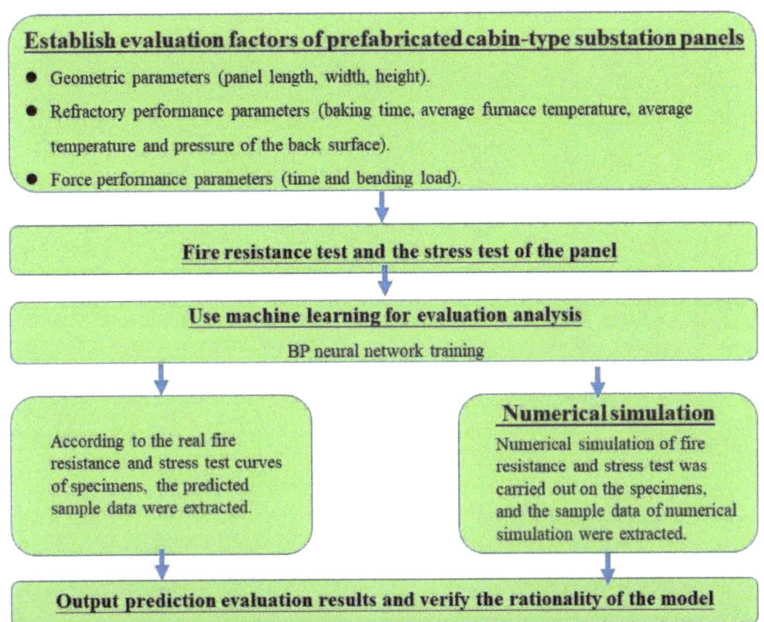

Figure 1. Research process of thermal–mechanical coupling evaluation of panel performance.

2.2. Thermal–Mechanical Coupling Evaluation Model of the Panel Performance Based on BP Neural Networks

2.2.1. Establishment of Evaluation Factors

In theory, the performance state of the prefabricated substation panel can be better described by the more comprehensive evaluation indexes. However, in practical engineering, on the one hand, it is very difficult to collect data. On the other hand, the more indexes there are, the more complex the nonlinear relationship of the thermal–mechanical coupling evaluation of the prefabricated substation panel performance is. Therefore, the determination of evaluation indexes cannot be simply generalized but should be analyzed in specific cases. As a complex system, the thermal–mechanical coupling evaluation of panel performance is affected by many factors. This study, adhering to the principles of representativeness, integrity, and desirability, takes the geometric parameters, mechanical performance, and fire resistance performance of the panel as evaluation factors of the thermal–mechanical coupling evaluation of the panel's performance.

1. The geometric parameters of the panel include length, width, and height.
2. The fire resistance performance parameters of the panel include the heating time, average furnace temperature, average temperature of the backfire surface, and pressure parameters.
3. The mechanical performance parameters of the panel include time and bending load.

2.2.2. Construction of BP Neural Network

The BP neural network as a method of machine learning is suitable for addressing complex nonlinear problems, such as the nonlinear relationship between the mechanical performance and the fire resistance performance of substation panels. The research process of the BP neural network model for the thermal–mechanical coupling evaluation of substation panel performance is shown in Figure 2. Firstly, the data parameters are input into the BP neural network for training. Secondly, the thermal–mechanical coupling evaluation results of the panel performance can be obtained through the model after model training. After that, we carried out numerical simulation of fire resistance test and stress test on the panel. We used the curve data of numerical simulation as sample data to predict the sample of numerical simulation. Finally, the correctness of the model is verified by comparing the real results with the numerical simulation results and the neural network prediction results.

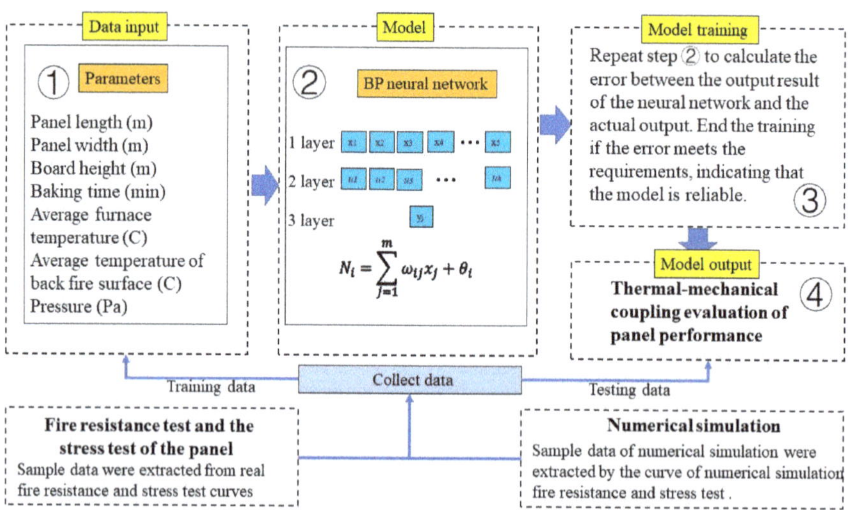

Figure 2. Research process of the BP neural network model in the thermal–mechanical coupling evaluation of prefabricated substation panel performance. x_1, x_2, ..., x_5, respectively, represents input layer parameters of neural network; u_1, u_2, ..., u_k represent hidden layer parameters of the neural network, respectively; y_j represents output layer parameters of neural network; N_i represents output results of neural network; ω represents weights of neural network and θ represents thresholds of neural network.

As shown in Figure 3, the BP neural network used for the thermal–mechanical coupling evaluation training of the prefabricated cabin-type substation panel performance is composed of three layers, representing the input layer, hidden layer, and output layer, respectively.

The input layer has seven impact indicators corresponding to the identification indicators, which are the length, width, height, heating time, average furnace temperature, average temperature, and pressure of the backfire surface. The output layer represents time and bending load. Therefore, there are seven input layer nodes in this model, six hidden layer nodes, and two output nodes. Each node is a specific output function, and each connection between two nodes represents a weighted value (weight) for the signal passing through the connection. The learning rate determines the amount of weight change generated in each cycle. The fixed learning rate in this research is 0.1, the training target is 0.00001, and the maximum number of learning iterations is 100. Through repeated iterative calculations, the correlation coefficient and threshold are determined. After that, the learning and training process ends, which means the model is successfully established. After the BP neural network model training, the actual value is compared with the predicted value. In order to solve the problem of inconsistency in the units and magnitudes of the input variables in the BP neural network, normalization is used to control the sample data to 0–1.

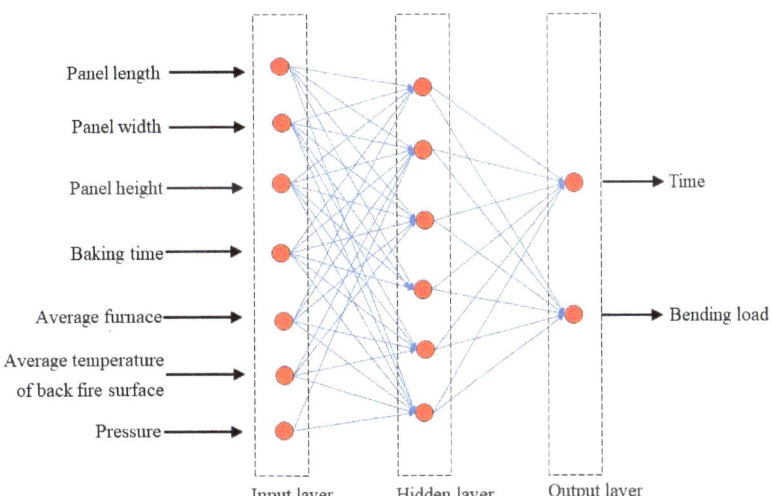

Figure 3. Application of the BP neural network in the thermal–mechanical coupling evaluation of substation panel performance.

The normalization formula is as follows:

$$Y_i = \frac{X_i - X_{min}}{X_i - X_{max}} \alpha + \beta \tag{1}$$

In the formula, X_i and Y_i represent the variables before and after normalization, respectively; X_{min} and X_{max} are the minimum and maximum values of X_i, respectively; α is a parameter with a value between 0–1, and $\beta = 1 - \frac{\alpha}{2}$.

3. Case Application Analysis

3.1. Substation Panel

3.1.1. Fire Resistance Test of Panel

The fire resistance test of panel refer to the requirements of GB/T 9978.1-2008 "Fire resistance Test Methods for Building Components part 1: General Requirements [39]" and GB/T 9978.8-2008 "Fire resistance Test Methods for Building Components Part 8: Characteristics of non-load-bearing vertical dividers [40]", as shown in Table 1. The test conditions and test plan were formulated according to the requirements of GB/T 9978.1-2008 [39] and GB/T 9978.8-2008 [40].

The length (m) width (m) × height (m) of the special panel for a box-type substation is 2.0 × 1.0 × 0.12. Ten temperature measurement points are set on the backfire surface of the panel with the vertical side on a free side, as shown in Figure 4.

According to the test requirements, the test uses vertical component fire test furnace device in Beijing Gequ fire test laboratory. The device can meet the requirements of the furnace temperature and pressure in Table 1. This device also can measure the temperature and pressure change value of the panel specimen. The data changes during the test can be visually displayed on the display screen of the equipment.

Table 1. Reference standards for fire resistance.

Test Items		Standard Clause	Judgment Criteria
Fire resistance	Completeness	GB/T 9978.8-2008 Article 10 GB/T 9978.1-2008 Article 10.2.2 Article 8.4	The duration of the test specimen's continuous fire resistance performance in the fire test. Any one of the following limited conditions of the test specimen shall be considered as a loss of integrity: (a) A cotton pad test is conducted, and the cotton pad is ignited. (b) A gap probe of 6 mm penetrates the specimen into the furnace and moves 150 mm along the length of the crack; a gap probe of 25 mm penetrates the specimen into the furnace. (c) A flame appears on the backfire surface and lasts for more than 10 s.
	Thermal insulation	GB/T 9978.8-2008 Article 10 GB/T 9978.1-2008 Article 10.2.3	If the duration of the fire resistance and heat insulation performance of the test specimen in the fire test as well as the temperature rise of the backfire surface of the test specimen exceeds any of the following limits, it is considered to have lost the heat insulation: (a) The average temperature rise exceeds the initial average temperature of 140 °C. (b) The temperature rise at any point exceeds the initial temperature (including the moving thermocouple) by 180 °C (the initial temperature should be the initial average temperature of the back surface at the beginning of the test).
		GB/T 9978.1-2008 Article 12.2.2	If the "integrity" of the test specimen does not meet the requirements, it is considered that the "heat insulation" of the test specimen does not meet the requirements.

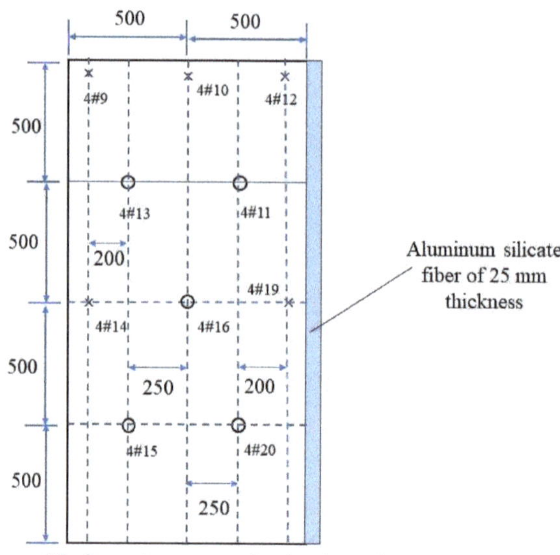

Figure 4. Schematic diagram of the measuring point layout on the backfire surface of the test specimen.

The experiment was terminated at 181 min. The test process was observed and recorded. The test phenomena are shown in Table 2.

Table 2. Test phenomena.

Time	Observation Record
0	Test start.
30	No significant change from the previous stage.
60	No significant change from the previous stage.
90	No significant change from the previous stage.
120	Concave deformation.
150	No significant change from the previous stage.
181	Integrity and thermal insulation are undamaged; test is stopped.

The fire resistance data of the panels are shown in Figures 5 and 6.

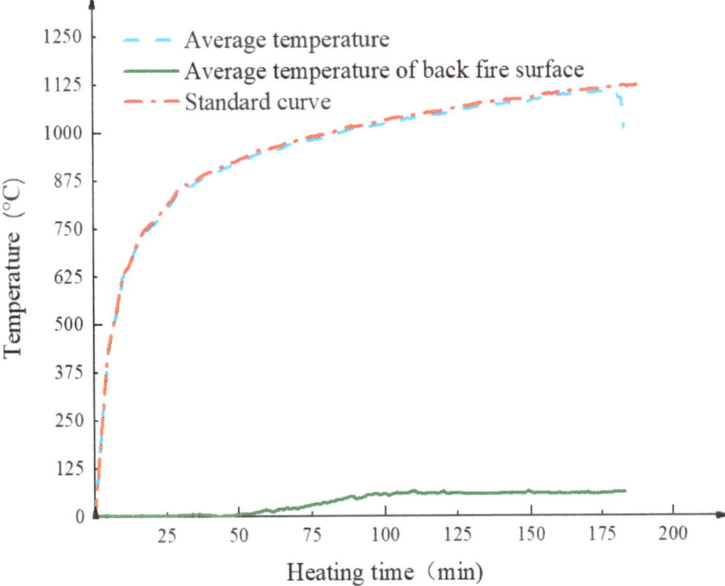

Figure 5. Temperature rise curve.

3.1.2. The Stress Test of the Panel

The same panel specimen as Section 3.1.1 was used in this experiment. Static loading is carried out by force control. A hydraulic jack was used for loading. During the test, the load is acted on the mid-span position of the panel through the actuating head. Once the specimen was destroyed, the test was over. The data of the stress test of the panel are shown in Figure 7.

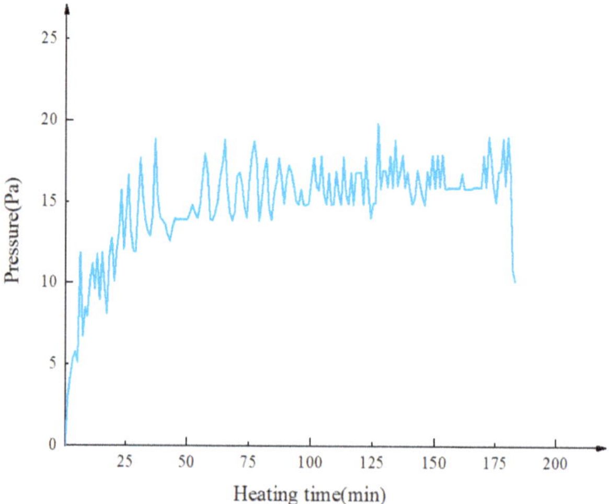

Figure 6. Pressure curve at 500 mm below the furnace roof.

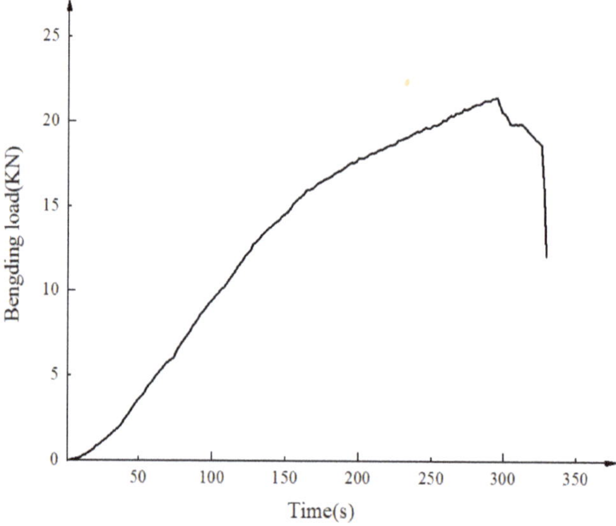

Figure 7. Stress curve of the panel strength test. Bending load refer to a load that causes bending deformation of a panel during a fixed strength test.

3.2. Thermal–Mechanical Coupling Evaluation of Panel Performance

The values were recorded every minute from the origin of the coordinates. Figures 5 and 6 show that the test specimen was damaged when heated to the 183rd minute. Figure 7 shows that the test specimen was damaged under stress at 329.052 s. The time from loading to failure was divided into 183 segments for the values recorded every 1.798 s. The fire resistance and stress performance data of the panel are shown in Appendix A. It should be emphasized that the temperature measured in Table A1 has subtract the ambient temperature. The data of columns 1 represent the number of samples; the data of columns 2 represent the heating time of panel; the data of columns 6 represent the load time of the panel.

According to the BP neural network structure constructed in Section 2.2, the thermal–mechanical coupling evaluation model of the panel performance was learned and trained:

1. Initialize the BP neural network. We randomly selected 100 sets of data from Table A1 as the input node data of the training sample, and the remaining 84 sets of data in Table A1 were used as prediction samples. Then, the weights and offsets of the neural network were initialized. Finally, the sample data were normalized.
2. Train the BP neural network. The BP neural network was used to train 100 sets of training sample data until the calculations at the end of the network training. The thermal–mechanical coupling evaluation model of the panel performance based on the BP neural network was obtained when the BP neural network converged after learning and training.
3. Predict the BP neural network. The randomly selected 84 sets of test sample data were predicted through the trained BP neural network to finally obtain the prediction result output. The graph is drawn as shown in Figure 8.

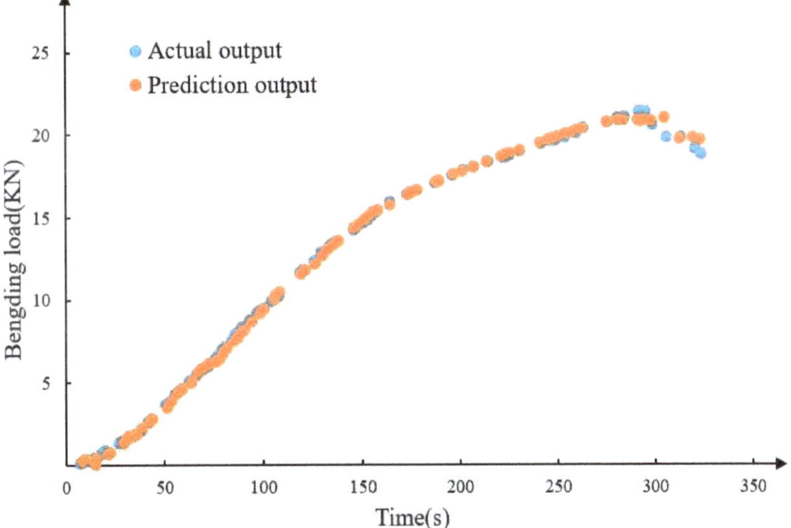

Figure 8. Comparison of sample predicted output and actual output.

It can be seen from Figure 8 that the predicted output values of the 84 groups of predicted samples fit well with the actual output values for the trend of the sample points showing basically the same, which indicates that the thermal–mechanical coupling evaluation model of panel performance based on a BP neural network is reasonable and accurate.

The mechanical performance data of the panel corresponding to the heating time of the 162nd minute to the 183rd minute were collected, as shown in Figure 9.

It can be seen from Figure 9 that, when the test specimen reaches the maximum bending load of 21.443 KN, the corresponding stress time of the substation plate is 294.888 s. When the time is 325.456 s, the bending load drops sharply from 18.664 KN, which means the material is damaged at this time. The prediction sample data of the fire resistance performance of the substation are input into the thermal–mechanical coupling evaluation model of the panel performance. The corresponding panel performance parameters can then be obtained. The test specimen reaches the maximum bending load of 21.128 KN when the predicted value of the neural network is displayed for 297.147 s. The bending load drops sharply from 18.683 KN for the material being damaged at the time of 323.658 s. By comparing the predicted value and actual value of the time and bending load, it is found that the maximum bending load and the corresponding stress time from the thermal–mechanical coupling evaluation model and actual test is very close, and the two values essentially satisfy the error requirements. This further demonstrates the accuracy and reliability of the thermal–mechanical coupling evaluation model of the panel performance.

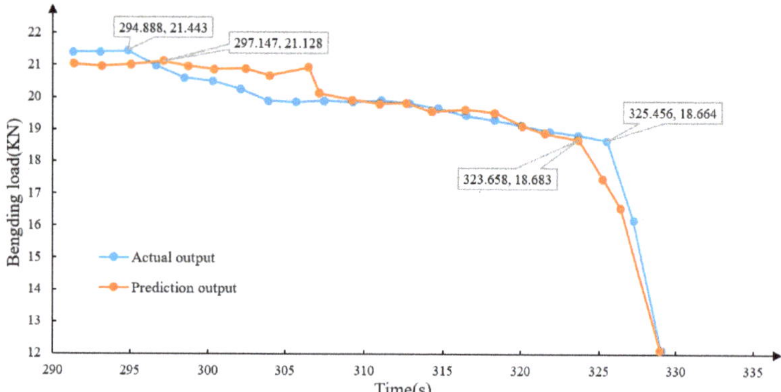

Figure 9. Sample result output of panel performance prediction.

3.3. Numerical Simulation

In order to verify the results of neural network calculation, we carried out numerical simulation on the specimen. The length (m) × width (m) × height (m) of the special panel for numerical simulation is 2.0 × 1.0 × 0.12, as shown in Figure 10. The fire resistance test and pressure test of numerical simulation model are consistent with the actual situation in Section 3.1.

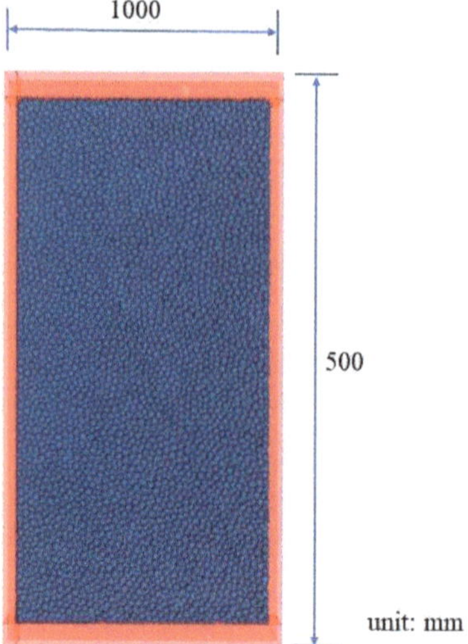

Figure 10. Numerical simulation model.

The numerical simulation results are shown in Figures 11 and 12.

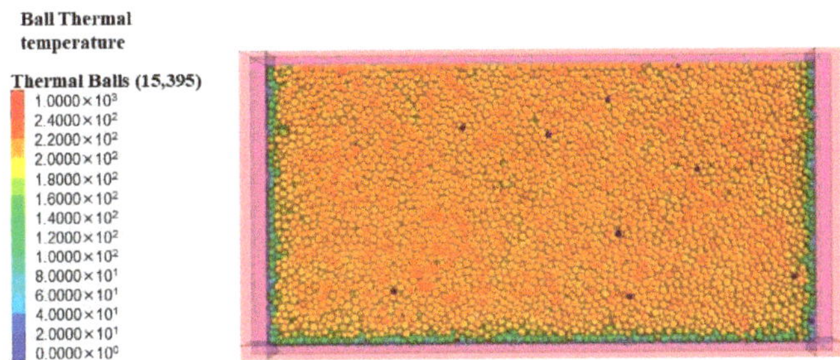

Figure 11. Numerical simulation of fire resistance test.

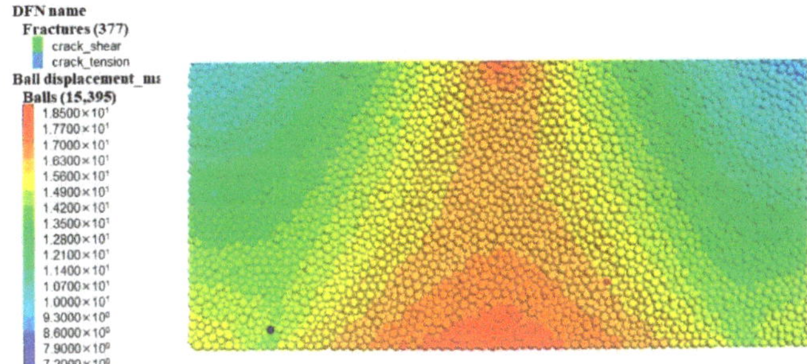

Figure 12. Numerical simulation of stress test.

The curve of the fire resistance test and pressure test parameters for the panel is shown in Figures 13 and 14. Each step in the diagram represents a unit of time.

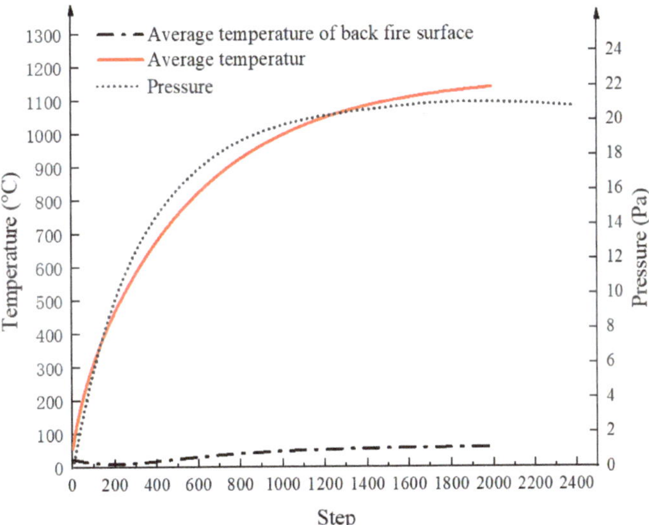

Figure 13. The curve of the fire resistance test.

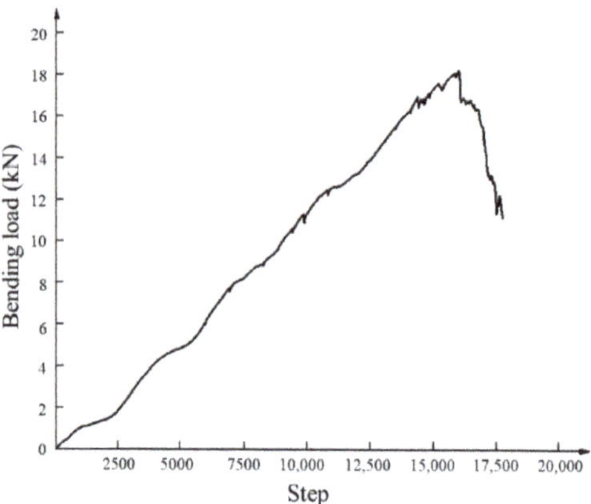

Figure 14. Stress curve of the panel samples. Bending load refer to a load that causes bending deformation of a panel during a fixed strength test.

The failure time step of numerical simulation corresponds to the failure time of fire resistance test and pressure test in real time, and the simulated result curve is also divided into 183 sections. Corresponding values are recorded in each section and 184 sample data of numerical simulation can be obtained.

According to the BP neural network structure trained in Section 3.2, we conduct neural network learning, training and prediction using the sample data of numerical simulation. According to the sample data of numerical simulation, the prediction results of numerical simulation are obtained. By converting the failure time of the real stress curve into the corresponding time step, we plotted the prediction curve of the neural network, the prediction curve of the numerical simulation and the real stress test curve in the same figure, as shown in Figure 15.

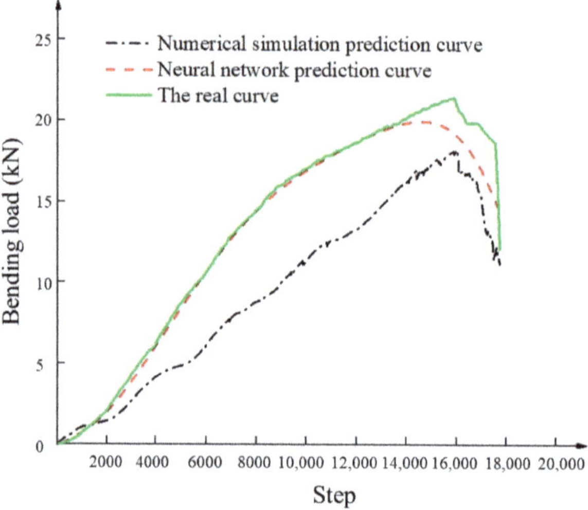

Figure 15. The stress test curve.

It can be seen from Figure 15 that the curve of prediction result of neural network and array simulation is basically consistent with the curve of real pressure test. The force increases gradually and decreases rapidly after reaching the peak value. Numerical simulation results show that when the time step is 15,850, the maximum bending load is 18.11064 kN. The neural network prediction results show that when the time step is 14,687, the bending load reaches the maximum value of 19.963 KN. The actual test results show that when the time step is 15,889, the bending load reaches the maximum value of 21.443 kN. Compared with the results of numerical simulation, the prediction curve of neural network is closer to the real pressure curve. The percentage error of the maximum bending load calculated by numerical simulation is 15.5%, the percentage error of the maximum bending load calculated by neural network prediction is 6.9%, and the error of neural network prediction is about half smaller than that of numerical simulation. The prediction result of neural network is better than that of numerical simulation. Thus, the accuracy and rationality of the neural network prediction model can be proved.

3.4. The Functional Relationship between Fire Resistance and Stress Resistance

The relationship between the parameters of fire resistance and stress resistance can be obtained by deriving the training parameters of the neural network, as shown in Equations (2)–(5):

$$\alpha_h = \sum_{i=1}^{M} v_{ih} x_i + r_h \tag{2}$$

$$b_h = f(\alpha_h) \tag{3}$$

$$y_j = \sum_{h=1}^{q} w_{hj} b_h + \theta_j \tag{4}$$

$$f(x) = \frac{1}{1 + e^{-x}} \tag{5}$$

M refers to the number of nodes in the input layer, $M = 7$; x_i $(i = 1, 2, \ldots \ldots, M)$ refers to length (m), width (m), height (m), heating time (min), average furnace temperature (°C), average temperature of backfire surface (°C), and pressure parameter (Pa); h refers to the number of hidden layer nodes, $h = 6$; q is the number of nodes in the output layer, $q = 2$; y_j $(j = 1, 2)$ refers to the values of the time (s) and bending load (KPa), respectively. v refers to weight parameters from input layer to hidden layer of neural network; r_h refers to threshold parameters from input layer to hidden layer of neural network; W refers to weight parameters from hidden layer to output layer of neural network; θ_j refers to threshold parameters from hidden layer to output layer of neural network.

$$v = \begin{bmatrix} 0 & 0 & 0 & 4.4136 & -1.6295 & -0.0460 & -0.1070 \\ 0 & 0 & 0 & 0.8386 & -0.7978 & -2.0962 & 0.1305 \\ 0 & 0 & 0 & -1.4522 & -1.0707 & -0.3682 & -0.0326 \\ 0 & 0 & 0 & 0.1633 & -0.3356 & -0.6002 & -0.0347 \\ 0 & 0 & 0 & 1.0436 & -0.0877 & -0.4381 & -0.0416 \\ 0 & 0 & 0 & 1.1642 & -0.2854 & -0.3663 & 0.1397 \end{bmatrix}$$

$$r_h = \begin{bmatrix} -3.3526 & -1.0328 & -0.2724 & 0.6572 & -0.1576 & 0.9250 \end{bmatrix}^{\mathrm{T}}$$

$$\theta_j = \begin{bmatrix} 0.2891 & -1.0308 \end{bmatrix}^{\mathrm{T}}$$

$$w = \begin{bmatrix} 0.0798 & -0.1080 & -0.0044 & -0.7329 & 0.9450 & 0.2738 \\ -1.4042 & 0.0712 & -0.6059 & -0.4502 & 0.7257 & 0.0036 \end{bmatrix}$$

4. Conclusions

Based on the evaluation factors such as the geometric data of the substation panel, the stress performance, the fire resistance performance data, etc., a BP neural network, a method of machine learning, was used to establish the nonlinear relationship between panel performance stress and fire resistance under impact of fire. This model can quickly predict the performance of the substation panel under fire. The prediction of the thermal–mechanical coupling evaluation model is very close to the actual test, and satisfy the error requirements. Additionally, the specimen was verified by numerical simulation. Comparing the neural network with numerical simulation, the result indicates the error of neural network prediction is about half smaller than that of numerical simulation, the prediction result of neural network is better than that of numerical simulation. The correctness and reliability of the thermal–mechanical coupling performance evaluation model is verified. If meeting the requirements of the test itself and the amount of data required by the structure of the neural network, the thermal–mechanical coupling evaluation model constructed in this study can be directly used for similar models. It does not need to conduct additional tests. As the types and quantities of data for training become richer, the models we build will become more and more refined. Therefore, this can provide a reference for exploring more thermal coupling evaluation models and complex functional relationships of materials based on neural networks under different loading modes in the future.

Author Contributions: Conceptualization, methodology, data curation, formal analysis; writing—review and editing, project administration, funding acquisition, Z.L.; conceptualization, methodology, supervision, project administration, funding acquisition, C.Z.; data curation, formal analysis, writing—original draft, preparation and editing, J.O.; data Curation, X.L.; data curation, Y.W.; data curation, X.W.; data curation, X.Z.; data curation, F.C.; data Curation, C.X., J.O. and X.L. contributed equally to this work and they are co-first authors of this article. All authors have read and agreed to the published version of the manuscript.

Funding: This research was funded by Science and Technology Project of Guangdong Power Grid Co., Ltd (037700KK52190022), the National Natural Science Foundation of China (NSFC) (Grant No.41977230), the National Key Research and Development Project (Grant No. 2017YFC1501203, No. 2017YFC1501201), the Special Fund Key Project of Applied Science and Technology Research and Development in Guangdong (Grant No. 2015B090925016, No. 2016B010124007).

Institutional Review Board Statement: The study did not require ethical approval.

Informed Consent Statement: Not applicable.

Data Availability Statement: Not applicable.

Acknowledgments: The authors would like to thank the anonymous reviewers for their very constructive and helpful comments.

Conflicts of Interest: The authors declare that they have no conflict of interest.

Appendix A

Table A1. Fire resistance and stress performance data of the panel.

Sample	Heating Time (min)	Average Furnace Temperature (°C)	Average Temperature of Backfire Surface (°C)	Pressure (Pa)	Time (s)	Bending Load (KN)
1	0	0	0	0	0.000	0
2	1	79.84	3.29	3.0577	1.798	0.0152
3	2	173.4	3.06	4.349	3.596	0.0619
4	3	267.91	3.12	5.4032	5.394	0.0949
5	4	360.6	4.9	5.8117	7.192	0.1097
6	5	433.88	2.36	5.1342	8.991	0.1888
7	6	468.52	3	11.858	10.789	0.287
8	7	504.38	2.5	6.7311	12.587	0.3562
9	8	535.65	1.09	8.499	14.385	0.4535
10	9	577.35	3.45	7.9913	16.183	0.5464

Table A1. Cont.

Sample	Heating Time (min)	Average Furnace Temperature (°C)	Average Temperature of Backfire Surface (°C)	Pressure (Pa)	Time (s)	Bending Load (KN)
11	10	625	2.94	10.133	17.981	0.6664
12	11	634.44	2.04	11.221	19.779	0.8527
13	12	643.43	1.94	9.6595	21.577	0.9194
14	13	664.89	1.28	11.8	23.375	1.0552
15	14	690.58	1.9	9.017	25.173	1.1762
16	15	705.34	3.27	11.87	26.972	1.2974
17	16	721.01	3.77	9.6982	28.770	1.4346
18	17	734.37	3.24	8.1376	30.568	1.5541
19	18	742.72	2.88	11.704	32.366	1.7039
20	19	750.87	2.65	12.759	34.164	1.8248
21	20	757.59	3.6	10.145	35.962	1.9602
22	21	763.69	3.13	12.014	37.760	2.114
23	22	773	2.27	13.136	39.558	2.3759
24	23	780.88	2.33	15.751	41.356	2.5796
25	24	794.49	3.52	12.086	43.154	2.7826
26	25	802.42	2.67	14.159	44.953	2.9971
27	26	809.7	2.96	16.672	46.751	3.2553
28	27	817.29	2.32	13.244	48.549	3.4676
29	28	828.68	1.68	11.989	50.347	3.659
30	29	840.37	4.26	11.958	52.145	3.8044
31	30	844.53	3.45	14.878	53.943	4.0017
32	31	851.16	3.73	17.732	55.741	4.3081
33	32	857.39	4.34	15.323	57.539	4.5063
34	33	854.14	5.18	13.932	59.337	4.6692
35	34	861.78	6.84	13.22	61.135	4.9182
36	35	867.05	3.7	12.95	62.934	5.1096
37	36	871.16	6.4	14.14	64.732	5.3131
38	37	876.67	5.2	18.894	66.530	5.4892
39	38	881.12	5.24	15.398	68.328	5.7004
40	39	884.83	4.83	14.108	70.126	5.8432
41	40	887.87	4.21	13.907	71.924	5.9585
42	41	889.97	2.81	13.67	73.722	6.0633
43	42	893	1.73	12.992	75.520	6.4429
44	43	898.57	1.2	12.621	77.318	6.692
45	44	900.57	1.55	13.437	79.116	6.9692
46	45	905.35	1.64	14.015	80.915	7.1593
47	46	908.14	3.61	13.948	82.713	7.3998
48	47	910.72	2.79	13.949	84.511	7.6056
49	48	914.22	2.12	13.951	86.309	7.9198

Table A1. *Cont.*

Sample	Heating Time (min)	Average Furnace Temperature (°C)	Average Temperature of Backfire Surface (°C)	Pressure (Pa)	Time (s)	Bending Load (KN)
50	49	914.51	4.92	13.954	88.107	8.1175
51	50	919.4	3.94	13.989	89.905	8.3583
52	51	922.69	6.19	14.364	91.703	8.634
53	52	925.51	6.38	14.84	93.501	8.8009
54	53	929.22	6.44	14.399	95.299	9.002
55	54	933.64	7.51	14.027	97.097	9.2243
56	55	938.97	9.24	14.843	98.896	9.3998
57	56	942.03	11.24	16.543	100.694	9.5576
58	57	944.37	12.45	18.004	102.492	9.7716
59	58	947.88	14.54	16.851	104.290	9.937
60	59	944.27	17.79	13.966	106.088	10.116
61	60	947.25	15.6	13.934	107.886	10.237
62	61	949.48	13.49	14.342	109.684	10.439
63	62	952.08	13.32	15.022	111.482	10.679
64	63	954.15	16.91	16.586	113.280	10.9
65	64	955.83	19.37	17.402	115.078	11.103
66	65	958.84	18.06	18.863	116.877	11.379
67	66	962.18	21.68	15.877	118.675	11.6
68	67	966.24	20.39	14.384	120.473	11.794
69	68	969.58	20.66	13.91	122.271	12.07
70	69	967.65	21.86	14.387	124.069	12.231
71	70	970.98	19.67	16.629	125.867	12.374
72	71	973.61	23.63	16.834	127.665	12.742
73	72	976.41	24.6	16.088	129.463	12.884
74	73	978.48	26.5	14.731	131.261	13.016
75	74	978.69	28.38	14.053	133.059	13.21
76	75	981.34	29.56	16.432	134.858	13.381
77	76	982.55	31.64	18.029	136.656	13.548
78	77	983.33	31.82	18.777	138.454	13.693
79	78	986	32.33	17.387	140.252	13.835
80	79	985.67	37.18	13.891	142.050	13.985
81	80	986.19	34.77	15.217	143.848	14.102
82	81	987.83	37.34	16.949	145.646	14.218
83	82	989.37	38.31	17.731	147.444	14.448
84	83	992.76	40.21	14.677	149.242	14.555
85	84	997.82	43.91	13.965	151.040	14.666
86	85	999.42	42.37	15.563	152.839	14.86
87	86	1001.8	47.47	16.311	154.637	15.123
88	87	1004	47.86	17.738	156.435	15.303

Table A1. Cont.

Sample	Heating Time (min)	Average Furnace Temperature (°C)	Average Temperature of Backfire Surface (°C)	Pressure (Pa)	Time (s)	Bending Load (KN)
89	88	1005.8	50.86	16.551	158.233	15.469
90	89	1009.2	52.06	14.923	160.031	15.585
91	90	1006.1	45.67	16.52	161.829	15.71
92	91	1007.6	46.3	17.301	163.627	15.968
93	92	1008.8	48.17	16.862	165.425	16.002
94	93	1011.3	50.79	16.047	167.223	16.039
95	94	1013	55.77	15.029	169.021	16.167
96	95	1014.3	55.85	14.896	170.820	16.339
97	96	1016.1	57.01	15.814	172.618	16.428
98	97	1017.8	58.17	14.865	174.416	16.512
99	98	1020.6	56.41	14.865	176.214	16.596
100	99	1020.3	56.48	14.935	178.012	16.681
101	100	1022.5	58.42	16.329	179.810	16.74
102	101	1025.1	53.55	17.756	181.608	16.865
103	102	1026.3	56.35	15.992	183.406	16.986
104	103	1028.8	58.47	15.757	185.204	17.071
105	104	1029.3	61.61	17.863	187.002	17.123
106	105	1031.7	58.83	15.589	188.801	17.229
107	106	1031.5	60.25	14.911	190.599	17.369
108	107	1034.3	60.88	16.814	192.397	17.487
109	108	1036.5	60.93	14.881	194.195	17.605
110	109	1035.1	64.07	14.881	195.993	17.577
111	110	1037.6	65.42	16.92	197.791	17.662
112	111	1037.8	60	15.529	199.589	17.877
113	112	1038.9	58.92	14.885	201.387	17.853
114	113	1040.1	58.82	17.807	203.185	17.876
115	114	1042.2	58.58	15.499	204.983	17.968
116	115	1042.9	60.23	14.923	206.782	18.081
117	116	1043.3	59.86	16.86	208.580	18.181
118	117	1045.1	59.06	14.858	210.378	18.182
119	118	1046.5	58.54	16.83	212.176	18.27
120	119	1045.9	60.74	16.83	213.974	18.366
121	120	1047.3	64.13	16.83	215.772	18.392
122	121	1048.9	58.24	14.897	217.570	18.488
123	122	1051	58.63	17.819	219.368	18.58
124	123	1052.8	59.49	15.749	221.166	18.639
125	124	1054.6	59.2	14.052	222.964	18.655
126	125	1055.7	60.53	14.97	224.763	18.733
127	126	1057.5	60.28	15.006	226.561	18.916

Table A1. *Cont.*

Sample	Heating Time (min)	Average Furnace Temperature (°C)	Average Temperature of Backfire Surface (°C)	Pressure (Pa)	Time (s)	Bending Load (KN)
128	127	1058.9	58.28	19.863	228.359	18.961
129	128	1060.6	57.76	15.857	230.157	18.998
130	129	1060.7	57	16.98	231.955	19.101
131	130	1063.2	57.95	16.98	233.753	19.139
132	131	1064.8	58.31	15.963	235.551	19.271
133	132	1066.1	57.56	17.866	237.349	19.345
134	133	1068.7	58.53	15.933	239.147	19.469
135	134	1066.8	58.26	18.854	240.945	19.456
136	135	1069.3	58.71	16.037	242.744	19.544
137	136	1070.3	58.68	16.853	244.542	19.677
138	137	1070.7	58.49	17.941	246.340	19.674
139	138	1072	58.11	15.973	248.138	19.672
140	139	1071.6	58.5	16.891	249.936	19.815
141	140	1073.3	59.79	15.738	251.734	19.827
142	141	1074	60.42	14.925	253.532	19.85
143	142	1077.2	59.75	15.367	255.330	19.961
144	143	1075.3	58.78	16.999	257.128	20.132
145	144	1077.2	59.56	16.185	258.926	20.111
146	145	1077.1	59.34	15.405	260.725	20.281
147	146	1077.6	60.26	14.863	262.523	20.4
148	147	1078	58.89	16.935	264.321	20.36
149	148	1077.2	60.51	15.918	266.119	20.547
150	149	1079.2	63.73	17.923	267.917	20.555
151	150	1081.2	61.49	15.921	269.715	20.593
152	151	1083.1	58.71	17.925	271.513	20.787
153	152	1086	58.99	15.991	273.311	20.758
154	153	1085.5	59.37	17.962	275.109	20.805
155	154	1087.8	59.31	15.892	276.907	20.912
156	155	1090	59.96	15.895	278.706	21.004
157	156	1090.9	59.23	15.895	280.504	21.061
158	157	1092.7	58.71	15.895	282.302	21.038
159	158	1089.6	58.42	15.93	284.100	21.135
160	159	1093.5	58.03	15.933	285.898	21.264
161	160	1095.2	60.64	15.933	287.696	21.275
162	161	1096.7	58.66	16.818	289.494	21.257
163	162	1098.2	58.48	15.868	291.292	21.402
164	163	1096	57.96	15.868	293.090	21.405
165	164	1098.5	58.82	15.868	294.888	21.443

Table A1. *Cont.*

Sample	Heating Time (min)	Average Furnace Temperature (°C)	Average Temperature of Backfire Surface (°C)	Pressure (Pa)	Time (s)	Bending Load (KN)
166	165	1099.2	60.62	15.868	296.687	20.986
167	166	1099.9	59.08	15.943	298.485	20.603
168	167	1101.3	58.55	15.946	300.283	20.502
169	168	1099	59.65	15.946	302.081	20.263
170	169	1101	58.56	15.946	303.879	19.9
171	170	1101.8	61.27	17.916	305.677	19.858
172	171	1102.2	57.09	15.982	307.475	19.903
173	172	1102.4	56.92	19.039	309.273	19.868
174	173	1102.4	57.55	17.783	311.071	19.913
175	174	1104.5	58.75	15.985	312.869	19.827
176	175	1105	58.93	15.037	314.668	19.665
177	176	1106.2	60.01	16.871	316.466	19.445
178	177	1107	60.76	16.975	318.264	19.301
179	178	1107.5	60.35	18.946	320.062	19.131
180	179	1107.9	61.09	16.06	321.860	18.953
181	180	1108.2	61.39	19.016	323.658	18.834
182	181	1089.3	61.3	16.809	325.456	18.664
183	182	1089.2	61.44	10.868	327.254	16.174
184	183	1010.7	61.69	10.19	329.052	12.114

References

1. Hazel, T.; Norris, A.; Barbizet, M.; Et, A. Designing prefabricated substation buildings according to GOST standards; Record of Conference Papers; Industry Applications Society; Forty-Ninth Annual Conference. In Proceedings of the 2002 Petroleum and Chemical Industry Technical Conference, New Orleans, LA, USA, 23–25 September 2002; pp. 251–259.
2. Zhengmao, F.; Xiuhua, S.; Hongzhi, C.; Et, A. Optimization design of box structure for prefabricated substation. *Int. J. Res. Eng. Technol.* **2018**, *7*, 85–90.
3. Zou, P.L. Comparative analysis of traditional civil construction new energy substation and modular prefabricated cabin substation. *Mech. Electr. Inf.* **2020**, *38*, 9.
4. Gerges, M.; Demian, P.; Adamu, Z. Customising Evacuation Instructions for High-Rise Residential Occupants to Expedite Fire Egress: Results from Agent-Based Simulation. *Fire* **2021**, *4*, 21. [CrossRef]
5. Ghodrat, M.; Shakeriaski, F.; Nelson, D.J.; Simeoni, A. Existing Improvements in Simulation of Fire–Wind Interaction and Its Effects on Structures. *Fire* **2021**, *4*, 27. [CrossRef]
6. Ali, F.; Nadjai, A.; Silcock, G.; Et, A. Outcomes of a major research on fire resistance of concrete columns. *Fire Saf. J.* **2004**, *39*, 433–445. [CrossRef]
7. Kodur, V.K.R.; Dwaikat, M.M.S.; Dwaikat, M.B. High-temperature properties of concrete for fire resistance modeling of structures. *ACI Mater. J.* **2008**, *105*, 517–527.
8. Ran, L.; Zhao, H.; Huang, W.; Li, X.; Wang, Y.; Hu, Y. Fire resistance analysis of door and wall composite components. *Fire Sci. Technol.* **2014**, *33*, 1031–1033.
9. Serrano, R.; Cobo, A.; Prieto, M.I.; Et, A. Analysis of fire resistance of concrete with polypropylene or steel fibers. *Constr. Build. Mater.* **2016**, *122*, 302–309. [CrossRef]
10. Tian, J.; Zhu, P.; Qu, W. Study on fire resistance time of hybrid reinforced concrete beams. *Struct. Concr.* **2019**, *20*, 1941–1954. [CrossRef]
11. Naser, M.Z.; Kodur, V.K.R. Comparative fire behavior of composite girders under flexural and shear loading. *Thin-Walled Struct.* **2017**, *116*, 82–90. [CrossRef]
12. Hawileh, R.A.; Naser, M.Z. Thermal-stress analysis of RC beams reinforced with GFRP bars. *Compos. Part B Eng.* **2012**, *43*, 2135–2142. [CrossRef]

13. Hawileh, R.A.; Naser, M.; Zaidan, W.; Al, E. Modeling of insulated CFRP-strengthened reinforced concrete T-beam exposed to fire. *Eng. Struct.* **2009**, *31*, 3072–3079. [CrossRef]

14. Hawileh, R.A.; Naser, M.; Zaidan, W.; Al, E. Transient Thermal-Stress Finite Element Analysis of CFRP Strengthened RC beams Exposed to different Fire Scenarios. *Mech. Adv. Mater. Struc.* **2011**, *18*, 172–180. [CrossRef]

15. Aguado, J.V.; Albero, V.; Espinos, A.; Al, E. A 3D finite element model for predicting the fire behavior of hollow-core slabs. *Eng. Struct.* **2016**, *108*, 12–27. [CrossRef]

16. Faridmehr, I.; Nikoo, M.; Baghban, M.H.; Pucinotti, R. Hybrid Krill Herd-ANN Model for Prediction Strength and Stiffness of Bolted Connections. *Buildings* **2021**, *11*, 229. [CrossRef]

17. Avossa, A.M.; Picozzi, V.; Ricciardelli, F. Load-Carrying Capacity of Compressed Wall-Like RC Columns Strengthened with FRP. *Buildings* **2021**, *11*, 285. [CrossRef]

18. Abd-Elhamed, A.; Shaban, Y.; Mahmoud, S. Predicting Dynamic Response of Structures under Earthquake Loads Using Logical Analysis of Data. *Buildings* **2018**, *8*, 61. [CrossRef]

19. Mishra, P.; Samui, P.; Mahmoudi, E. Probabilistic Design of Retaining Wall Using Machine Learning Methods. *Appl. Sci.* **2021**, *11*, 5411. [CrossRef]

20. Jain, N.; Bansal, V.; Virmani, D.; Gupta, V.; Salas-Morera, L.; Garcia-Hernandez, L. An Enhanced Deep Convolutional Neural Network for Classifying Indian Classical Dance Forms. *Appl. Sci.* **2021**, *11*, 6253. [CrossRef]

21. Wu, M.; Wang, J. Estimating Contact Force Chains Using Artificial Neural Network. *Appl. Sci.* **2021**, *11*, 6278. [CrossRef]

22. Jiao, Z.; Hu, P.; Xu, H.; Al, E. Machine learning and deep learning in chemical health and safety: A systematic review of techniques and applications. *ACS Chem. Health Saf.* **2020**, *27*, 316–334. [CrossRef]

23. Wang, W.; Kiik, M.; Peek, N.; Al, E. A systematic review of machine learning models for predicting outcomes of stroke with structured data. *PLoS ONE* **2020**, *15*, e234722.

24. Abuodeh, O.R.; Abdalla, J.A.; Hawileh, R.A. Prediction of shear strength and behavior of RC beams strengthened with externally bonded FRP sheets using machine learning techniques. *Compos. Struct.* **2020**, *234*, 111698. [CrossRef]

25. Abuodeh, O.; Abdalla, J.A.; Hawileh, R.A. Prediction of compressive strength of ultra-high performance concrete using SFS and ANN. In Proceedings of the 2019 8th International Conference on Modeling Simulation and Applied Optimization (ICMSAO), Sanya, China, 9–10 November 2019; pp. 1–5.

26. Liu, T.; Wang, Z.; Zeng, J.; Al, E. Machine-learning-based models to predict shear transfer strength of concrete joints. *Eng. Struct.* **2021**, *249*, 113253. [CrossRef]

27. Chen, C.S.; Tzeng, Y.M.; Hwang, J.C. The application of artificial neural networks to substation load forecasting. *Electr. Power Syst. Res.* **1996**, *38*, 153–160. [CrossRef]

28. Hsu, Y.Y.; Lu, F.C. A combined artificial neural network-fuzzy dynamic programming approach to reactive power/voltage control in a distribution substation. *IEEE Trans. Power Syst.* **1998**, *13*, 1265–1271.

29. Borkowski, D.; Wetula, A.; Bień, A. Contactless measurement of substation busbars voltages and waveforms reconstruction using electric field sensors and artificial neural network. *IEEE Trans. Smart Grid* **2014**, *6*, 1560–1569. [CrossRef]

30. Nguyen, B.N.; Quyen, A.H.; Nguyen, P.H.; Al, E. Wavelet-based Neural Network for recognition of faults at NHABE power substation of the Vietnam power system. In Proceedings of the 2017 International Conference on System Science and Engineering (ICSSE), Ho Chi Minh City, Vietnam, 21–23 July 2017; pp. 165–168.

31. Dudzik, M.; Jagiello, A.; Drapik, S.; Et, P.J. The selected real tramway substation overload analysis using the optimal structure of an artificial neural network. In Proceedings of the 2018 International Symposium on Power Electronics, Electrical Drives, Automation and Motion (SPEEDAM), Amalfi, Italy, 20–22 June 2018; pp. 413–417.

32. Da Silva, A.P.A.; Insfran, A.H.F.; Da Silveira, P.M.; Et, A. Neural networks for fault location in substations. *IEEE Trans. Power Deliv.* **1996**, *11*, 234–239. [CrossRef]

33. Wang, J.; You, Z.; Xiao, J.; Tan, Z. Deep learning based state recognition of substation switches. In Proceedings of the AIP Conference Proceedings, Kuala Lumpur, Malaysia, 24–26 July 2018; p. 1971.

34. Jiang, H.; Liu, S.; Zhou, J.; Zhu, G.; Wang, K.; Shi, Z. Adaptive Noise Reduction of Transformer in Substation Based on Genetic Wavelet Neural Network. *Electr. Power Sci. Eng.* **2020**, *36*, 25–31.

35. Oliveira, B.A.S.; Neto, A.P.D.F.; Fernandino, R.M.A.; Et, A. Automated Monitoring of Construction Sites of Electric Power Substations Using Deep Learning. *IEEE Access* **2021**, *9*, 19195–19207. [CrossRef]

36. Wang, L.; Zeng, Y.; Chen, T. Back propagation neural network with adaptive differential evolution algorithm for time series forecasting. *Expert Syst. Appl.* **2015**, *42*, 855–863. [CrossRef]

37. Li, J.; Cheng, J.; Shi, J.; Al, E. Brief introduction of back propagation (BP) neural network algorithm and its improvement. In *Advances in Computer Science and Information Engineering*; Springer: Berlin/Heidelberg, Germany, 2012.

38. Singh, A.K.; Kumar, B.; Singh, S.K.; Al, E. Multiple watermarking technique for securing online social network contents using back propagation neural network. *Future Gener. Comput. Syst.* **2018**, *86*, 926–939. [CrossRef]

39. Fire-Resistance Tests—Elements of Building Construction—Part 1: General Requirements (GB/T 9978.1-2008). Available online: https://gf.1190119.com/list-704.htm (accessed on 15 November 2021).

40. Fire-Resistance Tests—Elements of Building Construction—Part 8: Specific Requirements for Non-Loadbearing Vertical Separating Elements (GB/T 9978.8-2008). Available online: https://www.doc88.com/p-7798292250942.html (accessed on 15 November 2021).

MDPI
St. Alban-Anlage 66
4052 Basel
Switzerland
Tel. +41 61 683 77 34
Fax +41 61 302 89 18
www.mdpi.com

Fire Editorial Office
E-mail: fire@mdpi.com
www.mdpi.com/journal/fire